2022“一带一路”国际大学生数字建筑设计竞赛作品集

丁 帅 主编

中国建材工业出版社

图书在版编目（CIP）数据

2022"一带一路"国际大学生数字建筑设计竞赛作品集 / 丁帅主编. -- 北京：中国建材工业出版社，2023.6

ISBN 978-7-5160-3680-8

Ⅰ.①2… Ⅱ.①丁… Ⅲ.①数字技术－应用－建筑设计－作品集－世界－现代 Ⅳ.①TU206

中国国家版本馆CIP数据核字（2023）第003113号

内容简介

"一带一路"建筑类大学国际联盟秘书处组织编写的《2022"一带一路"国际大学生数字建筑设计竞赛作品集》共收录竞赛征集的94件作品，竞赛以"铁路——架起'丝路'新纽带"为主题，围绕铁路及设施建设与管理、城市更新、建筑遗产保护、环境监测等方向，设立了数字化建筑设计、数字化结构设计、地理场景建模与表达三个赛题方向。其中数字化建筑设计方向作品50件，数字化结构设计方向作品20件，地理场景建模与表达方向作品24件。

本次竞赛由中国宋庆龄基金会、中国丝路集团有限公司、北京国际和平文化基金会共同协办，得到联合国教科文组织、G-Global国际秘书处、国际摄影测量与遥感协会、丝路国际公益基金、北京和苑博物馆、中国企业文化促进会、中国交通运输协会"一带一路"物流分会、广联达科技股份有限公司、中国铁路设计集团有限公司、北京建大资产经营管理有限公司、中测国际地理信息有限公司、武汉大势智慧科技有限公司及"一带一路"建筑类大学国际联盟成员高校的大力支持和参与。同时，中国建筑设计研究院本土设计研究中心及北京未来城市设计高精尖创新中心对竞赛给予特别支持。

竞赛以"专业性、权威性、创新性"为核心理念，以"以赛促教、以赛促学、以赛促培、以赛促融"和有效提升联盟各成员高校学生的国际创新能力为宗旨，为"一带一路"建筑类大学搭建了科技创新与教育交流平台，积极助力国际化创新型专业人才的培养，引导、促进了相关专业的教育教学改革。

本书可供广大工科院校师生以及建筑设计师、结构工程师、城市设计研究者参考。

2022"一带一路"国际大学生数字建筑设计竞赛作品集

2022 "YIDAIYILU" GUOJI DAXUESHENG SHUZI JIANZHU SHEJI JINGSAI ZUOPINJI

丁　帅　主编

出版发行：中国建材工业出版社

地　　址：北京市海淀区三里河路11号

邮政编码：100831

经　　销：全国各地新华书店

印　　刷：北京天恒嘉业印刷有限公司

开　　本：889mm×1194mm　1/16

印　　张：22.25

字　　数：400千字

版　　次：2023年6月第1版

印　　次：2023年6月第1次

定　　价：198.00元

本社网址：www.jccbs.com，微信公众号：zgjcgycbs

请选用正版图书，采购、销售盗版图书属违法行为

本书如有印装质量问题，由我社市场营销部负责调换，联系电话：（010）57811387

本书编委会

主　　编　丁　帅
副 主 编　邓　扬
参　　编　李　洋　刘书青　王　璇
　　　　　何静涵　刘　星　刘　璐
组织编写　“一带一路”建筑类大学国际联盟秘书处

前言

• P R E F A C E •

为积极响应国家“一带一路”倡议，深入贯彻落实北京市《新时期北京教育对外开放工作规划（2016—2020）》《北京市“十三五”时期加强国际交往中心建设规划》和《北京市对接共建“一带一路”教育行动计划实施方案》等文件精神，北京建筑大学发起并于2017年10月10日成立了“一带一路”建筑类大学国际联盟（以下简称联盟）。

目前，来自俄罗斯、哈萨克斯坦、波兰、法国、美国、英国、亚美尼亚、保加利亚、捷克、韩国、马来西亚、希腊、尼泊尔、以色列等28个国家的74所大学成为联盟成员高校。北京建筑大学作为联盟主席单位，积极谋划，不断创新，与成员院校携手，推动“一带一路”建筑类大学国际联盟可持续发展。

为培养大学生的科技创新精神和实践能力，提高大学生科学素养和科研技能水平，增强各国大学生之间的学术和文化交流，促进“一带一路”建筑类大学交流合作，从而进一步提升教育教学国际化水平，2020年，由“一带一路”建筑类大学国际联盟发起举办首届“一带一路”国际大学生数字建筑设计竞赛。竞赛组委会积极探索常态化疫情防控条件下对外交往新模式，打造“云竞赛”“云展览”等系列品牌活动，以“线上+线下”相融合的形式，至今已成功举办2020年至2022年三届大赛。

2022“一带一路”国际大学生数字建筑设计竞赛以“铁路——架起‘丝路’新纽带”为主题，旨在引导各国大学生以“数字化”思维思考铁路及沿线城市的历史文脉延续、综合品质提升与可持续发展，围绕铁路及设施建设与管理、城市更新、建筑遗产保护、环境监测等方向，运用计算机辅助、虚拟现实、数字孪生、地理信息系统、遥感技术等新技术，设立了数字化建筑设计、数字化结构设计、地理场景建模与表达三个赛题方向。

本次竞赛由中国宋庆龄基金会、中国丝路集团有限公司、北京国际和平文化基金会共同协办，得到联合国教科文组织、G-Global国际秘书处、国际摄影测量与遥感协会、丝路国际公益基金、北京和苑博物馆、中国企业文化促进会、中国交通运输协会“一带一路”物流分会、广联达科技股份有限公司、中国铁路设计集团有限公司、北京建大资产经营管理有限公司、中测国际地理信息有限公司、武汉大势智慧科技有限公司及联盟各高校的大力支持和参与。同时，中国建筑设计研究院有限公司本土设计研究中心及北京未来城市设计高精尖创新中心对竞赛给予特别支持。组委会聘请中国工程院院士、全国工程勘察设计大师崔愷、挪威工程院院士贾军波、国际摄影测量与遥感学会第四技术委员会主席 Sisi Zlatanova 为特聘专家进行指导，共 11 个国家的 24 名专家组成三个方向的评审小组，最终评选出获奖作品。

本届竞赛参赛团队数量和作品提交数量再创历史新高，吸引了来自 11 个国家的 45 所高校、650 余名师生参与，提交参赛作品百余件。经过三年的探索与发展，竞赛的影响力、知名度、专业权威性及参与度与日俱增，已成为国际大学生之间的学术和文化交流盛会。

中国丝路集团有限公司董事长闫立金在贺信中写道:“‘一带一路’国际大学生数字建筑设计竞赛为推动线上线下融合式开展教育合作、助力丝绸之路沿线教育协同发展提出了一种新的可能。我们希望通过竞赛的持续举办，加强丝绸之路沿线各国人文交流、科技合作、产业发展，弘扬‘和平合作、开放包容、互学互鉴、互利共赢’的丝绸之路精神，推动丝绸之路沿线教育领域务实合作。未来，丝路集团将携手‘一带一路’建筑类大学国际联盟，为丝绸之路沿线各国大学生搭建多元化学术和实践平台，推动教育科技助力数字经济发展及人类可持续发展。”

在此，对积极参与和鼎力支持大赛的院校师生及社会各方人士表示衷心感谢！

编　者

2023.1

（书中部分图片做虚化处理，如有疑问，欢迎联系“一带一路”建筑类大学国际联盟秘书处。）

Preface

To actively respond to China's Belt and Road Initiative and *Beijing's Work Plan for Opening Up Education in the New Period (2016—2020), Beijing's Thirteenth Five Year Plan for Improving Its Role as An International Exchange Center, Beijing's Implementation Plan for Participating in Promoting Education Along the Belt and Road,* the Belt and Road Architectural University International Consortium (BRAUIC) was initiated by the Beijing University of Civil Engineering and Architecture (BUCEA) on October 10, 2017.

BRAUIC comprises 74 universities hailing from 28 countries, including Russia, Kazakhstan, Poland, France, the United States, the United Kingdom, Armenia, Bulgaria, the Czech Republic, South Korea, Malaysia, Greece, Nepal, and Israel. As the chairman's unit of the BRAUIC, BUCEA is committed to promoting the sustainable development of the BRAUIC through active planning and concerted cooperation with member universities.

In 2020, the First Belt and Road International Student Competition on Digital Architectural Design was launched by the Belt and Road Architectural University International Consortium (BRAUIC) with the aim to cultivate the scientific and technological innovation spirits and practical abilities of university students, improve their scientific literacy and research skills, promote the academic and cultural exchanges among university students from various countries and increase communication among architectural universities along the Belt and Road to further level up the international education and teaching. Through active innovation, the Organizing Committee has created a new mode for international communication and exchange under regular pandemic prevention and control, carrying out a series of brand activities including "cloud competition" and "cloud exhibition". So far, the competition has been successfully held for three sessions from 2020 to 2022 with an integrated "online + offline" mode.

Under the theme of *Railway: A New Bond Along the Silk Road*, the 2022 Belt and Road International Student Competition on Digital Architectural Design was devoted to cultivating all college students' digitalization-based thinking that explores how to preserve the historical features of railways and cities, improve overall quality, and achieve sustainable development along the Silk Road. Urban renewal, architectural heritage protection, environmental monitoring, and other topics related to the railways are included in the Competition. Centering around these topics, participants will use computer-assisted technology, virtual reality, digital twins, geographic information system, remote sensing technology, and other new technologies to contribute works under three categories: Architectural Design, Structural Design, Scene Modeling and Visualization.

The competition was co-organized by China Soong Ching Ling Foundation, China Silk Road Group Limited, and Beijing Peace Garden Museum; supported by United Nations Educational, Scientific and Cultural Organization (UNESCO), G-Global International Secretariat, International Society for Photogrammetry and Remote Sensing (ISPRS), Silk Road International Foundation, Beijing Peace Garden Museum, China Enterprise Culture Improvement Association, Belt and Road Logistics Branch of China Communications and Transportation Association, Glodon Company Limited, China Railway Design Corporation, Beijing Jianda Assets Management Co., Ltd., China International Geo-Information Corporation Limited, and Daspatial Technology Co., Ltd.; and widely participated by member universities. Meanwhile, the competition was offered special support by China Architecture Design Group Land-based Rationalism D. R. C and Beijing Advanced Innovation Center for Future Urban Design. The Organizing Committee invited Cui Kai, Academician of Chinese Academy of Engineering and National Master of Engineering Survey and Design, Jia Junbo, Elected Member of Norwegian Academy of Technological Sciences, and Sisi Zlatanova, President of Technical Commission Ⅳ 'Spatial Information Science' of ISPRS as honorary members of the scientific committee. A total of 24 experts from 11 countries formed the jury in three categories to select award-winning works.

This year's competition has witnessed a record number of participating teams and submitted entries, attracting more than 650 students and teachers from 45 universities and institutions in 11 countries and receiving over 100 submissions. After three years of exploration and development, the competition has become a welcoming event for academic and cultural exchanges for international university students with its influence, popularity, professional authority and participation rate greatly increased.

Mr. Yan Lijin, Chairman of China Silk Road Group Limited, said in his congratulation letter: "The Belt and Road International Student Competition on Digital Architectural Design opens up new possibilities for integrated educational cooperation via both online and offline methods and coordinated development of education along the Silk Road. It is our shared expectation that through hosting the event consecutively, we will strengthen people-to-people and cultural exchanges, scientific and technological cooperation, and industrial development among countries along the Silk Road, carry forward the Silk Road Spirit featuring 'peace and cooperation, openness and inclusiveness, mutual learning, and mutual benefit', and promote practical cooperation in the field of education along the Silk Road. In the future, China Silk Road Group Limited will join hands with the BRAUIC to build diversified academic and practical platforms for college students from countries along the Silk Road and promote educational technologies to aid the development of the digital economy and sustainable development of mankind."

Taking this opportunity, we would like to extend our sincere gratitude to the students and teachers of the BRAUIC members and the community for their active participation and extensive support!

Compiler

January, 2023

(Some feather pictures are permitted in this book, any question, please contact BRAUIC.)

目录

CONTENTS

数字化建筑设计 Category A: Architectural Design

一等奖 First Prize

二等奖 Second Prize

三等奖 Third Prize

优秀奖 Honorable Mention

其他作品 Other Works

数字化结构设计 Category B: Structural Design

一等奖 First Prize

二等奖 Second Prize

204 荔樟丝梦
北京建筑大学
Libo Dream-Silk Bridge
Beijing University of Civil Engineering and Architecture, China

207 丝绸之桥
苏州科技大学
Bridge of Silk
Suzhou University of Science and Technology, China

三等奖 Third Prize

210 冰雪之缘
河北建筑工程学院
Frozen
Hebei University of Architecture, China

213 STS 大桥
河南城建学院
STS-Bridge
Henan University of Urban Construction, China

216 中俄首座“国家团结—人民友谊”客运铁路桥
俄罗斯莫斯科国立建筑大学
The First Passenger Railway Bridge Russia-China “Unity of Countries-Friendship of People”
Moscow State University of Civil Engineering, Russia

219 丝路红环
沈阳建筑大学
Red Ring Bridge
Shenyang Jianzhu University, China

222 丝路之脉
苏州科技大学
The Vein of the Silk Road
Suzhou University of Science and Technology, China

225 伦敦格罗夫纳新铁路桥
英国东伦敦大学
New Grosvenor Railway Bridge London
University of East London, the UK

优秀奖 Honorable Mention

地理场景建模与表达 Category C: Scene Modeling and Visualization

一等奖 First Prize

二等奖 Second Prize

三等奖 Third Prize

优秀奖 Honorable Mention

293 基于风格修复理念的老合肥火车站历史空间场景复原的数字孪生
安徽建筑大学
The Digital Twin of the Restoration of the Historical Space Scene of the Old Hefei Railway Station Based on the Concept of Style Restoration
Anhui Jianzhu University, China

296 徐州高铁站
河南城建学院
Xuzhou High-speed Railway Station
Henan University of Urban Construction, China

299 新丝绸之路——郑州航空港站场景设计
河南城建学院
New Silk Road—Scene Design of the Zhengzhou Airport Station
Henan University of Urban Construction, China

301 绸缎
吉林建筑大学
Silk
Jilin Jianzhu University, China

303 一翩新驿
吉林建筑大学
A New Station
Jilin Jianzhu University, China

306 “一带一路”铁路生态园景观设计
山东建筑大学
The Design of the Belt and Road Railway Ecological Park
Shandong Jianzhu University, China

309 高铁啤酒博物馆还原设计
山东建筑大学
Restoration Design for the High-Speed Railway Beer Museum
Shandong Jianzhu University, China

312 古韵雁塔——基于海量点云数据的大雁塔三维重建
西安科技大学
Dayan Pagoda
Xi’an University of Science and Technology

其他作品 Other Works

数字化建筑设计

Category A: Architectural Design

设有客运枢纽站的索尔塔瓦拉多功能社区中心

Multifunctional Community Center with a Passenger Terminal in Sortavala

参赛学校 University/College
俄罗斯圣彼得堡国立建筑大学
St. Petersburg State University of Architecture and Civil Engineering, Russia

指导教师 Supervisor(s)
Olga Kokorina
Dmitrii Zinenkov

参赛学生 Participant(s)
Veronika Andriyenko

简介 Description

索尔塔瓦拉是俄罗斯卡累利阿共和国的一个城市，位于拉多加湖北岸，距离圣彼得堡 270 公里，距离芬兰不到 60 公里。这座城市位于 16 到 19 世纪的贸易路线沿线，是与圣彼得堡和芬兰交流的大型古代贸易中心。1893 年，连接索尔塔瓦拉、维堡和约恩苏的铁路开通，自此，大量游客开始涌入索尔塔瓦拉。

铁路是索尔塔瓦拉发展的强大动力，但铁路分隔了城市区域，因而在城市结构中成为“异物”。

建筑项目的地点位于历史城区街道网中，四通八达的街道汇聚在一起时，一个三角形的广场便形成了，这是索尔塔瓦拉的城市特色。三角形广场因而成为重要的城市中心。我们根据道路的延展方向，使铁路划分而成的城市各个部分形成一个包含市中心、路堤和公园的综合区域。

综合区域能实现以下三种功能：交通（包括售票处、行政场所、行李寄存处、通往站台的道路、候车室）、休闲（包括餐厅、健身中心）和工作（包括联合办公地点、图书馆、会议室）。

Sortavala is a city in the Republic of Karelia of the Russian Federation, located on the northern shore of Lake Ladoga, 270km from St. Petersburg and less than 60km from Finland. Located along the trade routes of the 16th-19th centuries, the city was a large ancient trade center for communication with St. Petersburg and Finland. The influx of tourists was facilitated by the opening in 1893 of a railway link between Sortavala, Vyborg and Joensuu.

The railways became a powerful engine for the subsequent development of the city, however, they took the place of “foreign bodies” in its structure, came to be the dividers of the city.

The design site is located within the historic street grid. When the directions of the streets were combined, a triangular square which is characteristic of the city of Sortavala was formed. The triangular square becomes an important urban core. By means of tracing the pedestrian directions, parts divided by the railway of the city form a unified system consisting of the city center, the embankment, and the park.

The function of the complex is divided into 3 groups: transport (ticket offices, administrative premises, left-luggage offices, transition to the platform, waiting rooms), leisure (restaurants, fitness centers), work (coworking places, a library, conference rooms).

MULTIFUNCTIONAL COMMUNITY CENTER

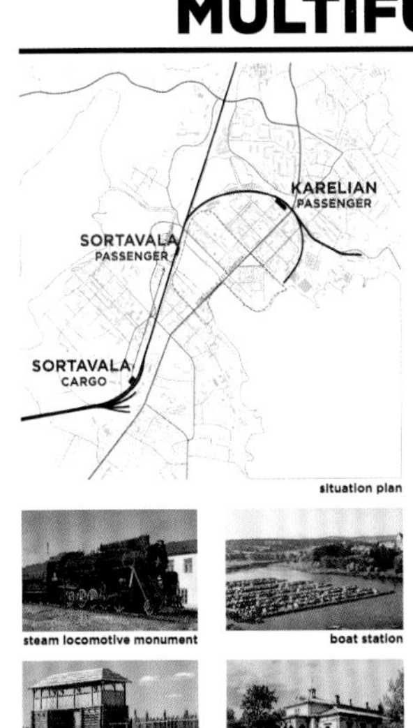

situation plan

Sortavala is a city in the Republic of Karelia of the Russian Federation, located on the northern shore of Lake Ladoga, 270 km from St. Petersburg and less than 60 km from Finland. The first mention of the settlement dates back to 1468.

Located along the trade routes of the 16th-19th centuries, the city was a large ancient trade center for communication with St. Petersburg and Finland. The influx of tourists was facilitated by the opening in 1893 of a railway link that connected the city of Sortavala with the city of Vyborg and the city of Joensuu.

The railways became a powerful engine for the subsequent development of the city, however, they took the place of "foreign bodies" in its structure, the dividers of the previously unified city, creating problems for the city for many subsequent decades.

Sortavala station is a double-track passenger railway station, carrying out more than 20 passenger flights daily (more than 3,000 passengers). The locomotive depo and cargo station (daily loading and unloading of 50 trains or more) are a little further from the passenger station.

The transport infrastructure, deployed mainly in 1950-1970, is largely worn out, physically and morally obsolete. All this requires the urgent implementation of systemic measures for its reconstruction, modernization and development.

Also, the weaknesses of the city of sortavala can be called the low popularity of the Sortavala brand, the lack of landscaped areas, the aging of the population, the lack of modern community centers.

cultural heritage

steam locomotive monument | boat station | Vakkolahti bay | Vakkosalmi park

historical park «Bastion» | town hall | Sortavala museum | fire station

church of John the Evangelist | publishing house | lyceum | Sortavala Forestry

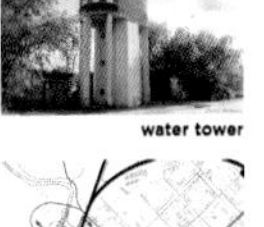

house of merchant siytonen | church of St. Nicholas | house of Leander | water tower

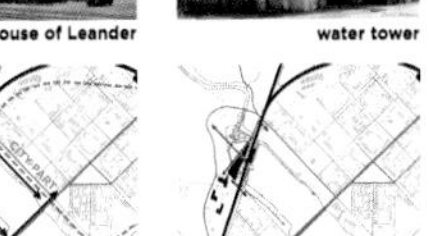
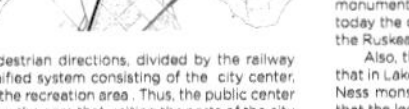

The design site is located within the historic street grid. When the directions of the streets were combined, a triangular square which is characteristic of the city of Sortavala was formed. The triangular square, outgoing parties to the Vakkolahti bay, to the Vakkosalmi park, to the recreation center, becomes an important urban core.

By means of tracing the pedestrian directions, divided by the railway forms parts of the city form a unified system consisting of the city center, the embankment of the bay, and the recreation area . Thus, the public center with a passenger terminal becomes the core that uniting the parts of the city fill the lack of buildings for sports, public and transport purposes

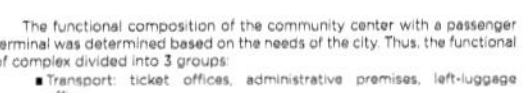

master plan of the development area 1:2500

The functional composition of the community center with a passenger terminal was determined based on the needs of the city. Thus, the functional of complex divided into 3 groups:

- Transport: ticket offices, administrative premises, left-luggage offices,
  transition to the platform, waiting rooms
- Leisure: restaurant, fitness center
- Work: coworking, library, conference room

The town of Sortavala is known for its sights. One of them is a monument to a locomotive. It is the same model of locomotive, which is also today the only daily locomotive-powered train in Russia, that makes trips to the Ruskeala mountain park in Karelia. The city of Sortavala its final stop.

Also, the people of Karelia have a rich folklore. One of the legends says that in Lake Ladoga near the city of Sortavala in Vakkolahti bay live the Loch Ness monster - the water dragon. Ethnographers, on the contrary,believe that the legendary Sadko of the painting by the famous Russian artist Repin visited the Ladoga underwater kingdom, where he met the girl Volkhova.

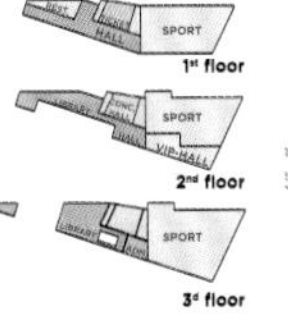

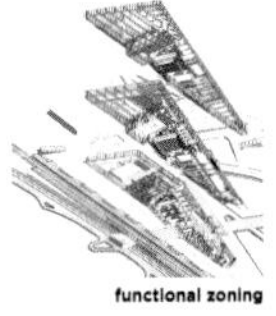

functional zoning

WITH A PASSENGER TERMINAL IN SORTAVALA

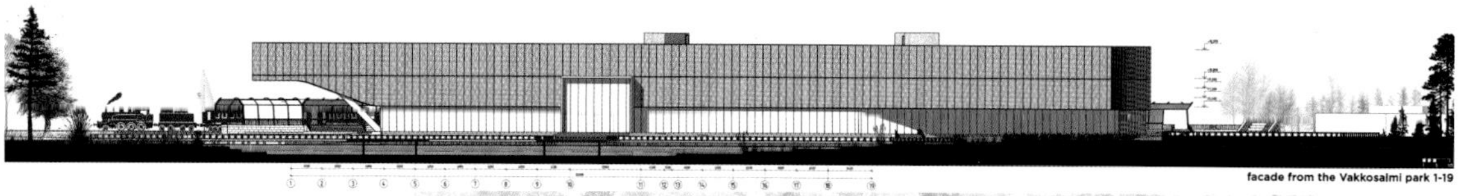

facade from the Vakkosalmi park 1-19

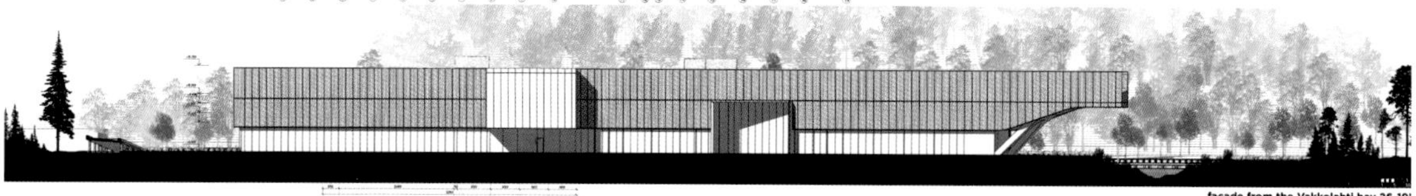

facade from the Vakkolahti bay 26-19'

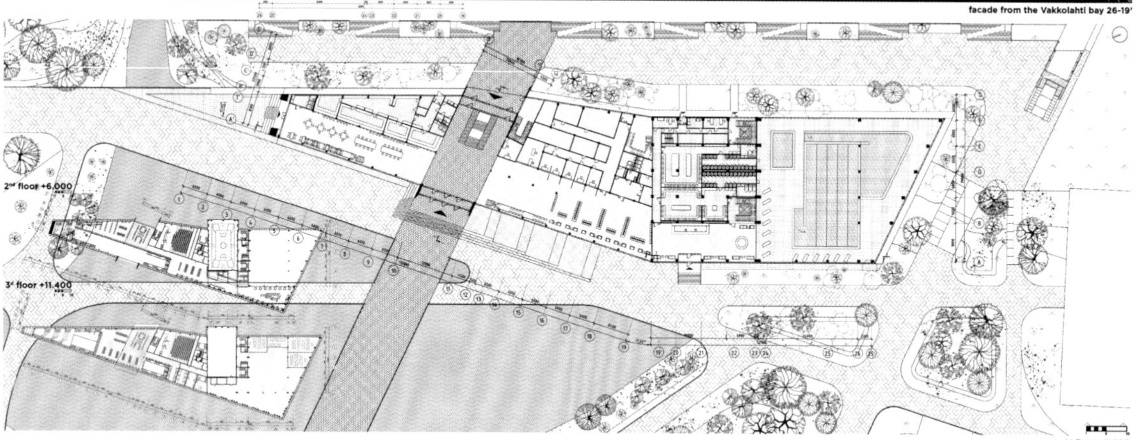

2nd floor +6.000

3rd floor +11.400

1st floor +0.600

Design features:

- Floors: 3 floors
- Structures: frame-panel system, large-span structures (steel/wooden/truss trusses), curtain wall system.
- Construction materials: reinforced concrete, steel, wood, glass.
- Color scheme: red mesh metal hinged panels (the context of the place, the environmental qualities of the environment were taken into account)
- Accessibility for visitors with limited mobility: ensured by the presence of elevators, lifts, ramps.
- Environmental justification: maximum preservation of the existing landscape and landscaping.

During the design, special attention was paid to the interiors of public areas.

In interior decoration was used typical for the city of Sortavala colors and materials - brick red color used in the solution of the hinged facade, light wood, and accent black color that was used to give the interior a graphic look.

Also, paying tribute to the Karelian-Finnish epic, was developed an art object - a red water dragon, which became a logical conclusion of the stairs leading to the observation deck with a view of the city.

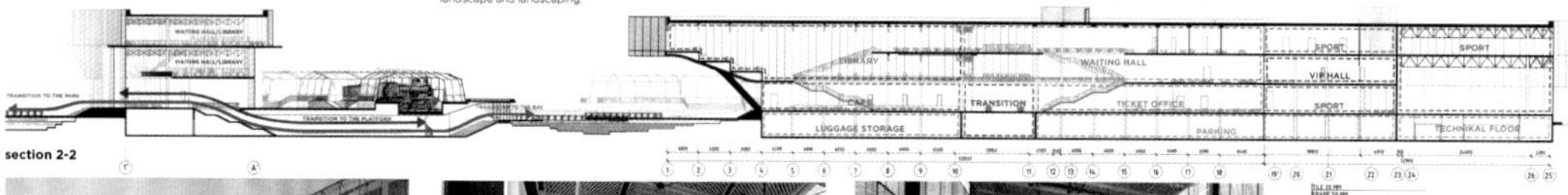

WAITING HALL/LIBRARY
TRANSITION TO THE PARK
TRANSITION TO THE PLATFORM
TRANSITION TO THE BAY
LIBRARY
WAITING HALL
SPORT
VIP HALL
CAFE
TRANSITION
TICKET OFFICE
LUGGAGE STORAGE
PARKING
TECHNIKAL FLOOR

section 2-2

section 1-1

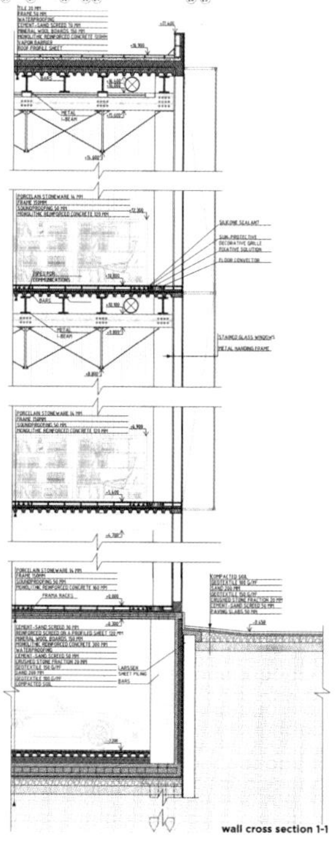

wall cross section 1-1

圣彼得堡市库罗特尼区佩索赫尼村的火车站设计

Design Project of the Railway Station in the Village of Pesochny in the Kurotny District of St.Petersburg

参赛学校 University/College
俄罗斯圣彼得堡国立建筑大学
St. Petersburg State University of Architecture and Civil Engineering, Russia

指导教师 Supervisor(s)
李晓东 LI Xiaodong
Vladimir Kuzmich

参赛学生 Participant(s)
宋万里 SONG Wanli
王林玉 WANG Linyu
李彦波 LI Yanbo
胡国庆 HU Guoqing
王子维 WANG Ziwei

简介 Description

原车站位于圣彼得堡市库洛特尼区佩索赫尼村，因为村庄的发展，原车站已无法满足人们的使用需求，需要重新设计。整个建筑以丝带为灵感来源，从视觉上连接了南北两个区域。通过抬升建筑体量从而留出更多的广场和灰空间供人们使用。

地下通道连接了南北广场，村民也可以通过地下通道方便地到达森林。基地附近的区域缺少公共空间，于是在场地东侧设计了景观广场。建筑表皮通过参数化开窗，让候车的人们更好地观赏两边景色。

The original station is located in Pesohny village, Kulotny District, St. Petersburg in Russia. Due to the development of the village, the original station can no longer meet the needs of people, so a renovation became necessary. The entire building was designed based on the image of a ribbon, the north and south areas are visually connected. By raising the volume of the building, squares are available and gray spaces are left for people to use. The underground passage connects the north and south squares, through which villagers can easily reach the forest. The area near the site lacks public space, so a plaza landscape was designed on the east side. Parametric windows on the surface of the building allow people to appreciate the scenery on both sides while waiting for the trains.

圣彼得堡市库罗特尼区佩索赫尼村的火车站设计

Design Project of the Railway Station in the Village of Pesochny in the Kurotny District of St.Petersburg

项目概况 Project summary

原车站 "Pesochny" (Песочная станция) 位于俄罗斯联邦城市圣彼得堡Kurortny区的村镇Pesochny (Песочная) 内，该村镇成立于1902年。时至今日，Pesochny村是一个维护良好的别墅式村庄，一个有着悠久文化历史的村庄，是圣彼得堡人们的度假休息的良好选择。

经过车站的列车全部都是圣彼得堡芬兰火车站和维堡站之间往返的列车。原车站现已无法完全满足人们的一些功能需求，因此，我们在原车站的位置重新设计加工，以便更好的服务于人们日常的工作及休行。

The original station "Pesochny" is located in the city of St. Petersburg, Russian Federation In Kurortny district Pesochny (Песочная станция), which was founded in 1902. Today, the village of Pesochny is a well-maintained villa-style village, a village with a long cultural history and a good choice for a holiday break for the people of St. Petersburg.

Go through The trains at the station are all round-trip trains between St. Petersburg Finnish Railway Station and Vyborg Station. The original station cannot fully meet some of the functional needs of people. Therefore, we redesigned and processed the location of the original station to better serve people's daily work and rest.

村镇基础情况介绍 Town introduction

原车站场地分析 Analysis of the original station site

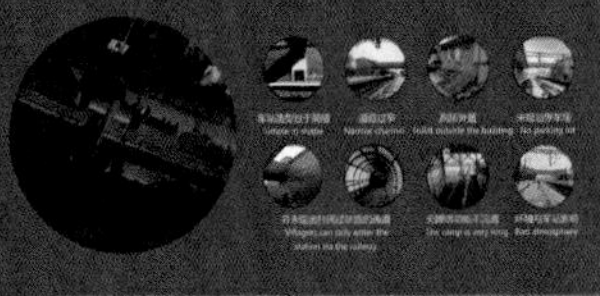

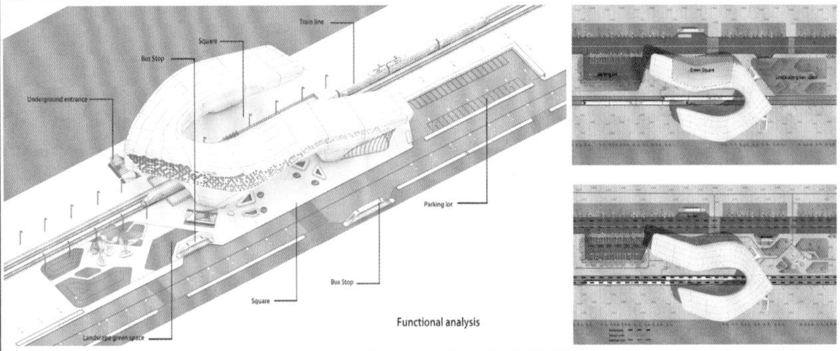

场地红线与道路规划 Building red lines and road planning

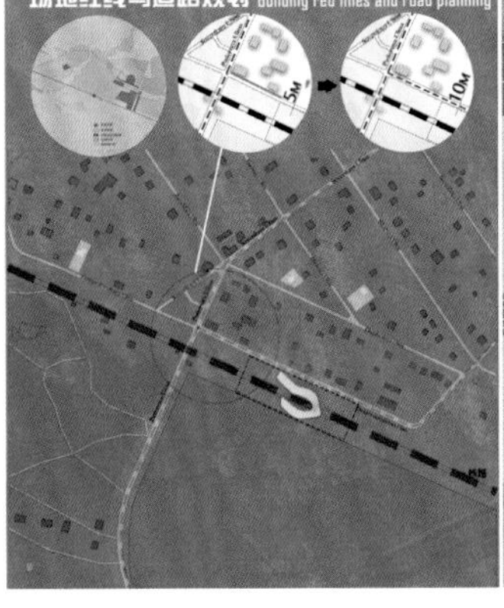

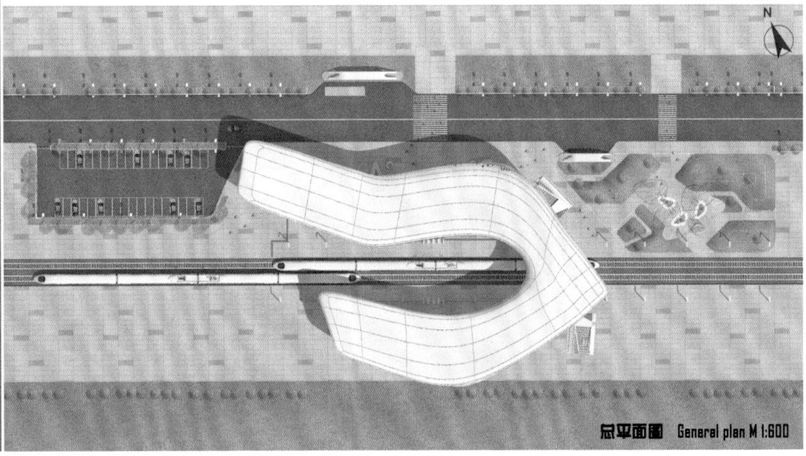

场地功能示意 Site function analysis

Functional analysis

总平面图 General plan M 1:600

圣彼得堡市库罗特尼区佩索赫尼村的火车站设计

Design Project of the Railway Station in the Village of Pesochny in the Kurotny District of St.Petersburg

设计说明 Design description

整个建筑以丝带的意向开始设计，从视觉上连接了南北两个区域。通过抬升建筑体量从而留出更多的广场和灰空间供人们使用。地下通道连接了南北广场，村民也可以通过地下通道方便的到达森林。基地附近的区域缺少公共空间，于是在场地东侧设计了景观广场。建筑表皮通过参数化开窗，让候车的人们更好的观赏两边景色。

The entire building is designed with the intention of a ribbon, which visually connects the north and south areas. By raising the building volume, more plazas and grey spaces are left for people to use. The underground passage connects the north and south squares, and villagers can also easily reach the forest through the underground passage. The area near the site lacks public space, so a landscape plaza is designed on the east side of the site. Through parametric window openings on the building skin, people waiting for the bus can better appreciate the scenery on both sides.

形体生成分析 Blok analysis

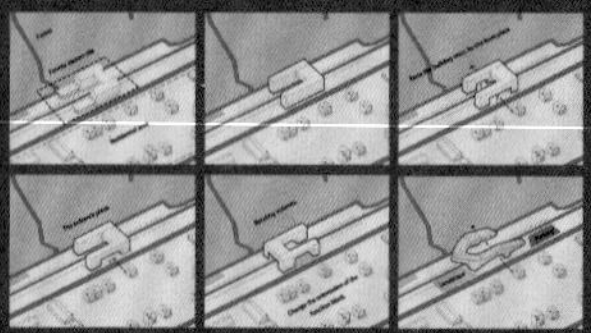

体块功能分析图 Functional analysis

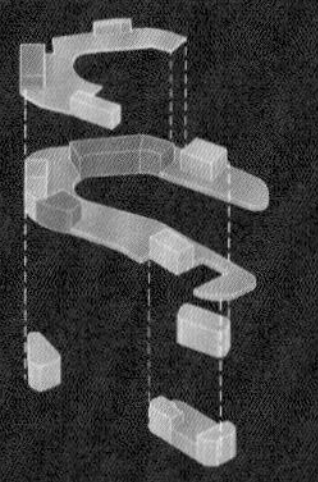

Traffic space
Waiting space
Dining space
Bathroom
Relaxation area
Office Space

建筑内流线分析图 Station Streamline analysis

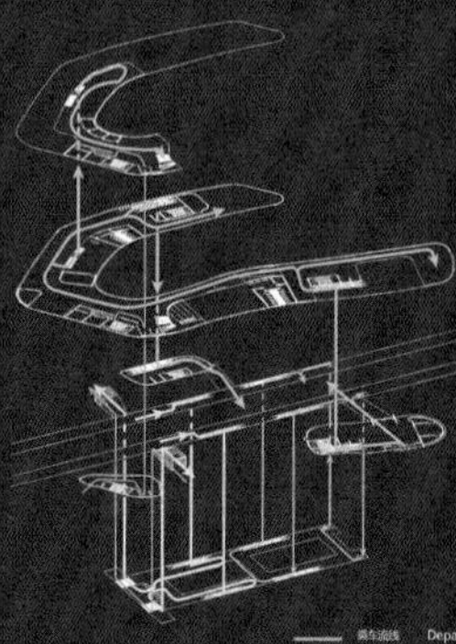

Departure
Arrival
Staff member
Only by rail

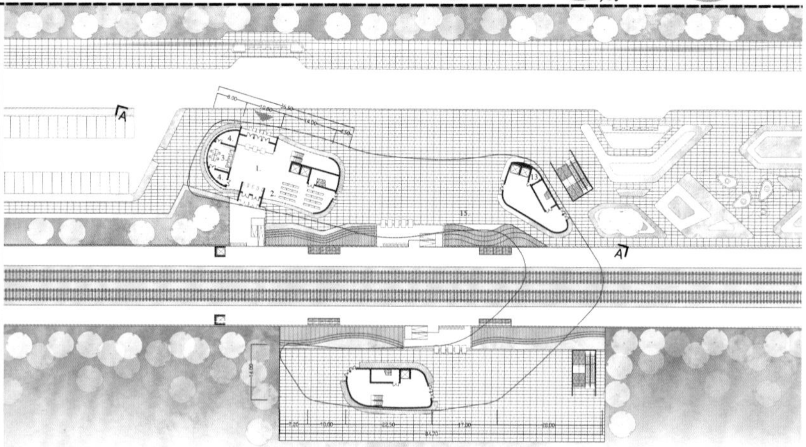

首层平面图 Ground plan M 1:500

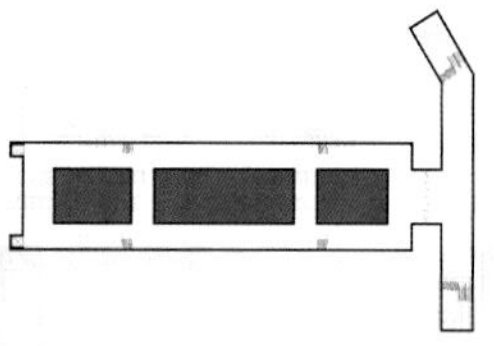

负一层平面图 -1 F plan M 1:750

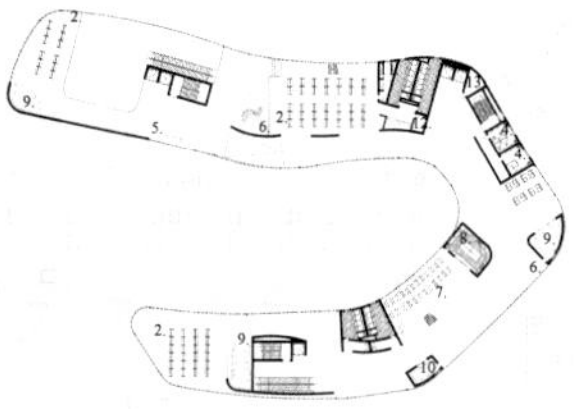

二层平面图 2 F plan M 1:750

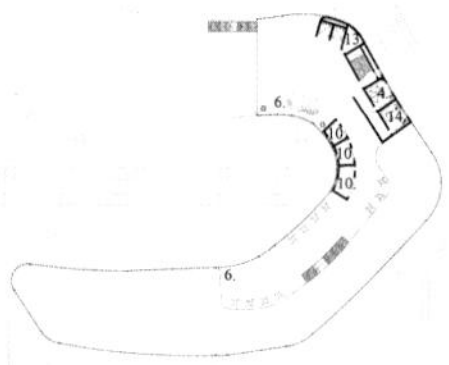

三层平面图 3 F plan M 1:750

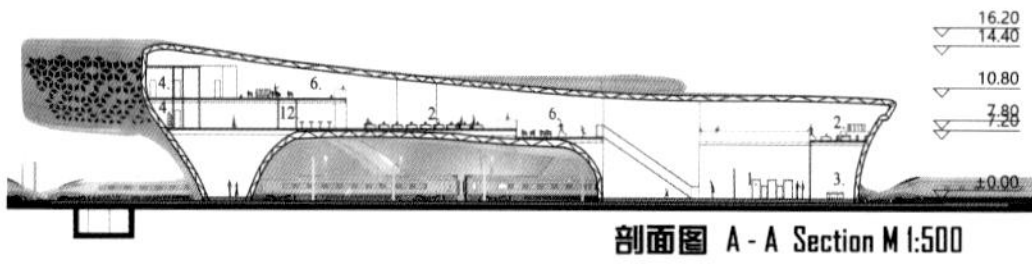

剖面图 A-A Section M 1:500

1.售票大厅 Ticket hall
2.候车室 Waiting Room
3.售票室 Box Office
4.办公室 Office
5.问讯处 Information desk
6.公共空间 Public space
7.餐厅 Restaurant
8.餐厅厨房 Dining room kitchen
9.咖啡厅 Cafe
10.小卖部 Commissary
11.母婴室 Mother and baby room
12.储藏室 Storage Room
13.配电室 Power distribution room
14.医疗室 Medical room
15.广场 Square
16.停车场 Parking lot

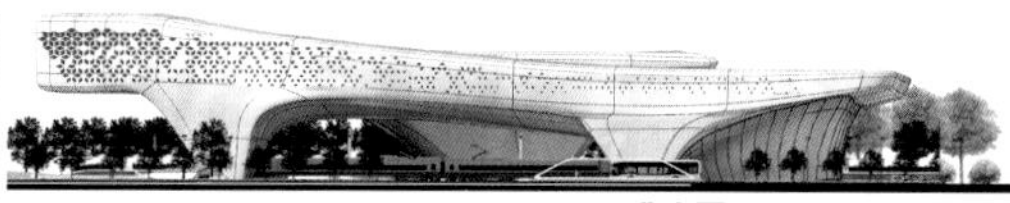

北立面 North Elevation

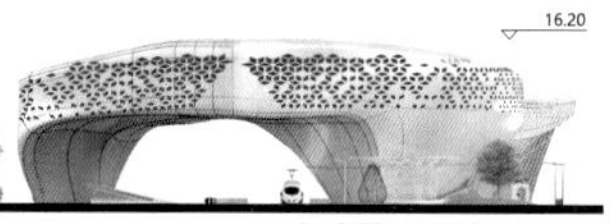

东立面 East Elevation

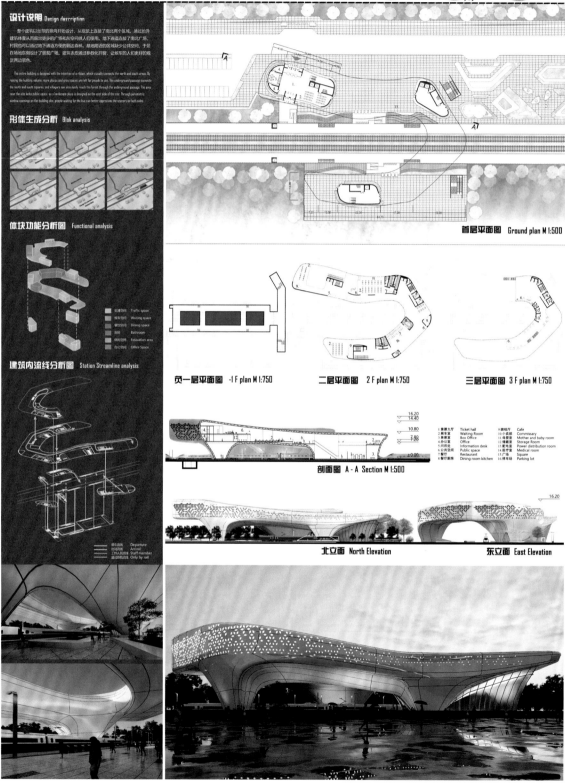

城有园 垣无向——基于民族石刻艺术的铁路客运站设计

The City has a Garden, and the Walls have no Direction—The Design of the Railway Station Based on the Art of Ethnic Stone Carvings

参赛学校 University/College
郑州大学
Zhengzhou University, China

指导教师 Supervisor(s)
段武俊 DUAN Wujun
刘芳芳 LIU Fangfang

参赛学生 Participant(s)
袁 亮 YUAN Liang
陈云祥 CHEN Yunxiang
刘吉龙 LIU Jilong
郭亚婷 GUO Yating
黄 蕊 HUANG Rui

简介 Description

“一带一路”对沿线各地区的经济发展和文化传承有着很强的推动作用，因此我们选取具有优质自然景观和文化底蕴的新密市米村为基地。米村作为玉石的生产地而广为人知，因此以玉石作为设计的出发点，建筑形体采用轴线和几何元素进行玉石切割一般的纯粹操作；内部空间采用空间句法约束、激发各个空间的行为、视线、效率等内在关系；最后采用参数化设计进行站台雨棚设计，雨棚形态既形似"伞"，体现其功能性，又形似"花"，寓意"一带一路，一路生花"。

The "Belt and Road" plays a significant role in promoting the economic development and cultural inheritance of the regions along the route, so we chose the site of the Micun Village, Xinmi City in China with the high-quality natural landscape and cultural heritage. the Micun Village is widely known as an abundant source of jade. Therefore, jade became the starting point of the design. The architectural form contains an axis and geometric elements to present the pure operation of jade cutting; the interior space uses spatial syntax constraints to stimulate the relationship between the behavior, sight and efficiency of each space. Finally, the parametric design was used in the platform canopy. The shape of the canopy is like an "umbrella" and also a "flower" to reflect its functionality and carry the connotation of "Flowers bloom alongside the Belt and Road".

Background 背景

城有园 垣无向

Design description 设计说明

Digital application analytics 石刻与数字化

Behavior simulation 行为模拟分析

总平面图 1: 1500

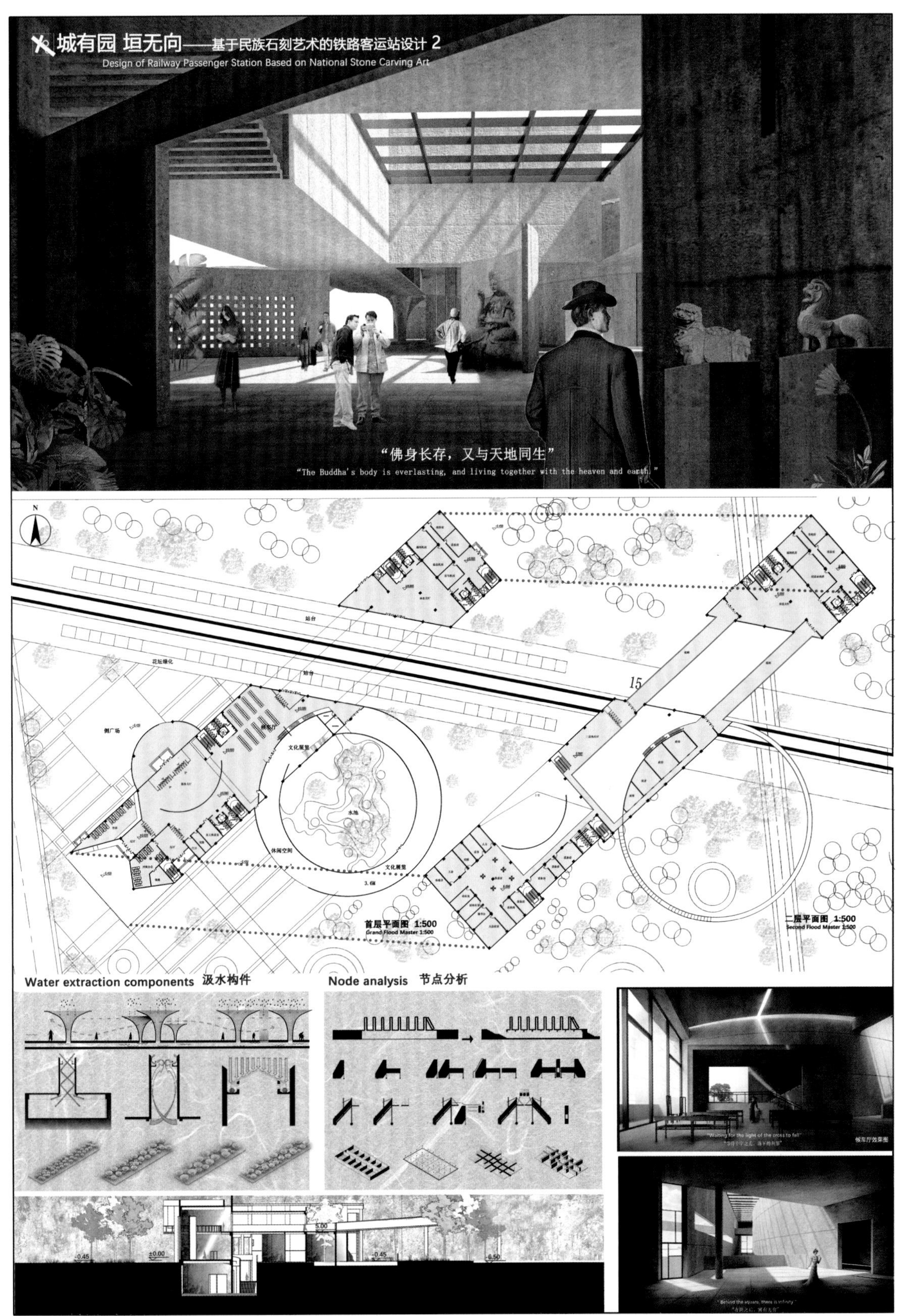
城有园 垣无向——基于民族石刻艺术的铁路客运站设计 2
Design of Railway Passenger Station Based on National Stone Carving Art
"佛身长存，又与天地同生"
"The Buddha's body is everlasting, and living together with the heaven and earth."
首层平面图 1:500
Grand Flood Master 1:500
二层平面图 1:500
Second Flood Master 1:500
Water extraction components 汲水构件
Node analysis 节点分析

三等奖 >> Third Prize

乡村引线

Rural Lead

参赛学校 University/College

北京建筑大学

Beijing University of Civil Engineering and Architecture, China

指导教师 Supervisor(s)

任中琦 REN Zhongqi

蒋　蔚 JIANG Wei

参赛学生 Participant(s)

高天石 GAO Tianshi

李梦瑶 LI Mengyao

刘世禹 LIU Shiyu

蔡依葵 CAI Yikui

简介 Description

从城市设计的角度出发，以构建车站与乡村的和谐关系为目标，采用分散式的体块策略，避免车站的巨大体量对乡村造成不利影响，并承载乡村中丰富多彩但往往无处安放的公共活动。涵盖不同功能的体块散布在村落之间，将乡村振兴的希望送入每家每户。

From the perspective of urban design, our goal is to develop a harmonious relationship between the station and the countryside by taking the strategy of decentralized blocks. This helps to avoid the adverse impact of the huge volume of the station on the countryside and accommodate those colorful activities enjoyed by the villagers which are often nowhere conducted. Blocks with different functions are scattered in the nearby villages and the hope of rural revitalization is therefore conveyed to every household.

乡村引线II
Rural Lead

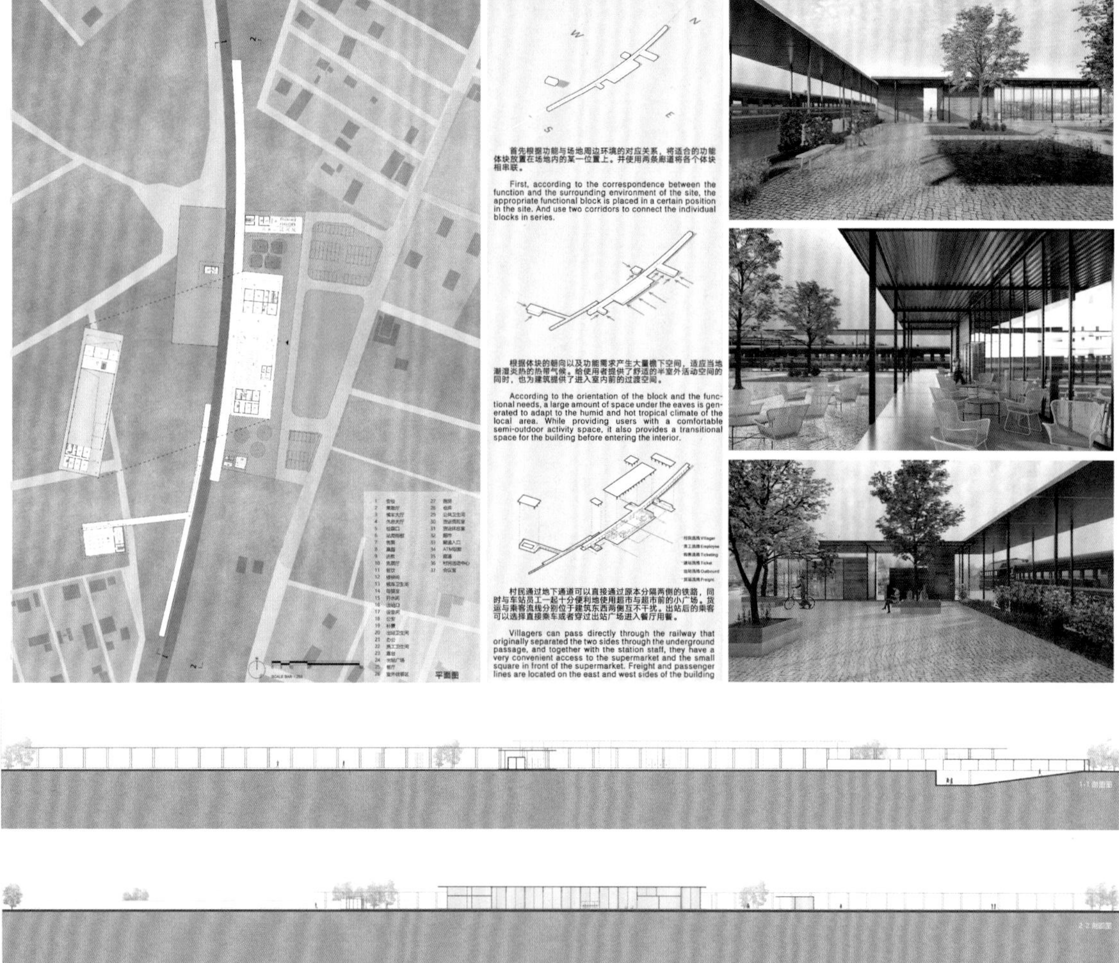
W
N
S
E
首先根据功能与场地周边环境的对应关系，将适合的功能体块放置在场地内的某一位置上。并使用两条廊道将各个体块相串联。
First, according to the correspondence between the function and the surrounding environment of the site, the appropriate functional block is placed in a certain position in the site. And use two corridors to connect the individual blocks in series.
根据体块的朝向以及功能需求产生大量檐下空间，适应当地潮湿炎热的热带气候。给使用者提供了舒适的半室外活动空间的同时，也为建筑提供了进入室内前的过渡空间。
According to the orientation of the block and the functional needs, a large amount of space under the eaves is generated to adapt to the humid and hot tropical climate of the local area. While providing users with a comfortable semi-outdoor activity space, it also provides a transitional space for the building before entering the interior.
村民通过地下通道可以直接通过原本分隔两侧的铁路，同时与车站员工一起十分便利地使用超市与超市前的小广场。货运与乘客流线分别位于建筑东西两侧互不干扰。出站后的乘客可以选择直接乘车或者穿过出站广场进入餐厅用餐。
Villagers can pass directly through the railway that originally separated the two sides through the underground passage, and together with the station staff, they have a very convenient access to the supermarket and the small square in front of the supermarket. Freight and passenger lines are located on the east and west sides of the building
平面图

双循环下 IOE

IOE under Double Circulation

参赛学校 University/College

长春建筑学院

Changchun University of Architecture and Civil Engineering, China

指导教师 Supervisor(s)

何　岩　HE Yan

张　蕾　ZHANG Lei

参赛学生 Participant(s)

高　京　GAO Jing

平亚宁　PING Yaning

计新远　JI Xinyuan

王立淇　WANG Liqi

肖　扬　XIAO Yang

简介 Description

项目位于山西省运城市（古称“河东”）盐湖区解州镇，因其地理环境特殊，孕育出“盐运文化”“关公文化”。在高速发展的信息时代，这里曾经凭借地域文化来延续地方生命力，但因地方建设无法承载未来地域发展需求，在一次次的文化红利上错失良机。基于对地方文化、旅游资源、产业结构等内容研究，我们引入“站城融合”理念，提出“双循环下 IOE”的品质提升策略，希望在城镇品质提升后建成“智慧解县”，迎接未来。

The project is located in Haizhou Town, Yanhu District, Yuncheng City (formerly known as "Hedong"), Shanxi Province in China. The special geographical environment serves as the cradle of "salt transportation culture" and "Guangong culture", which used to be the source of the local growth. However, in the fast-developing information era, the infrastructure of this region fails to satisfy the needs of development, losing opportunities that could have been seized with its profound cultural heritage again and again. Based on our research on local cultures, tourism resources, industrial structure, etc., we introduced the concept of "station-city integration" and put forward the strategy of "double cycle-IOE" for the quality updating work of the county. With the newly designed station, Haizhou will become a "smart town" that can better adapt to the future.

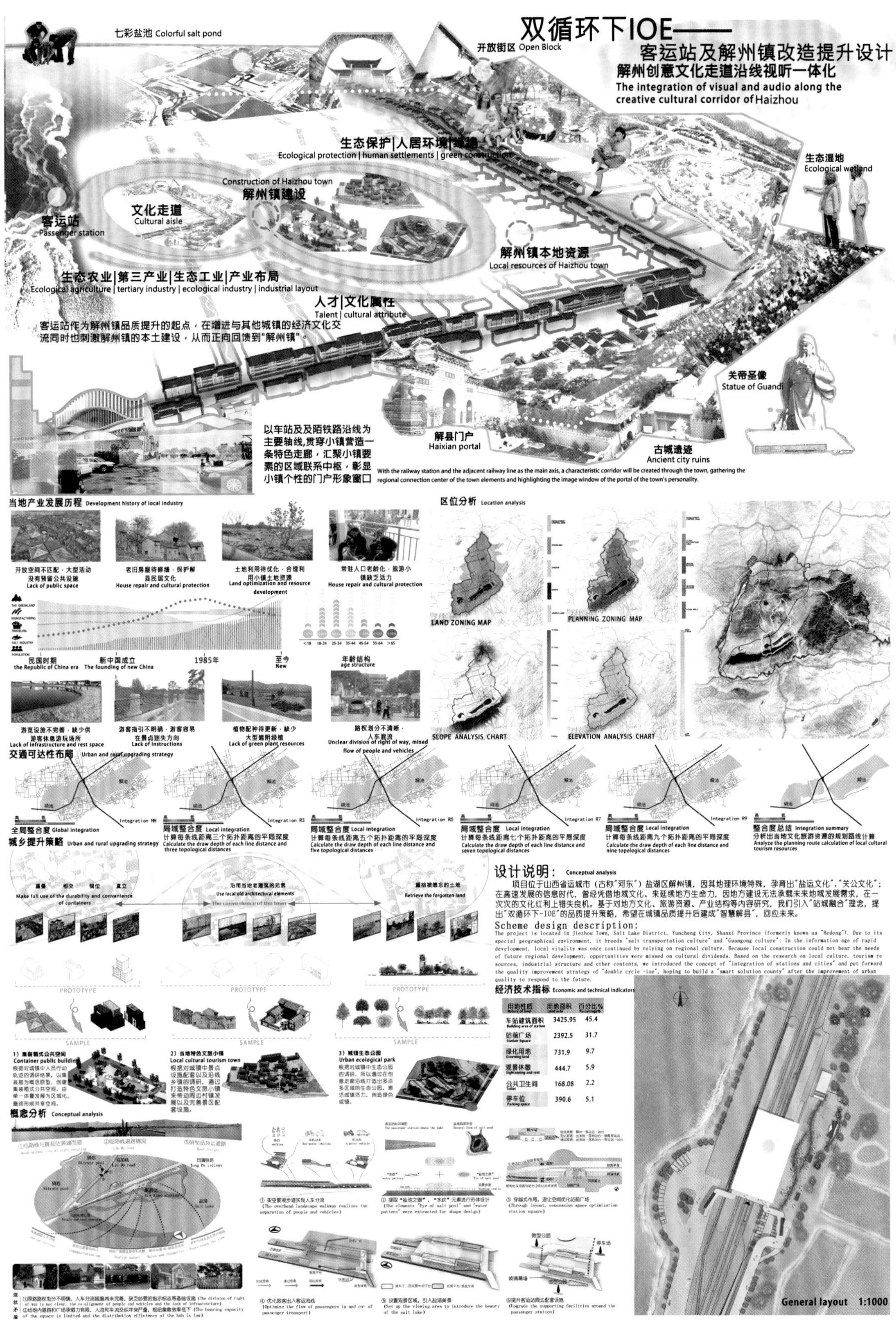

用地性质 Nature of land	用地面积 Land area	百分比% Percentage%
车站建筑面积 Building area of station	3425.95	45.4
站前广场 Station Square	2392.5	31.7
绿化用地 Greening land	731.9	9.7
观景休憩 Sightseeing and rest	444.7	5.9
公共卫生间 Toilet	168.08	2.2
停车位 Parking space	390.6	5.1

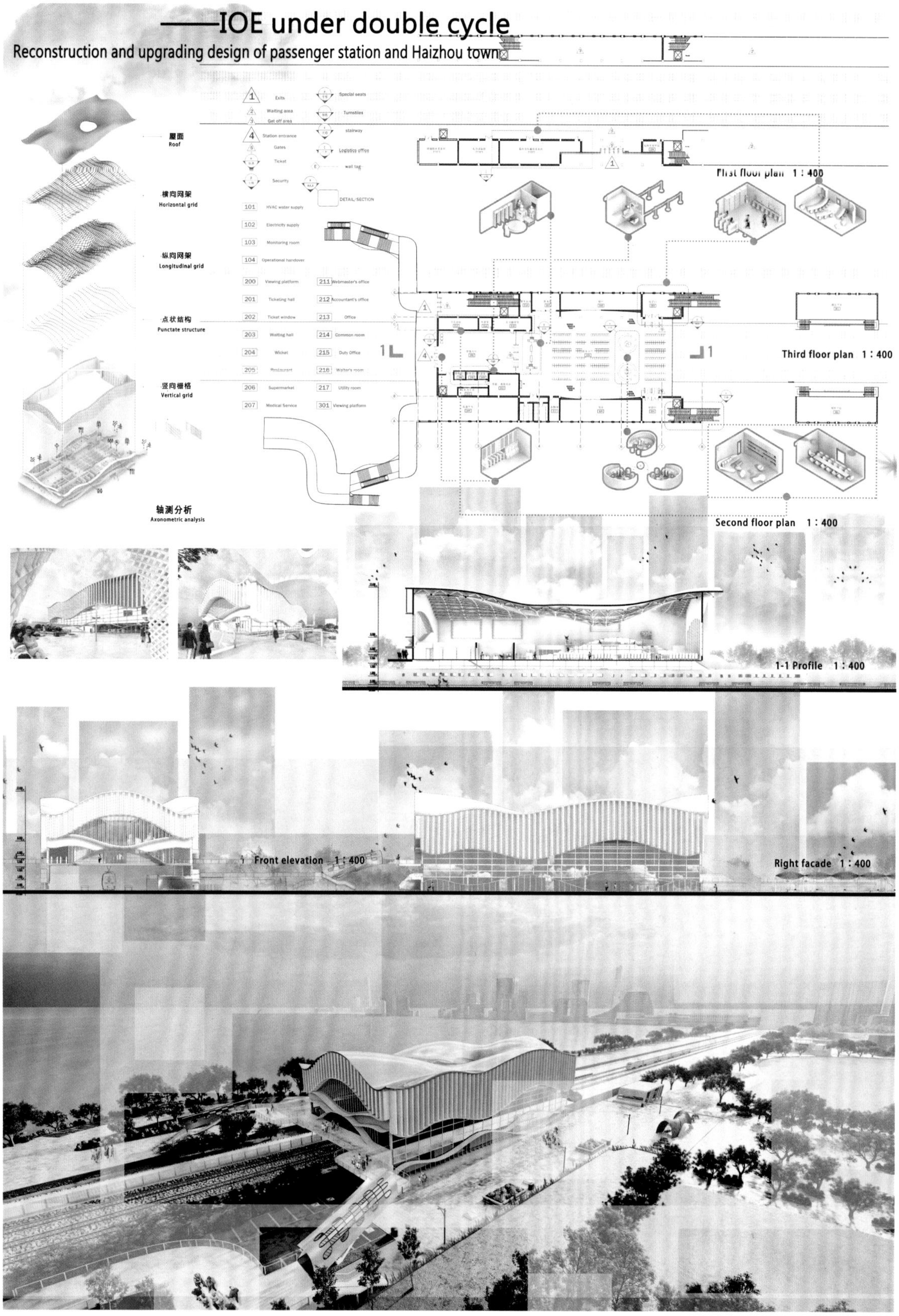

——IOE under double cycle

Reconstruction and upgrading design of passenger station and Haizhou town

屋面
Roof

横向网架
Horizontal grid

纵向网架
Longitudinal grid

点状结构
Punctate structure

竖向栅格
Vertical grid

轴测分析
Axonometric analysis

1 Exits
2 Waiting area
3 Get off area
4 Station entrance
5 Gates
Ticket
Security
Special seats
Turnstiles
stairway
Logistics office
wall tag
DETAIL/SECTION

101	HVAC water supply		
102	Electricity supply		
103	Monitoring room		
104	Operational handover		
200	Viewing platform	211	Webmaster's office
201	Ticketing hall	212	Accountant's office
202	Ticket window	213	Office
203	Waiting hall	214	Common room
204	Wicket	215	Duty Office
205	Restaurant	216	Waiter's room
206	Supermarket	217	Utility room
207	Medical Service	301	Viewing platform

First floor plan 1∶400

Third floor plan 1∶400

Second floor plan 1∶400

1-1 Profile 1∶400

Front elevation 1∶400

Right facade 1∶400

丁丁旅行记——城市触媒视角下的布鲁塞尔 Etterbeek 火车站建筑与城市更新设计

The Adventures of Tintin—Architecture and Urban Renewal Design of Brussels Etterbeek Railway Station from the Perspective of Urban Catalyst

参赛学校 University/College

重庆交通大学

Chongqing Jiaotong University, China

指导教师 Supervisor(s)

许　可　XU Ke

温　泉　WEN Quan

参赛学生 Participant(s)

孙思可　SUN Sike

宋澄怡　SONG Chengyi

夏梦琴　XIA Mengqin

周福星　ZHOU Fuxing

郭书伶　GUO Shuling

简介 Description

设计选址位于比利时布鲁塞尔自由大学片区，主要内容为以 Etterbeek 火车站设计为核心，并进行车站周边的城市更新设计以复兴城市历史街区。设计概念以比利时漫画“丁丁历险记”作为文化主线，实现现代与历史的融合。在火车站设计中，提取当地哥特建筑的尖顶、扶壁以及红砖元素，并以高线公园缝合新旧街区。在城市设计中，以“一轴 · 三线 · 四区”为结构，重新整合该片区的城市功能。在这里以丁丁为缘，冒险无处不在，艺术永远年轻。

The site of the design is located in the area near the Free University of Brussels in Belgium. The design is centered around the Etterbeek Railway Station, also including the urban renewal of its surroundings to revitalize of historical districts of the city. The design drew the cultural element from the Belgium comic, the Adventures of Tintin to achieve an integration of the present with the past. In conceptualizing the design of the station, the elements of Belgian Gothic architecture, spires, flying buttresses, and red brick were all adopted and the High Line Park serves to close the gap between the new and old districts. The design of the city takes the structure of “one spindle, three lines, four sections” to re-integrate the urban functions of this region. Here, let us follow Tintin to enjoy adventures everywhere and keep the art of architecture young forever.

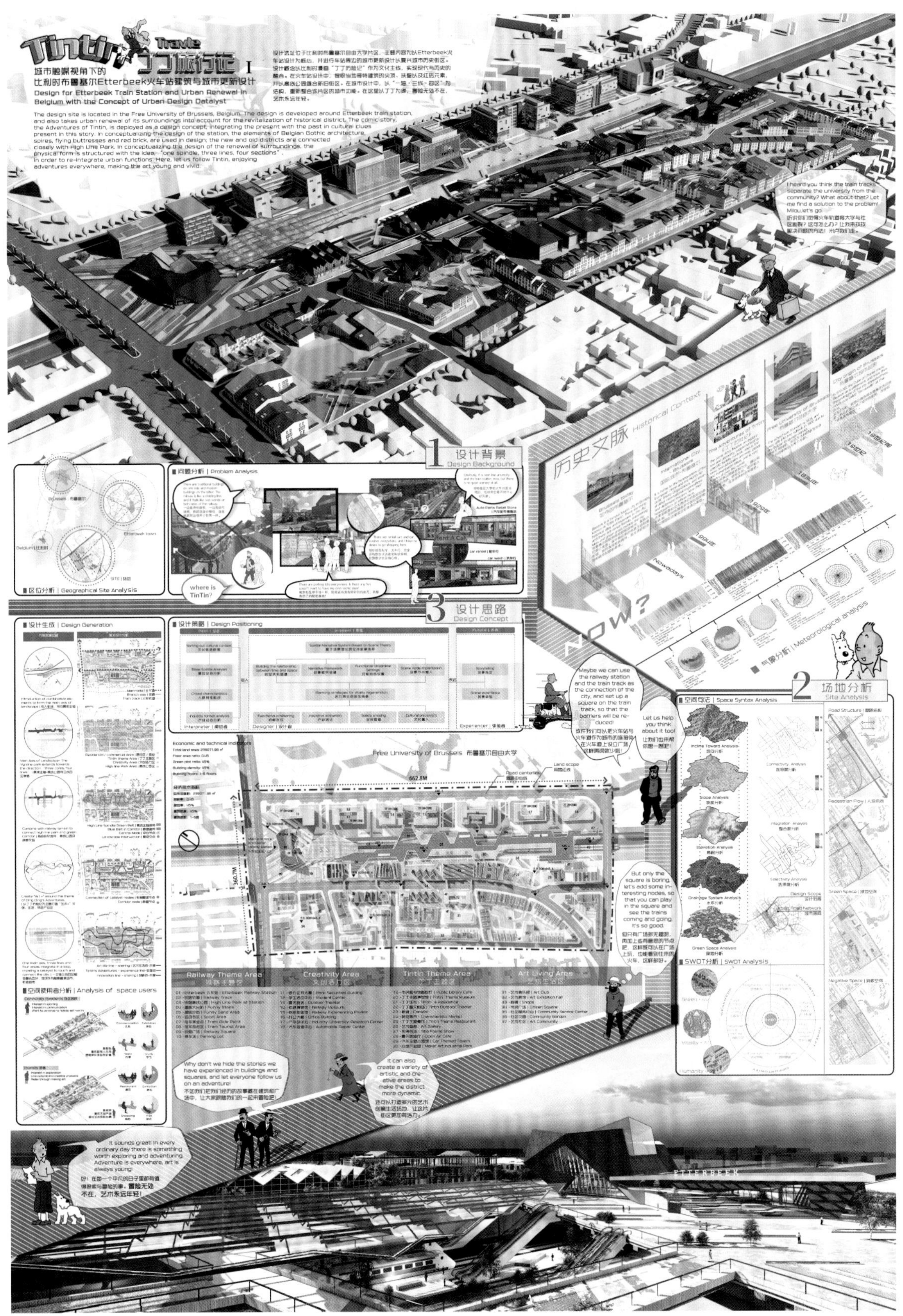
Tintin Travle 丁丁旅行记 I
城市触媒视角下的
比利时布鲁塞尔Etterbeek火车站建筑与城市更新设计
Design for Etterbeek Train Station and Urban Renewal in Belgium with the Concept of Urban Design Catalyst
设计场地位于比利时布鲁塞尔自由大学片区，主要内容为以Etterbeek火车站设计为核心，并进行车站周边的城市更新设计以复兴城市历史街区。设计概念以比利时漫画"丁丁历险记"作为文化主线，实现现代与历史的融合。在火车站设计中，提取哥特建筑的尖顶、扶壁以及红色元素，并以高线公园连接新旧街区。在城市设计中，以"一轴、三线、四区"为骨架，重新整合该片区的城市功能。在这里以丁丁为伴，冒险无处不在，艺术永远年轻。
The design site is located in the Free University of Brussels, Belgium. The design is developed around Etterbeek train station, and also takes urban renewal of its surroundings into account for the revitalization of historical district. The comic story, the Adventures of Tintin, is deployed as a design concept, integrating the present with the past in cultural clues present in this story. In conceptualizing the design of the station, the elements of Belgian Gothic architecture, spires, flying buttresses and red brick, are used in design; the new and old districts are connected closely with High Line Park. In conceptualizing the design of the renewal of surroundings, the physical form is structured with the idea: "one spindle, three lines, four sections", in order to re-integrate urban functions. Here, let us follow Tintin, enjoying adventures everywhere, making the art young and vivid.
I heard you think the train tracks separate the university from the community? What about that? Let me find a solution to the problem! Milou, let's go.
听说你们觉得火车轨道将大学与社区割裂？这可怎么办？让我来找找解决问题的方法！米卢我们走。
历史文脉 Historical Context
Nowadays
NOW?
气象分析 | Meteorological analysis
1 设计背景 Design Background
问题分析 | Problem Analysis
where is TinTin?
区位分析 | Geographical Site Analysis
3 设计思路 Design Concept
设计生成 | Design Generation
设计策略 | Design Positioning
Interpreter | 解说者
Designer | 设计者
Experiencer | 体验者
Maybe we can use the railway station and the train track as the connection of the city, and set up a square on the train track, so that the barriers will be reduced!
或许我们可以把火车站与火车道作为城市的连接体，在火车道上设立广场，这样隔阂就少啦！
Let us help you think about it too!
让我们也来帮你想一想吧！
2 场地分析 Site Analysis
空间句法 | Space Syntax Analysis
Incline Toward Analysis 坡向分析
Slope Analysis 坡度分析
Elevation Analysis 高程分析
Drainage System Analysis 水系分析
Green Space Analysis 绿地分析
Connectivity Analysis 连接度分析
Integration Analysis 整合度分析
Selectivity Analysis 选择度分析
Design Scope 设计范围
Urban Road Network 城市路网
Road Structure | 道路结构
Pedestrian Flow | 人流流向
Green Space | 绿地空间
Negative Space | 消极空间
SWOT分析 | SWOT Analysis
Free University of Brussels 布鲁塞尔自由大学
662.8M
Land scope 用地红线
Economic and technical indicators
But only the square is boring, let's add some interesting nodes, so that you can play in the square and see the trains coming and going. It's so good.
但只有广场多无趣呀，再加上些有意思的节点吧，这样既可以在广场上玩，也能看到往来的火车，这样多好。
Railway Theme Area 铁路主题区
Creativity Area 文创活力区
Tintin Theme Area 丁丁主题区
Art Living Area 艺术生活区
空间使用者分析 | Analysis of space users
Why don't we hide the stories we have experienced in buildings and squares, and let everyone follow us on an adventure!
不如我们把我们经历的故事藏在建筑和广场中，让大家跟随我们的一起来冒险吧！
It can also create a variety of artistic and creative areas to make the district more dynamic.
还可以打造多元的艺术创意生活场地，让这片街区更加有活力。
It sounds great! In every ordinary day there is something worth exploring and adventuring. Adventure is everywhere, art is always young!
好！在每一个平凡的日子里都有值得探索与冒险的事。冒险无处不在，艺术永远年轻！
ETTERBEEK

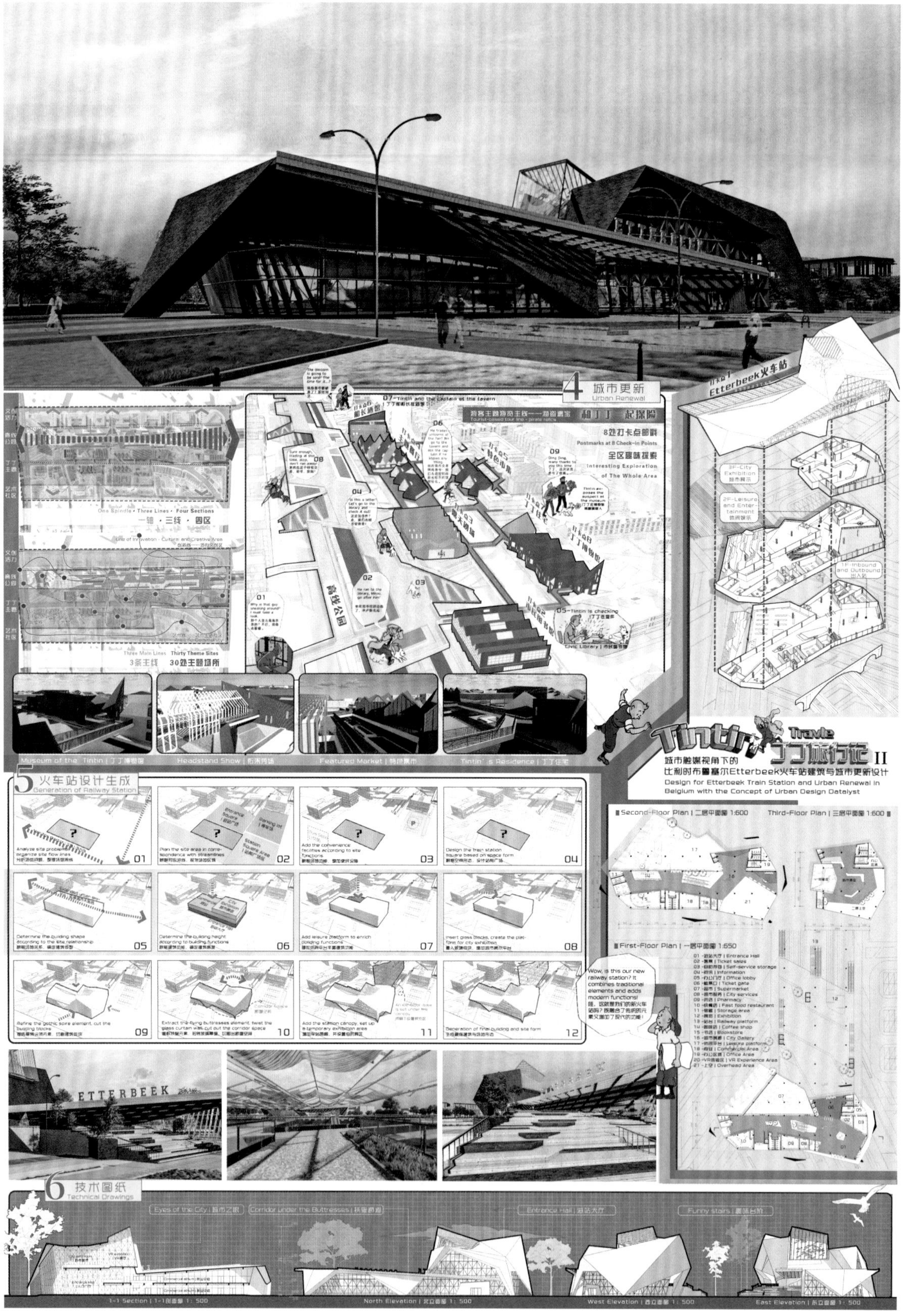

4 城市更新 Urban Renewal

游客主题游览主线——海盗遗宝 Tourist-based tour line - pirate relics

和丁丁一起探险

8处打卡点邮戳 Postmarks at 8 Check-in Points

全区趣味探索 Interesting Exploration of The Whole Area

One Spindle · Three Lines · Four Sections

一轴 · 三线 · 四区

Line of Innovation - Cultural and Creative Area

创新线——活力文创区

Three Main Lines　Thirty Theme Sites

3条主线　30处主题场所

07-Tintin and the captain at the tavern

05-Tintin is checking

Civic Library | 市民图书馆

Museum of the Tintin | 丁丁博物馆　Headstand Show | 街头秀场　Featured Market | 特色集市　Tintin's Residence | 丁丁住宅

Tintin Travle 丁丁旅行记 II

城市触媒视角下的
比利时布鲁塞尔Etterbeek火车站建筑与城市更新设计

Design for Etterbeek Train Station and Urban Renewal in Belgium with the Concept of Urban Design Datalyst

5 火车站设计生成 Generation of Railway Station

01 Analyze site problems, organize site flow lines

02 Plan the site area in correspondence with streamlines

03 Add the convenience facilities according to site functions

04 Design the train station square based on space form

05 Determine the building shape according to the site relationship

06 Determine the building height according to building functions

07 Add leisure platform to enrich building functions

08 Insert glass blocks, create the platform for city exhibition

09 Refine the gothic spire element, cut the building blocks

10 Extract the flying buttresses element, twist the glass curtain wall, cut out the corridor space

11 Add the station canopy, set up a temporary exhibition area

12 Generation of final building and site form

Second-Floor Plan | 二层平面图 1:600　Third-Floor Plan | 三层平面图 1:600

First-Floor Plan | 一层平面图 1:650

01 -进站大厅 | Entrance Hall
02 -售票 | Ticket sales
03 -自助存包 | Self-service storage
04 -问讯 | Information
05 -办公门厅 | Office lobby
06 -检票口 | Ticket gate
07 -超市 | Supermarket
08 -城市服务 | City services
09 -药店 | Pharmacy
10 -快餐店 | Fast food restaurant
11 -储藏 | Storage area
12 -展厅 | Exhibition
13 -站台 | Railway platform
14 -咖啡店 | Coffee shop
15 -书店 | Bookstore
16 -城市展廊 | City Gallery
17 -休闲平台 | Leisure platform
18 -商业 | Commercial Area
19 -办公区域 | Office Area
20 -VR体验区 | VR Experience Area
21 -上空 | Overhead Area

Wow, is this our new railway station? It combines traditional elements and adds modern functions!
哇，这就是我们的新火车站吗？既融合了传统的元素又增加了现代的功能！

6 技术图纸 Technical Drawings

Eyes of the City | 城市之眼　Corridor under the Buttresses | 扶壁廊道　Entrance Hall | 进站大厅　Funny stairs | 趣味台阶

1-1 Section | 1-1剖面图 1:500　North Elevation | 北立面图 1:500　West Elevation | 西立面图 1:500　East Elevation | 东立面图 1:500

融与荣

Integration and Prosperity

参赛学校 University/College

河北建筑工程学院

Hebei University of Architecture, China

指导教师 Supervisor(s)

班磊晶 BAN Leijing

曹迎春 CAO Yingchun

参赛学生 Participant(s)

曹　萍 CAO Ping

陈晓玲 CHEN Xiaoling

简介 Description

兰州，自古就是“座中连六”的丝路枢纽和商埠重镇，是一个多民族文化交融的城市。因此，火车站设计理念提取当地的古建筑文化，在建筑外形上，利用古代建筑的元素进行现代的转译，打造成具有地域特色的现代火车站。此外，复合功能空间满足了周边多人群的活动需求，为激活黄河滨江带景观创造了契机。建筑作为“城镇客厅”，将吸引更多的外地游客，带动周边经济、文化、生态共同发展。

“Sitting at the center and connected with six cities”, Lanzhou, a city in northwest China has been a hub on the Silk Road and an important town for business since ancient times, boasting diversified cultures of different ethnic groups. Therefore, the designed railway station extracts the ancient architectural culture of Lanzhou. The exterior of the building is a transcreation of the elements in ancient architecture so that the building is a modern one with local characteristics. In addition, a space with multiple functions meets the needs of different people in this area. Such a station serving as the “living room of the city” creates an opportunity for tapping into the landscape of the Yellow River-side belt, attracting more tourists from other cities and fueling the development of the economy, culture, and environment of this region.

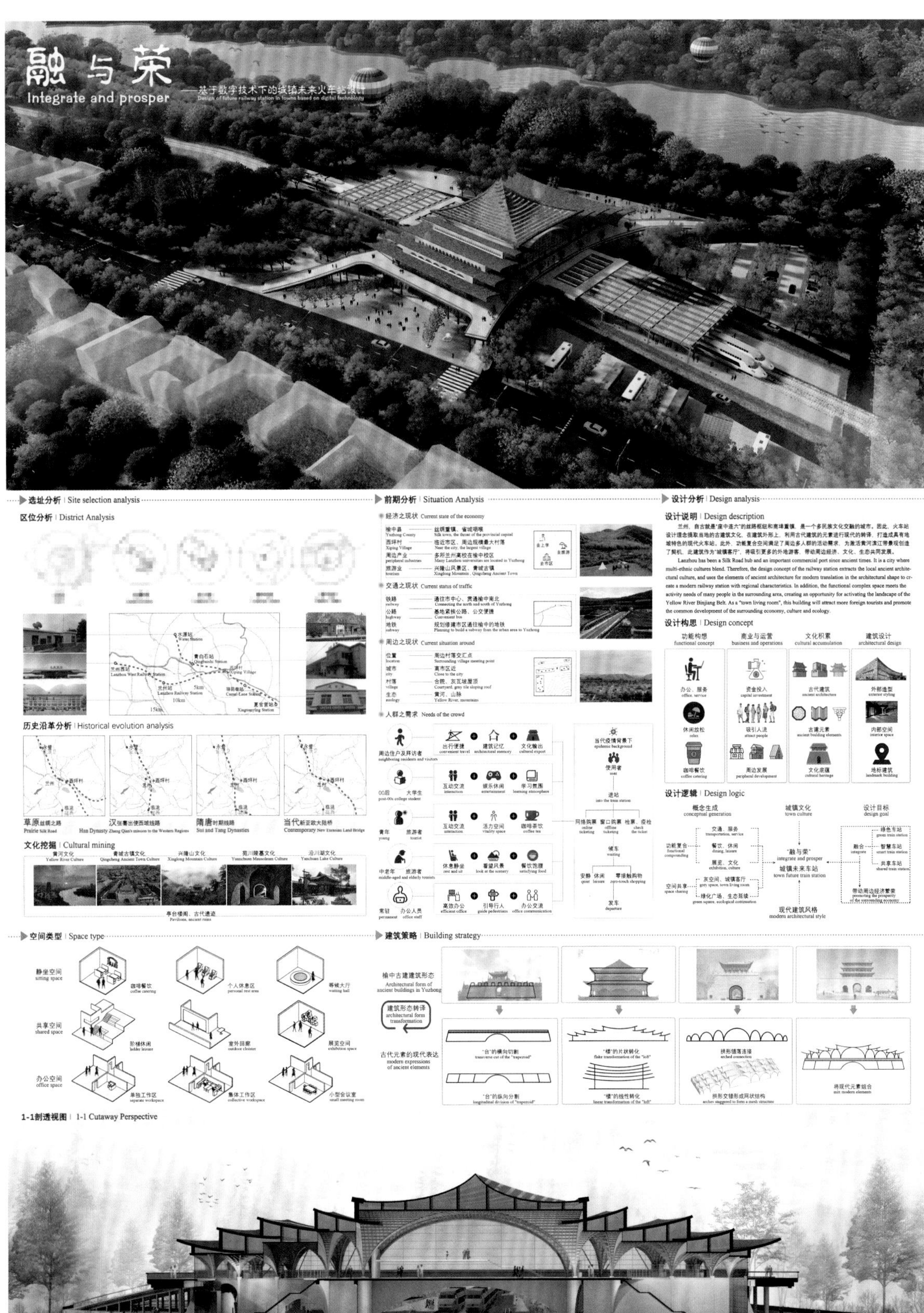
融与荣
Integrate and prosper
——基于数字技术下的城镇未来火车站设计
选址分析 | Site selection analysis
区位分析 | District Analysis
历史沿革分析 | Historical evolution analysis
文化挖掘 | Cultural mining
前期分析 | Situation Analysis
经济之现状 Current state of the economy
交通之现状 Current status of traffic
周边之现状 Current situation around
人群之需求 Needs of the crowd
设计分析 | Design analysis
设计说明 | Design description
设计构思 | Design concept
设计逻辑 | Design logic
空间类型 | Space type
建筑策略 | Building strategy
1-1剖透视图 | 1-1 Cutaway Perspective

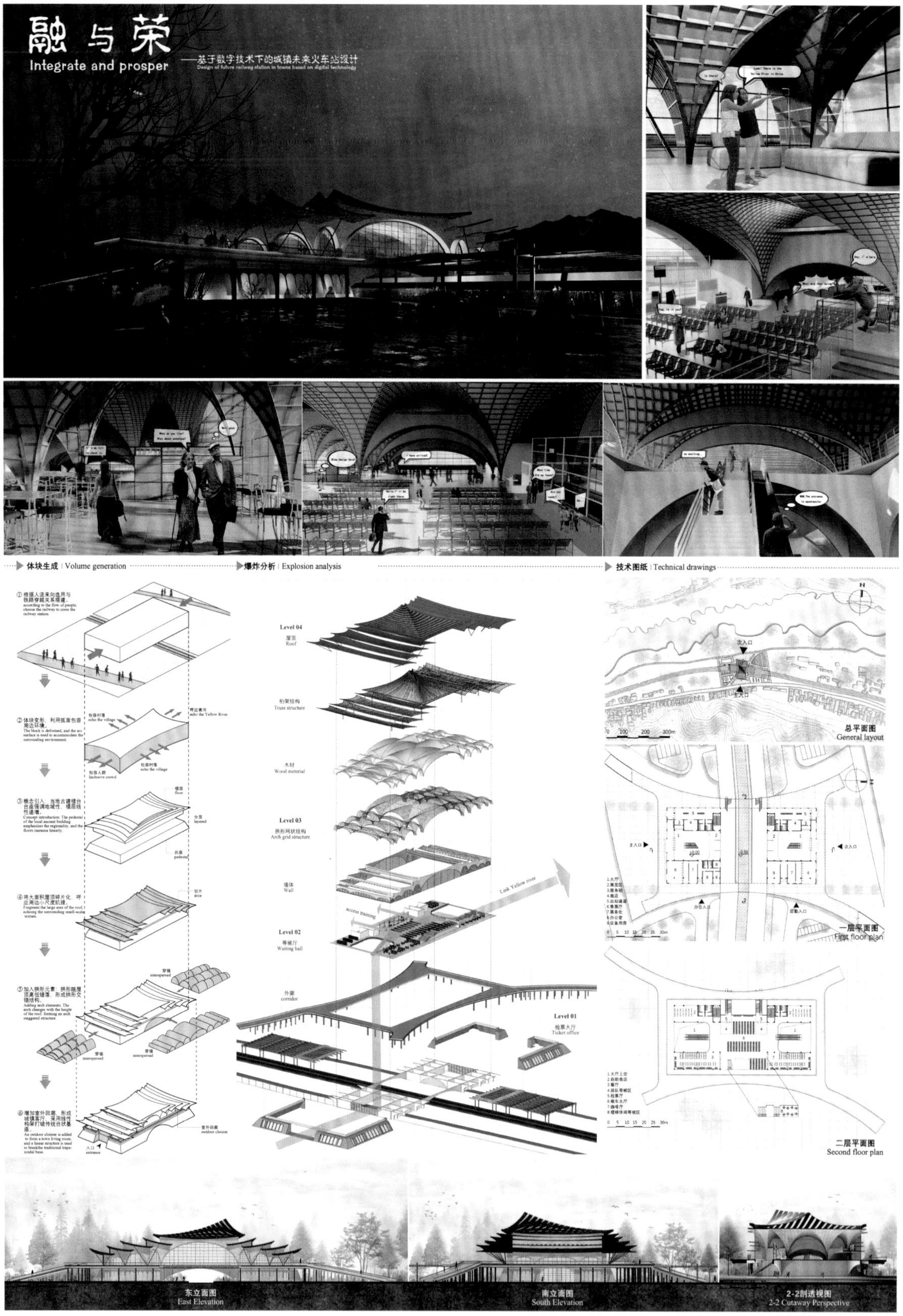

融与荣

Integrate and prosper

——基于数字技术下的城镇未来火车站设计

Design of future railway station in towns based on digital technology

体块生成 | Volume generation

① 根据人流来向选用与铁路穿越关系搭建。
according to the flow of people, choose the railway to cross the railway station.

② 体块变形，利用弧面包容周边环境。
The block is deformed, and the arc surface is used to accommodate the surrounding environment.

包容村落 echo the village
呼应黄河 echo the Yellow River
包容人群 Inclusive crowd
包容村落 echo the village

③ 概念引入：当地古建楼台台座强调地域性，楼层线性递增。
Concept introduction: The pedestal of the local ancient building emphasizes the regionality, and the floors increase linearly.

楼层 floor
分层 layered
台座 pedestal

④ 将大面积屋顶碎片化，呼应周边小尺度肌理。
Fragment the large area of the roof, echoing the surrounding small-scale texture.

切片 slice

⑤ 加入拱形元素：拱形随屋顶高低错落，形成拱形交错结构。
Adding arch elements: The arch changes with the height of the roof, forming an arch staggered structure.

穿插 interspersed
穿插 interspersed
穿插 interspersed

⑥ 增加室外回廊，形成城镇客厅，采用线性构架打破传统台状基座。
An outdoor cloister is added to form a town living room, and a linear structure is used to breakthe traditional trapezoidal base.

室外回廊 outdoor cloister
入口 entrance

爆炸分析 | Explosion analysis

Level 04
屋顶 Roof
桁架结构 Truss structure
木材 Wood meterial
Level 03
拱形网状结构 Arch grid structure
墙体 Wall
Link Yellow river
access training
Level 02
等候厅 Waiting hall
外廊 corridor
Level 01
检票大厅 Ticket office

技术图纸 | Technical drawings

次入口
主入口
0 100 200 300m

总平面图
General layout

主入口
次入口
办公入口
后勤入口
±0.00
-0.50

1.大厅
2.展览区
3.服务站
4.商店
5.出站通道
6.售票厅
7.票务处
8.办公室
9.设备用房

0 5 10 15 20 25 30m

一层平面图
First floor plan

1.大厅上空
2.自助商店
3.餐厅
4.排队等候区
5.检票厅
6.候车大厅
7.咖啡厅
8.楼梯休闲等候区

0 5 10 15 20 25 30m

二层平面图
Second floor plan

东立面图
East Elevation

南立面图
South Elevation

2-2剖透视图
2-2 Cutaway Perspective

外贝加尔湖地区一体化设计

Weave of Transbaikalia

参赛学校 University/College
俄罗斯莫斯科国立建筑大学
Moscow State University of Civil Engineering, Russia

指导教师 Supervisor(s)
Saltykov Ivan

参赛学生 Participant(s)
Khlebnikov Grigoriy Alekseevich
Zhvakina Polina Andreevna
Sakulina Anastasia Valeryevna
Mamiy Sultan Oscarovich
Trushkina Anna Andreevna

简介 Description

20 世纪，外贝加尔湖地区开始呈现繁荣景象，这里在修建西伯利亚大铁路，采矿，与邻国开展贸易活动。时代在不断变化，今天的外贝加尔湖地区需要新理念和新发展。

“外贝加尔湖地区一体化（Weave of Transbaikalia）”是基于统一化和可持续发展的设计理念，将文化、铁路和时代的交织融合于同一个项目中，重振外贝加尔湖地区。该建筑项目运用了艺术家 V. V. Vereshchagin 的作品作为核心装饰。他用独特的图案表现自己对东方文化的热情。博尔贾新车站成为一道引人注目的风景线，同时仍维持交通枢纽的功能。

该建筑体现了可持续发展的理念，原材料以木材为主，承重结构由 CLT 面板制成，并设计悬臂楼梯以支撑屋顶。立面设计采用雕花面板。

旧火车站经改造成为西伯利亚大铁路博物馆，车站广场改建后成为休闲场所。

“外贝加尔湖地区一体化”是建设外贝加尔湖地区新面貌的第一步，该地区将成为连接各个城市和整个国家的纽带，促进旅游和贸易的发展。

In the 20th century, the Trans-Baikal region had begun to flourish. The Trans-Siberian Railway is being built, mining is being carried out, and trade is conducted with neighbours. Times are changing, so today, Trans-Baikal region needs new ideas and development.

“Weave of Transbaikalia” is a concept that based on ideas of unification and sustainable development. Interweaving of cultures, railways and ages are merged in one project to revive the Transbaikalia. The Centerpiece of the project is works of artist V. V. Vereshchagin, whose passion for oriental culture transformed into unique patterns. New station in the Borzya becomes a point of attraction for everyone, continuing to function as a transport hub.

The building embodies the idea of sustainability with wood as the main raw material, the load-bearing structures are made of CLT panels and have a cantilever flight to support the roof. The facade design uses carved panels.

The old railway station is being restored with the adaptation for the museum of the Trans-Siberian Railway. The station square becomes a recreational area.

“Weave of Transbaikalia” is the first step to a new look of the region. Becoming a link for cities and entire states, this area will entail the development of tourism and trade.

外贝加尔湖地区一体化设计

Weave of Transbaikalia

View of the new station with the improvement of the station square in Borzya

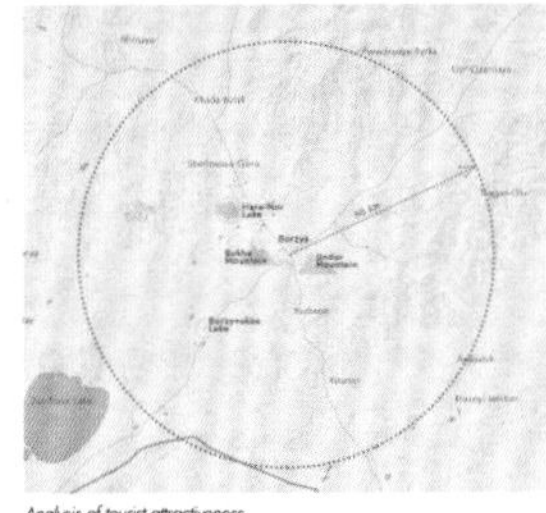

Analysis of tourist attractiveness

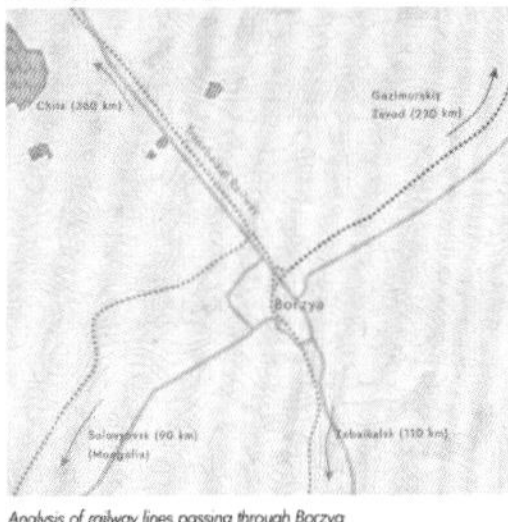

Analysis of railway lines passing through Borzya

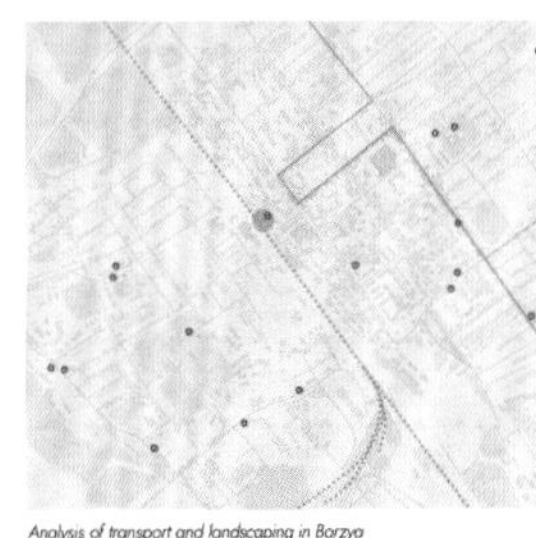

Analysis of transport and landscaping in Borzya

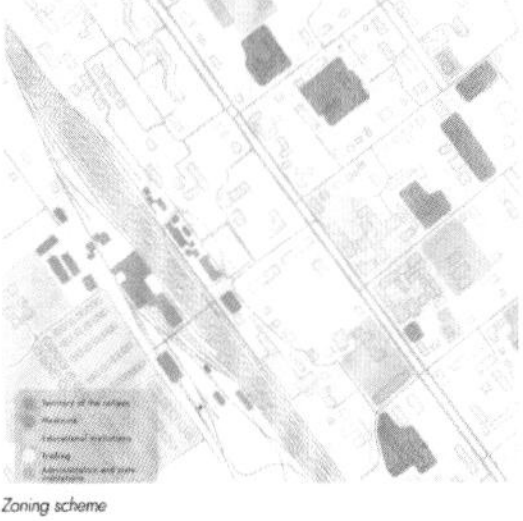

Zoning scheme

Weave of Transbaikalia

In the early 20th century, the Trans-Baikal region had begun to flourish. The Trans-Siberian Railway is being built, mining is being actively carried out, and trade is conducted with neighboring countries. Fortunately, or unfortunately, times are changing and priorities with them. Nowadays, Trans-Baikal region needs new ideas, meanings and intensive development. "Weave of Transbaikalia" – is a concept, that based on ideas of unification and sustainable development. Interweaving of cultures, railways and ages are merged in one project to breathe life into the Tranbaikalia. Centerpiece of project is works of Russian artist – V.V. Vereshchagin, whose passion of oriental culture transformed into unique patterns. New railway station in the Borzya becomes a point of attraction for citizens and tourists, continuing to function as a transport hub linking the Borzin district and its neighbors. The building reflects the idea of sustainability - wood is chosen as the main material, the load-bearing structures are created from CLT panels and have a cantilever flight to support the roof. In the design of the facades were used carved panels based on the works of Vereshchagin. The building of the old railway station is being restored with the adaptation for the museum of the history of the Trans-Siberian railway. The station square becomes a recreational area.
"Weave of Transbaikalia" is the first step to a new look of the region. Becoming a link not only for cities, but also for entire states, entails the development of tourism and increased trade.

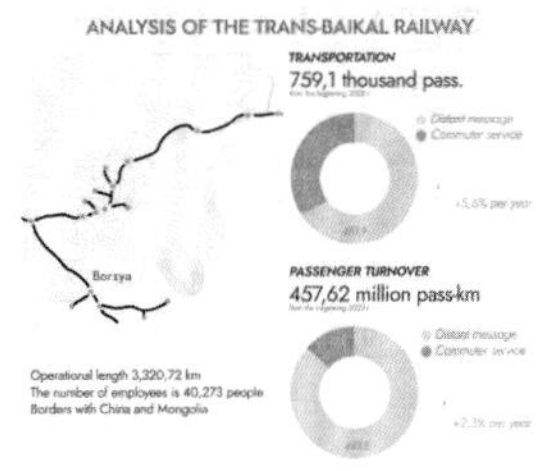

The interior of the reconstructed station from the hall

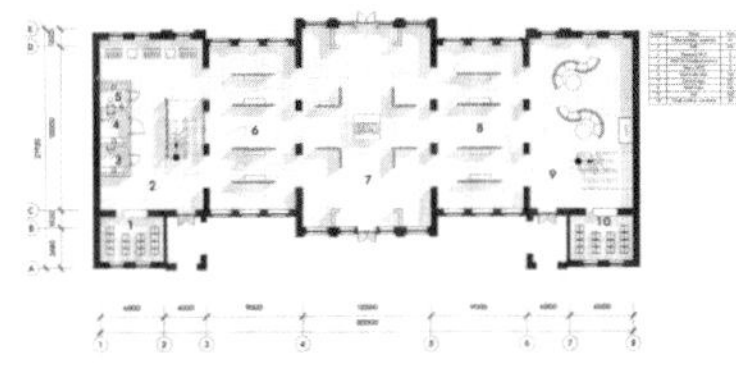

First floor

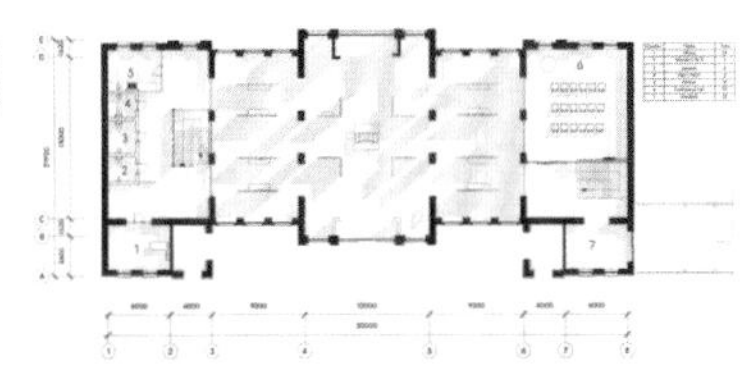

Second floor

Interior of the main exhibition hall

View of the new railway station with a transition from the platforms

外贝加尔湖地区一体化设计

Weave of Transbaikalia

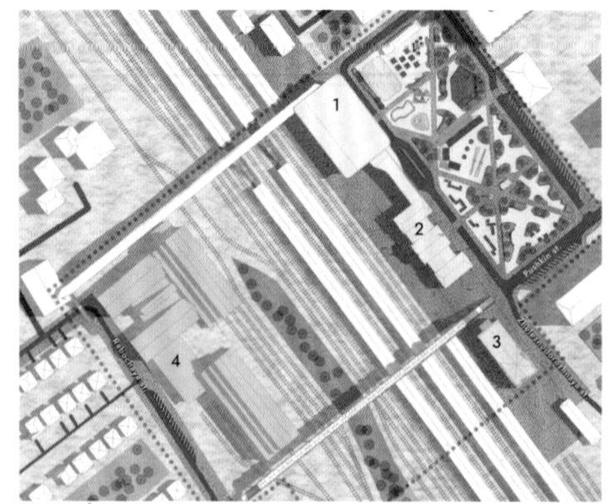

Master plan. 1 - New railway station. 2 - Old station/museum 3 - Administration. 4 - Locomotive depot.

View on the improvement of the station square

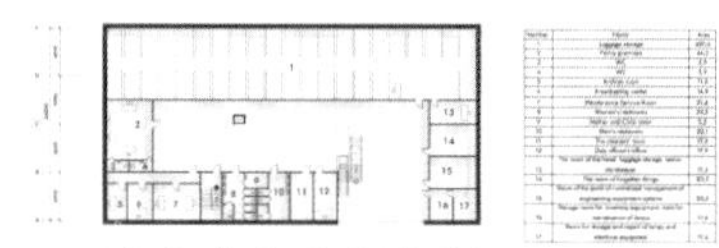

Ground floor

First floor

View from the square to the bridge between the old and new station buildings

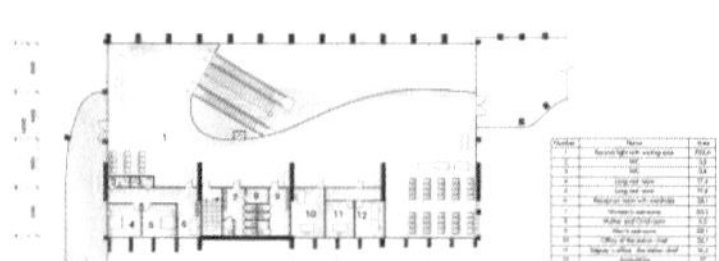

Second floor

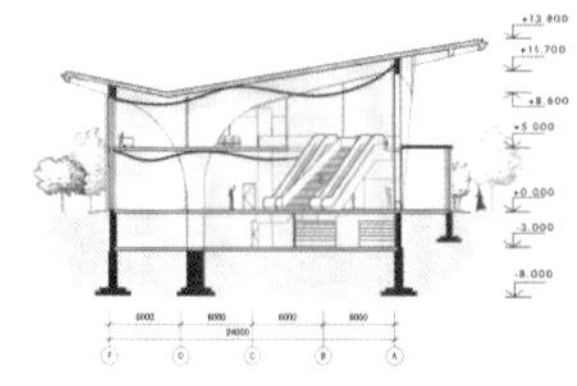

F-A section

View of the new railway station from the Zheleznodorozhnaya street

Interior of the 1st floor of the new station

Interior of the 2nd floor of the new station.

超级“C”元素

Hyper-C

参赛学校 | University/College
天津城建大学
Tianjin Chengjian University, China

指导教师 | Supervisor(s)
万　达　WAN Da
陈立镜　CHEN Lijing

参赛学生 | Participant(s)
张之新秀　ZHANG Zhixinxiu
王 菊 平　WANG Juping

简介 Description

该高铁站是基于时间观念引导下，整合了货物运输及交通枢纽的综合功能的铁路运输枢纽。运用 Hyperloop 的高速运输系统，同时充分利用其周边的铁路遗址和铁路运输线路，使历史遗迹再现出历史的交通景观，并且满足于市民活动需求，营造铁路遗址公园，是一个集休闲、乘车、换乘于一体的铁路客运站。

The high-speed railway station is a railway transportation hub that integrates the functions of freight transportation and transportation hub under the guidance of the concept of time. Using the high-speed transportation system of Hyperloop, the historical sites reproduce the scene of traffic in history by making full use of the railway relics and remaining railway transportation lines in the surroundings. To provide space for local citizens' activities, a Railway Heritage Park is also to be built. Therefore, the designed railway station enables relaxation, taking trains, and transfepping.

HYPER-C I 未来铁路客运站设计

CHAIN AND CONECT — CARRY — CIRCLE — CENTER — CUT —CONVENIENT

时间观念引导下——整合货物运输及交通枢纽——历史遗迹再现交通景观——集休闲乘车换乘与一体的铁路客运站

Status of Railway Passenger Station

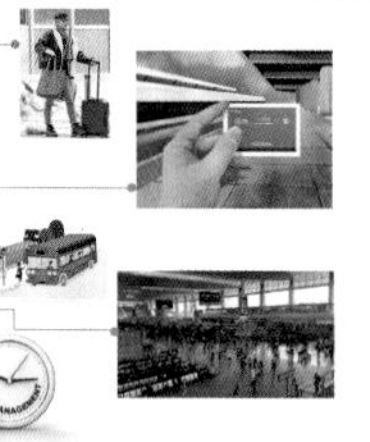

Baggage security check is heavy and inconvenient to carry.

The function of the passenger station is too strong, people who do not need to ride will not come to gather here, and the progress with ordinary citizens is low.

After getting off the bus, you need other slow transportation methods to transfer.

The waiting space is single, and the sense of scene is not strong.

The concept of time needs to be emphasized again.

Technology use

hyper-loop

Pack people or goods into a pod and transport them over long distances through pipes. Magnetic levitation and large vacuum pumps will eliminate friction and air resistance, allowing these bus-sized pods to travel at nearly 1,225.08 km/h.

类型	速度/km·h⁻¹		长度/m	总座位数/个	开行频次/列·h⁻¹	最短发车间隔/s	运输能力/人次（每小时每方向）
	最大速度	运行速度					
飞机	900	600	60~70	400	15	180	6 000
	900	400	40	200	20	180	4 000
高速列车	380	250	410	1 000	10	180	10 000
	250	150	200	450	12	180	5 400
Transrapid	500	225~250	125	438	12	300	5 250
SC Maglev	600	245		1 000	10	180	10 000
Hyperloop	1 200	1 000	25~30	28~40	12	300	336~480

How it works 超级高铁如何运作

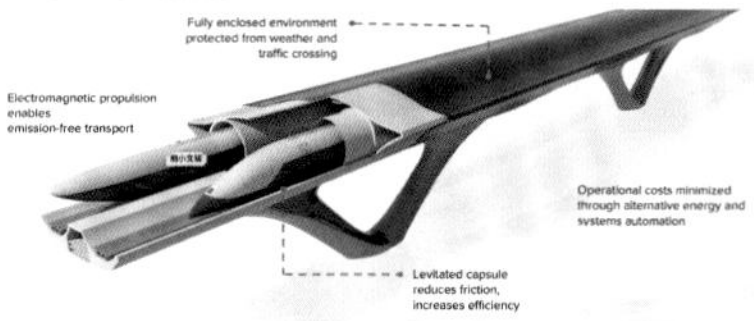

Fully enclosed environment protected from weather and traffic crossing

Electromagnetic propulsion enables emission-free transport

Operational costs minimized through alternative energy and systems automation

Levitated capsule reduces friction, increases efficiency

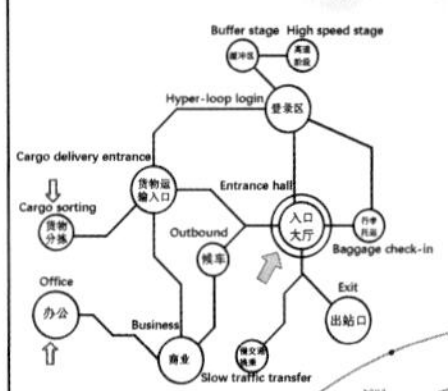

relationship diagram

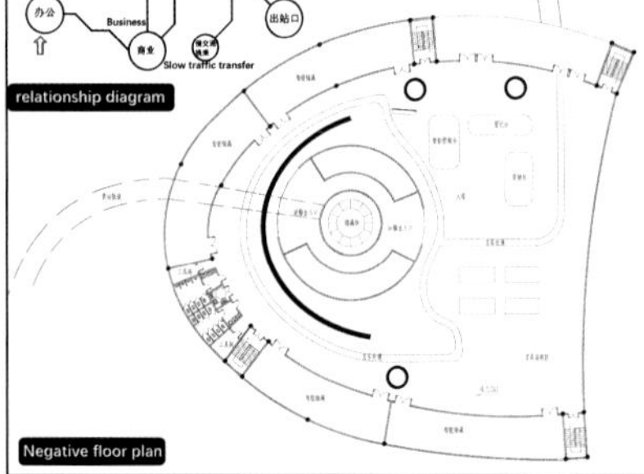

Negative floor plan

District Analysis

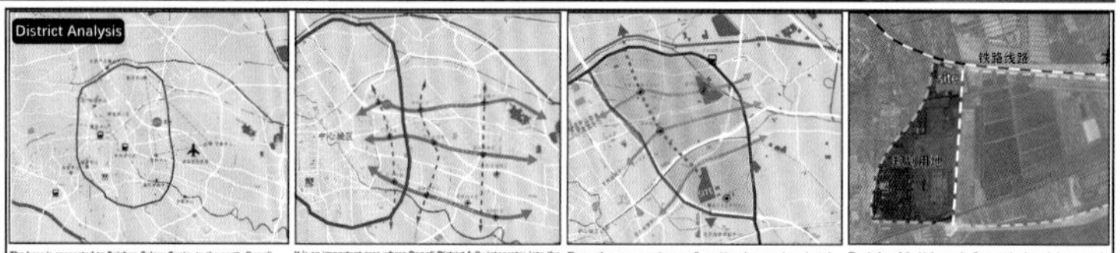

The base is connected to Beichen Galaxy Center to the north, Dongli Chenglin Center to the south, and Dongli Airport Center to the east. It is located at the intersection of the eastern secondary center.

It is an important area where Dongli District fully integrates into the main city, connects with industrial areas, and plays an important role in urban quality and potential release. It is an important bearing area for the overflowing population in the central city.

The northern new area is generally positioned as a service-oriented, technological and ecological new urban area. The base is located in Nandian Ecological Leisure Center.

The design of the high-speed railway station is carried out according to the geographical location of the base and the surrounding planning.

Base Planning and Location

The building base is located on the north side of the site, adjacent to railways and roads to the north, and residential and slow traffic in the planning area to the south. The base area is 15,000 square meters, and there is a new tower in the center of the site that echoes the old tower on the south side. A railway passenger station is proposed to be built in the center of the building site.

The new and old towers echo to form the central

Planning site building layout

Building site location

Thinking about the relationship between planning land use environment and base

Building Layout Analysis

Analyze the flow of people and set up the main entrance plaza and exit. And carry out the separation of people and goods, and the goods are transported from outside to this place for transit and sorting. The plot serves as both a railway passenger hub and cargo sorting and transportation. It is a complex functional requirement land.

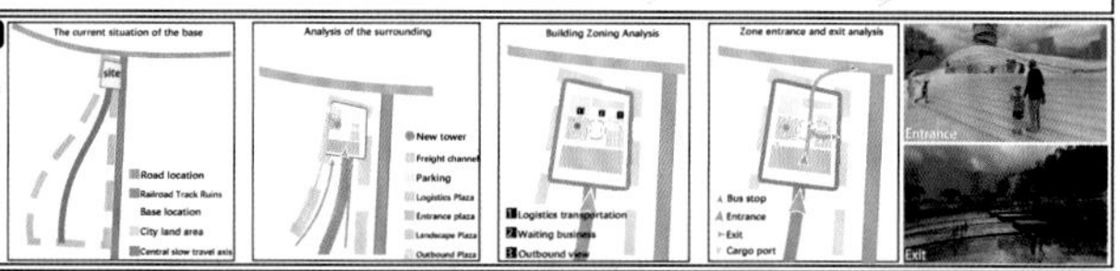

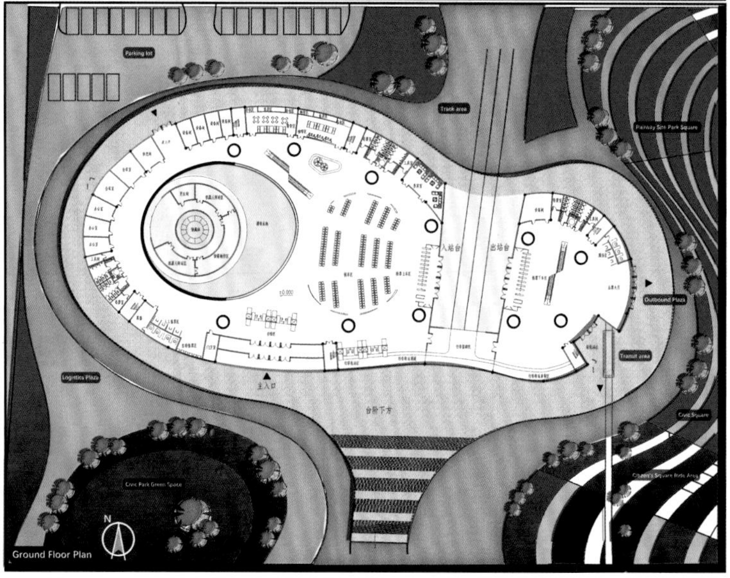

The current situation of the base

Analysis of the surrounding

Building Zoning Analysis

Zone entrance and exit analysis

Ground Floor Plan

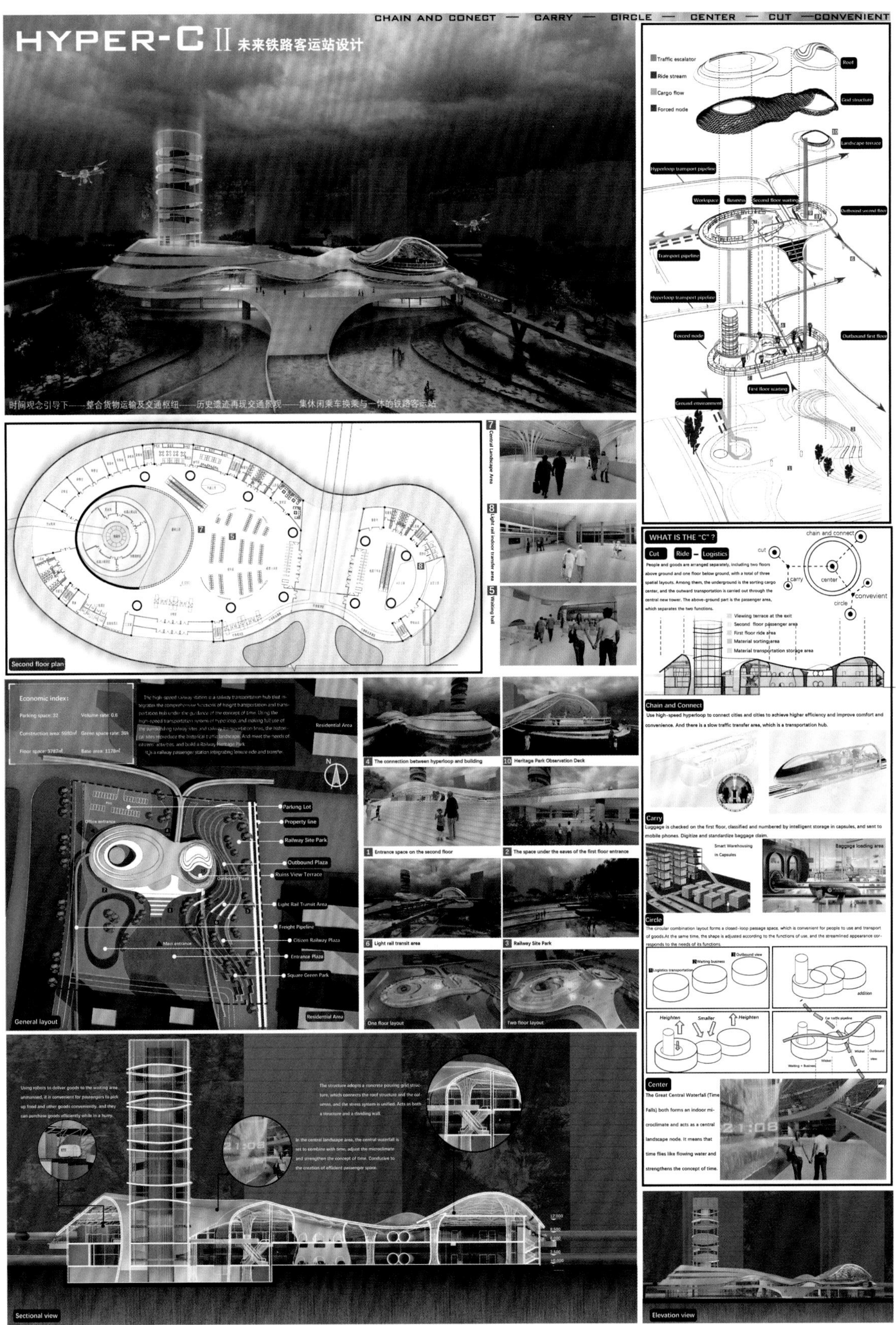
CHAIN AND CONECT — CARRY — CIRCLE — CENTER — CUT —CONVENIENT
HYPER-C II 未来铁路客运站设计
时间观念引导下——整合货物运输及交通枢纽——历史遗迹再现交通系观——集休闲乘车换乘与一体的铁路客运站
Second floor plan
General layout
Sectional view
Elevation view
WHAT IS THE "C"?
Chain and Connect
Carry
Circle
Center

曲转流觞 · 源缕塞鸣

Improve the Enthusiasm for Using Space in Desert Areas

参赛学校 University/College

中原工学院

Zhongyuan University of Technology, China

指导教师 Supervisor(s)

张淑润 ZHANG Shurun

马　骁 MA Xiao

参赛学生 Participant(s)

李　默 LI Mo

张健玮 ZHANG Jianwei

董恩惠 DONG Enhui

郭　瑶 GUO Yao

简介 Description

该设计从人群行为本身出发，不拘泥于传统的内部空间布局形式，探索未来客运站的发展形式，设置了复合多功能的TOD交通综合体。面对其独特的气候特征和场所感的缺失，我们从其本身的优势出发，通过人群行为分析出文化、交流、经济、展示等功能，再通过组合方式，使其以不同的形式在不同的时间为不同人群服务。从建筑形态上来说，在结合材料结构与功能属性的同时，转译传统元素“壁画丝带”“沙丘”“半穴居”“坎儿井”对周围景观及建筑进行回应。 该火车站解决地域特色同城镇发展的冲突，使建筑满足功能和需求的同时成为建筑的载体并激活经济。

Starting from crowd behavior itself, the design does not rigidly adhere to the traditional approaches of internal space layout, instead, it explores the development of the future railway station by setting up a transit-oriented development (TOD) complex with multiple functions. Considering its unique climate characteristics and a lack of sense of place, we we analyzed the functions of culture, communication, ecomomy and display through crowd behavior from its own advantages and then enabled it to serve the needs of different people at different times and in different forms through combination. In terms of architectural form, while considering both material structures and functional properties, we transcreated the traditional elements of “ribbons on murals”, “dune”, “semi cave”, and “Karez” to interact with the surrounding landscape and buildings. The design is devoted to resolving the conflict between regional characteristics and urban development so that the building can become a carrier and activate the local economy on top of fulfilling its functions and meeting expected needs.

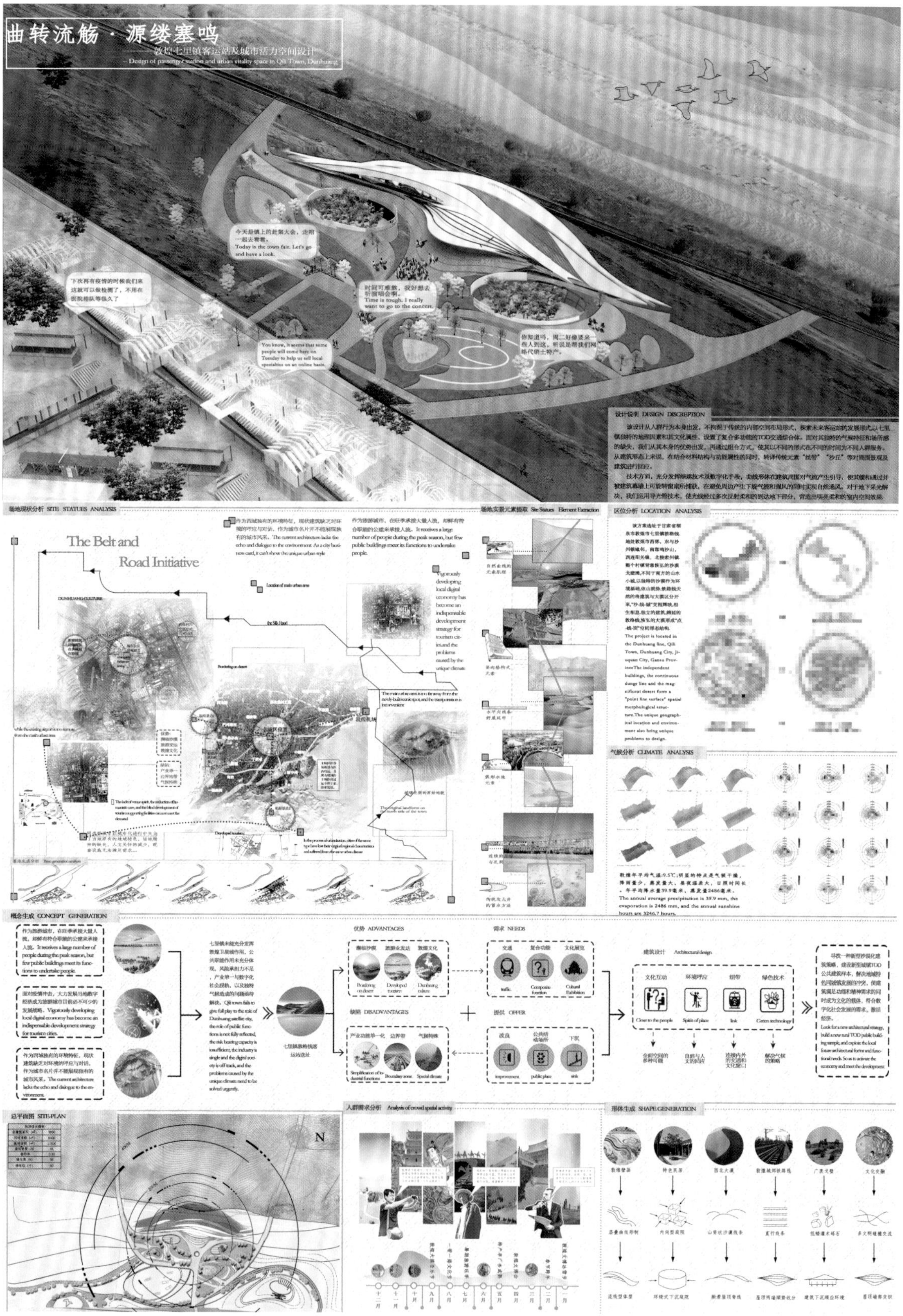

曲转流觞 · 源缕塞鸣

——敦煌七里镇客运站及城市活力空间设计

– Design of passenger station and urban vitality space in Qili Town, Dunhuang

设计说明 DESIGN DISCREPTION

该设计从人群行为本身出发，不拘泥于传统的内部空间布局形式，探索未来客运站的发展形式以七里镇独特的地理因素和其文化属性，设置了复合多功能的TOD交通综合体。面对其独特的气候特征和场所感的缺失，我们从其本身的优势出发，再通过组合方式，使其以不同的形式在不同的时间为不同人群服务。从建筑形态上来说，在结合材料结构与功能属性的同时，转译传统元素“丝带”“沙丘”等对周围景观及建筑进行回应。

技术方面，充分发挥绿建技术及数字化手段，曲线形体在建筑周围对气流产生引导，使其缓和通过并被建筑幕墙上可旋转窗扇所捕获，在避免周边产生下旋气流和强风的同时实现自然通风。对于地下采光解决，我们运用导光管技术，使光线经过多次反射柔和的到达地下部分，营造出明亮柔和的室内空间效果。

场地现状分析 SITE STATUES ANALYSIS

The Belt and Road Initiative

场地实景元素提取 Site Statues Element Extraction

区位分析 LOCATION ANALYSIS

气候分析 CLIMATE ANALYSIS

概念生成 CONCEPT GENERATION

总平面图 SITE-PLAN

人群需求分析 Analysis of crowd spatial activity

形体生成 SHAPE GENERATION

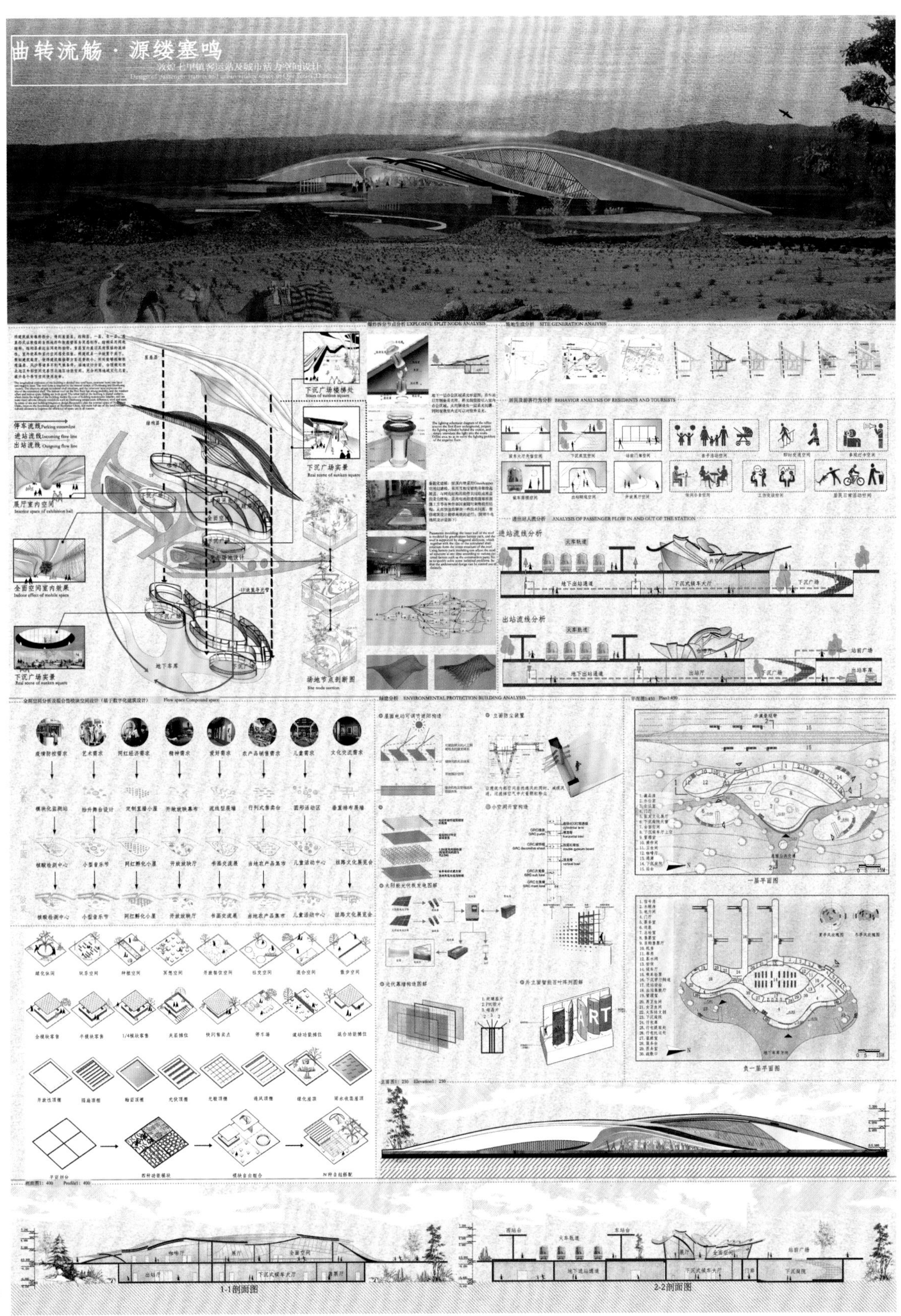
曲转流觞 · 源缕塞鸣
1-1剖面图
2-2剖面图
一层平面图
负一层平面图

汀木长林 耦合共生——低碳导向下垌炀河火车站改造设计探索

Coupling Symbiosis of Tingmu Long Forest—A Low Carbon-oriented Reconstruction Design of the Tongyang River Railway Station

参赛学校 University/College
安徽建筑大学
Anhui Jianzhu University, China

指导教师 Supervisor(s)
桂汪洋 GUI Wangyang
吴运法 WU Yunfa

参赛学生 Participant(s)
张 彤 ZHANG Tong
张 力 ZHANG Li
周 信 ZHOU Xin
张力群 ZHANG Liqun
张志伟 ZHANG Zhiwei

简介 Description

随着社会的发展，客运站作为单一的交通枢纽逐渐失去了以前的优势。垌炀河火车站是历史中不可或缺的一部分, 本次设计以“森林中的火车站”作为设计理念。以“木”为设计原型，以客运站为触媒，在不破坏原有肌理的情况下，解决该地区功能单一、无生活配套的问题，构建以铁路为依托的自给自足的微型城市循环系统，让居民、社区、自然、文化有更多方面的交流。通过这一途径来激活乡村活力，也让昔日的焖炀镇重新焕发新的活力。

With the development of society, railway stations are gradually losing their previous advantages as a mere transportation hub. As an indispensable part of history, the Tongyang River Railway Station is inspired by the concept of “a railway station in the forest”. Our design uses “wood” as the prototype and railway station as the presentation to solve the problem of single function and a lack of supportive living facilities in this area without destroying the original layout. The design aims to build a self-sufficient circulation system of a micro-city reliant on the railway where residents, communities, nature and culture can interact in more ways so that the local countryside and the Town of Tongyang could be vitalized.

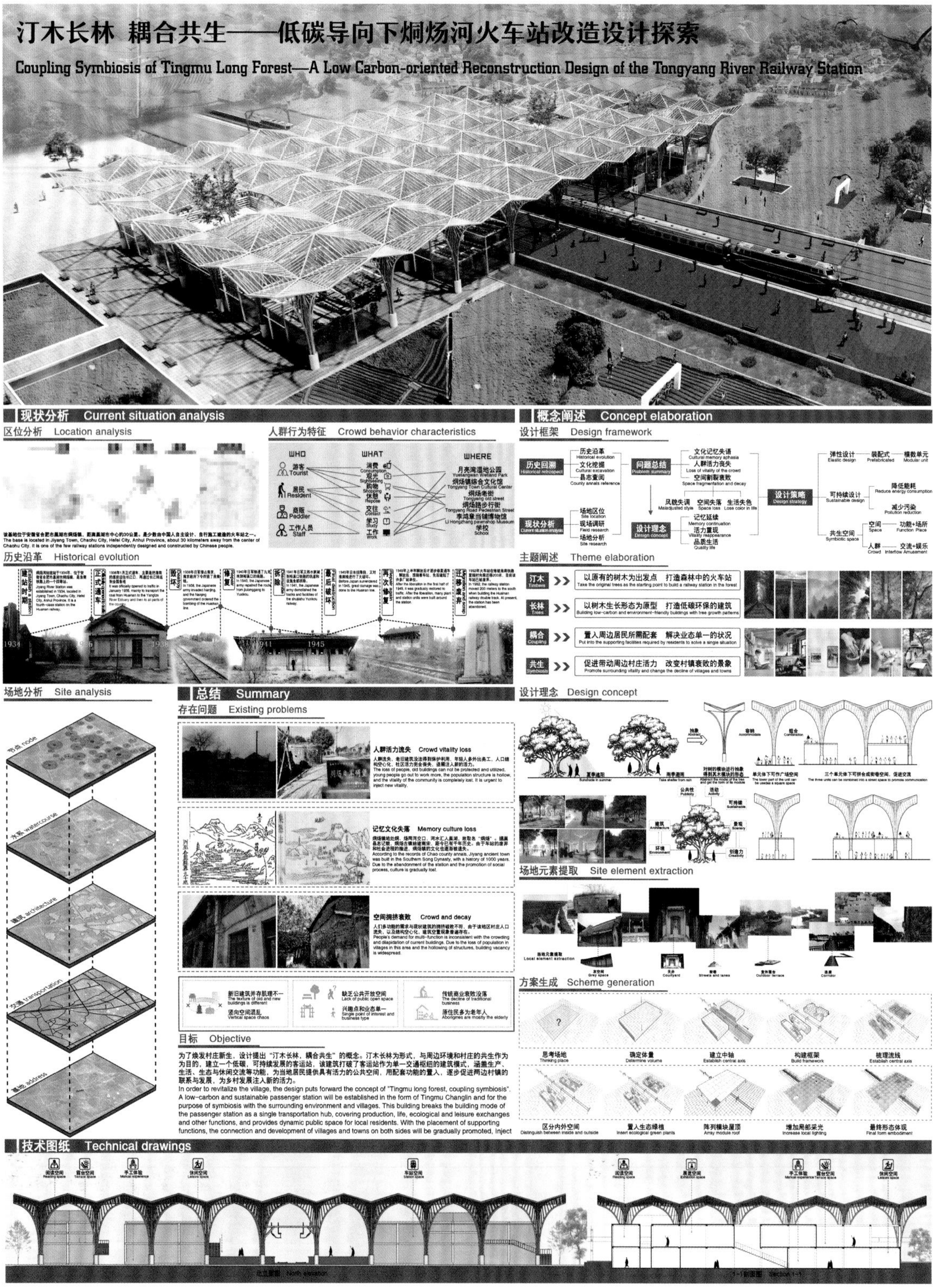

汀木长林 耦合共生——低碳导向下烔炀河火车站改造设计探索

Coupling Symbiosis of Tingmu Long Forest—A Low Carbon-oriented Reconstruction Design of the Tongyang River Railway Station

现状分析 Current situation analysis

区位分析 Location analysis

人群行为特征 Crowd behavior characteristics

历史沿革 Historical evolution

场地分析 Site analysis

总结 Summary

存在问题 Existing problems

人群活力流失 Crowd vitality loss

记忆文化失落 Memory culture loss

空间拥挤衰败 Crowd and decay

新旧建筑并存肌理不一 The texture of old and new buildings is different

竖向空间混乱 Vertical space chaos

缺乏公共开放空间 Lack of public open space

兴趣点和业态单一 Single point of interest and business type

传统商业衰败没落 The decline of traditional business

原住民多为老年人 Aborigines are mostly the elderly

目标 Objective

为了焕发村庄新生，设计提出"汀木长林，耦合共生"的概念。汀木长林为形式，与周边环境和村庄的共生作为为目的，建立一个低碳，可持续发展的客运站，该建筑打破了客运站作为单一交通枢纽的建筑模式，涵盖生产、生活、生态与休闲交流等功能，为当地居民提供具有活力的公共空间，用配套功能的置入，逐步促进两边村镇的联系与发展，为乡村发展注入新的活力。

In order to revitalize the village, the design puts forward the concept of "Tingmu long forest, coupling symbiosis". A low-carbon and sustainable passenger station will be established in the form of Tingmu Changlin and for the purpose of symbiosis with the surrounding environment and villages. This building breaks the building mode of the passenger station as a single transportation hub, covering production, life, ecological and leisure exchanges and other functions, and provides dynamic public space for local residents. With the placement of supporting functions, the connection and development of villages and towns on both sides will be gradually promoted, Inject

概念阐述 Concept elaboration

设计框架 Design framework

历史回溯 Historical retrospect — 历史沿革 Historical evolution；文化挖掘 Cultural excavation；县志查阅 County annals reference

现状分析 Current situation analysis — 场地区位 Site location；现场调研 Field research；场地分析 Site research

问题总结 Problem summary — 文化记忆失语 Cultural memory aphasia；人群活力丧失 Loss of vitality of the crowd；空间割裂衰败 Space fragmentation and decay；风貌失调 Maladjusted style；空间失落 Space loss；生活失色 Lose color in life

设计理念 Design concept — 记忆延续 Memory continuation；活力重现 Vitality reappearance；品质生活 Quality life

设计策略 Design strategy — 弹性设计 Elastic design；可持续设计 Sustainable design（降低能耗 Reduce energy consumption；减少污染 Pollution reduction）；共生空间 Symbiotic space（空间 Space — 功能+场所 Function Place；人群 Crowd — 交流+娱乐 Interflow Amusement）

主题阐述 Theme elaboration

汀木 Timbers >> 以原有的树木为出发点 打造森林中的火车站 Take the original trees as the starting point to build a railway station in the forest

长林 Trees >> 以树木生长形态为原型 打造低碳环保的建筑 Building low-carbon and environment-friendly buildings with tree growth patterns

耦合 Coupling >> 置入周边居民所需配套 解决业态单一的状况 Put into the supporting facilities required by residents to solve a single situation

共生 Symbiosis >> 促进带动周边村庄活力 改变村镇衰败的景象 Promote surrounding vitality and change the decline of villages and towns

设计理念 Design concept

场地元素提取 Site element extraction

方案生成 Scheme generation

技术图纸 Technical drawings

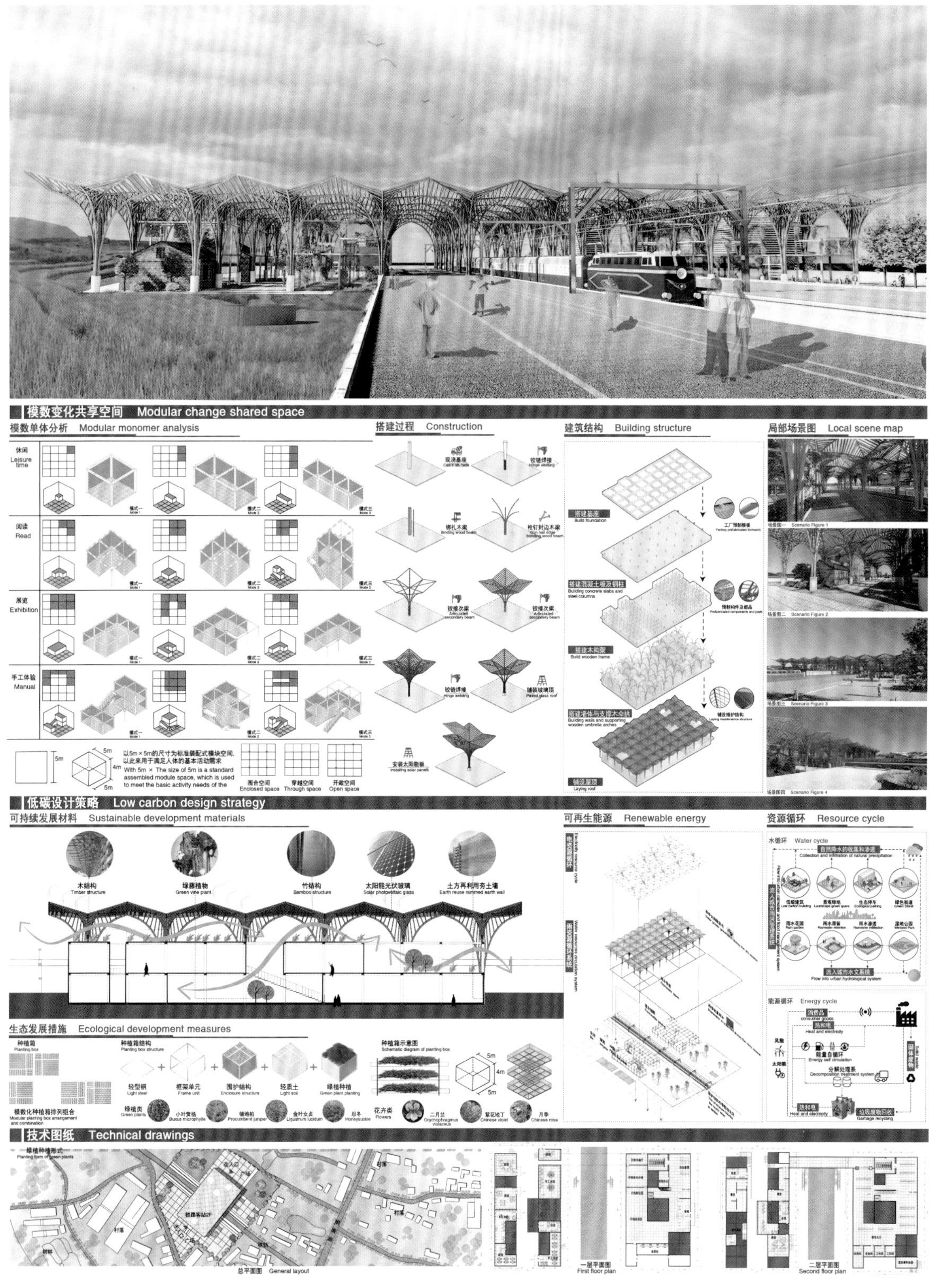
模数变化共享空间 Modular change shared space
模数单体分析 Modular monomer analysis
搭建过程 Construction
建筑结构 Building structure
局部场景图 Local scene map
低碳设计策略 Low carbon design strategy
可持续发展材料 Sustainable development materials
木结构
绿藤植物
竹结构
太阳能光伏玻璃
土方再利用夯土墙
可再生能源 Renewable energy
资源循环 Resource cycle
生态发展措施 Ecological development measures
技术图纸 Technical drawings
总平面图 General layout
一层平面图 First floor plan
二层平面图 Second floor plan

故驿新语——铜罐驿火车站片区更新改造设计

Renewal & Reconstruction Design of the Tongguanyi Railway Station Area

参赛学校 University/College
重庆交通大学
Chongqing Jiaotong University, China

指导教师 Supervisor(s)
温 泉 WEN Quan
刘亚南 LIU Yanan

参赛学生 Participant(s)
刘梓星 LIU Zixing
陈佳旭 CHEN Jiaxu
李悦鹏 LI Yuepeng
吕 峰 LYU Feng
王 瑜 WANG Yu

简介 Description

自明代以来，重庆长江之畔的铜罐驿即是著名水驿码头，新中国成立初修建的火车站更成为新中国建成的第一条铁路——成渝线重要的站点，而如今火车站片区面临产业分布不均、功能缺失、生态环境差、交通不便、智能化水平低等问题。设计以空间句法分析并改善片区交通可达性；通过 Grasshopper 及 DepthmapX 辅助设计改造闲置厂房，塑造自由集市、滨江公园等公共空间激活片区活力；以空间植入整合火车站与市民文化中心，构建文化、生态、交通一体的多维度交互空间，构建历史与未来互动共生的场景化城市客厅。

Sitting on the Bank of the Yangtze River of Chongqing, China, the Tongguan Station has been a famous wharf since the Ming Dynasty (1368—1644). The railway station built right after the founding of the People's Republic of China (PRC) has become an important station in the Chengdu-Chongqing Railway, the first railway built by the PRC. Nowadays, the area near this railway station is facing the problems of unbalanced industrial distribution, a lack of certain functions, a poor ecological environment, inconvenient transportation and a low level of intelligence. The design uses spatial syntax to analyze and improve the traffic accessibility of the area, Idle plants are to be transformed through Grasshopper and DepthmapX-assisted to activate public space in this area such as free markets and riverside parks. The railway station and the cultural center for citizens are integrated by implanting extra space to build a multi-dimensional interaction venue with culture, ecology and transportation, which is also the living room of the city that represents a scenario where the history and the future are interacting and coexisting.

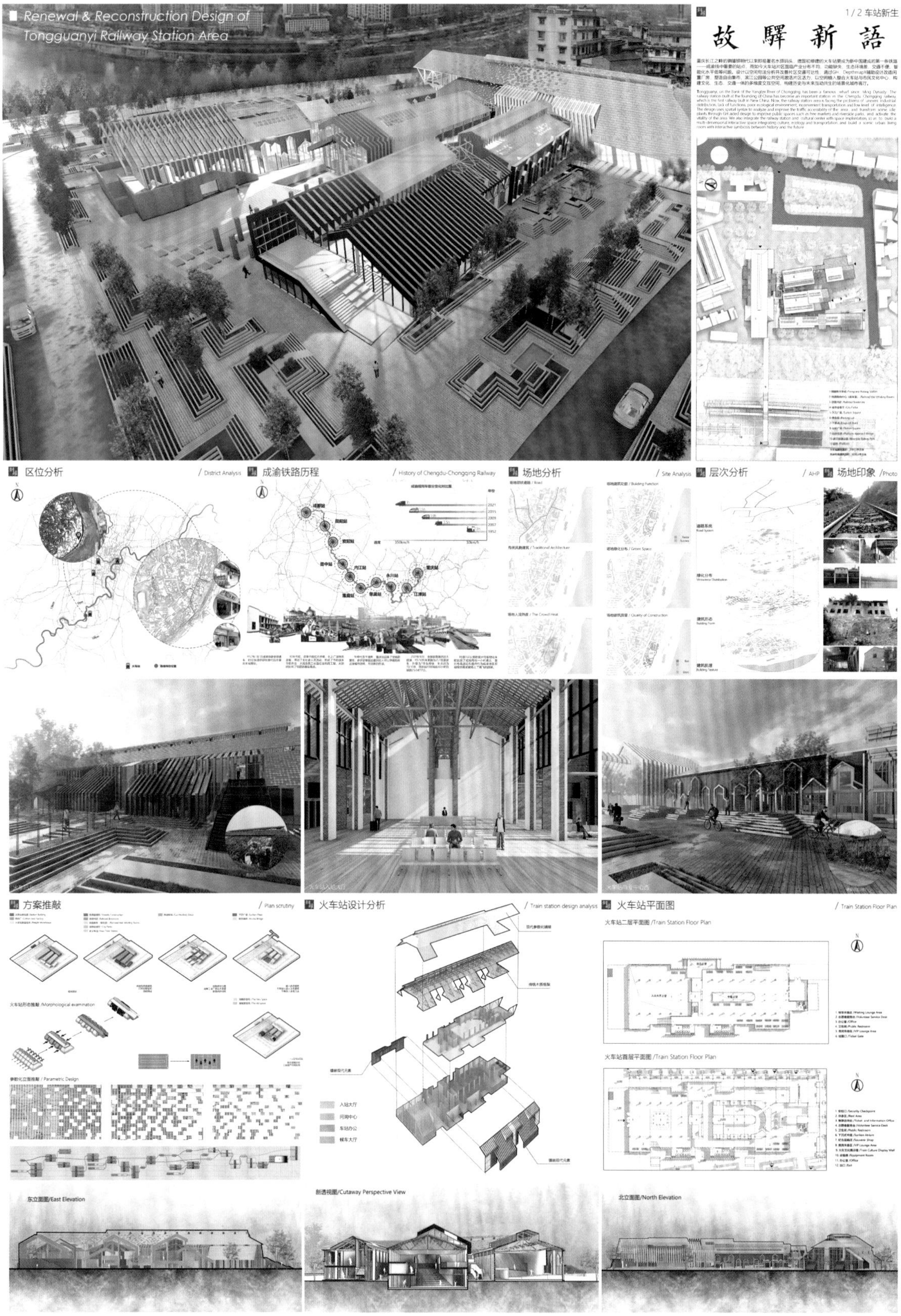

Renewal & Reconstruction Design of Tongguanyi Railway Station Area

1/2 车站新生

故 驛 新 語

重庆长江之畔的铜罐驿明代以来即是著名水驿码头，建国初修建的火车站更成为新中国建成的第一条铁路——成渝线中重要的站点，而如今火车站片区面临产业分布不均、功能缺失、生态环境差、交通不便、智能化水平低等问题。设计以空间句法分析并改善片区交通可达性，通过GH、DepthmapX辅助设计改造闲置厂房，塑造自由集市、滨江公园等公共空间激活片区活力，以空间植入整合火车站与市民文化中心，构建文化、生态、交通一体的多维度交互空间，构建历史与未来互动共生的场景化城市客厅。

Tongguanyi, on the Bank of the Yangtze River of Chongqing, has been a famous wharf since Ming Dynasty. The railway station built at the founding of China has become an important station in the Chengdu Chongqing railway, which is the first railway built in New China. Now, the railway station area is facing the problems of uneven industrial distribution, lack of functions, poor ecological environment, inconvenient transportation and low level of intelligence. The design uses spatial syntax to analyze and improve the traffic accessibility of the area, and transform some idle plants through GH aided design to improve public spaces such as free markets and riverside parks, and activate the vitality of the area. We also integrate the railway station and cultural center with space implantation, so as to build a multi-dimensional interactive space integrating culture, ecology and transportation, and build a scenic urban living room with interactive symbiosis between history and the future.

区位分析 / District Analysis

成渝铁路历程 / History of Chengdu-Chongqing Railway

场地分析 / Site Analysis

层次分析 / AHP

场地印象 / Photo

方案推敲 / Plan scrutiny

火车站形态推敲 /Morphological examination

参数化立面推敲 / Parametric Design

火车站设计分析 / Train station design analysis

火车站平面图 / Train Station Floor Plan

火车站二层平面图 /Train Station Floor Plan

火车站首层平面图 /Train Station Floor Plan

东立面图/East Elevation

剖透视图/Cutaway Perspective View

北立面图/North Elevation

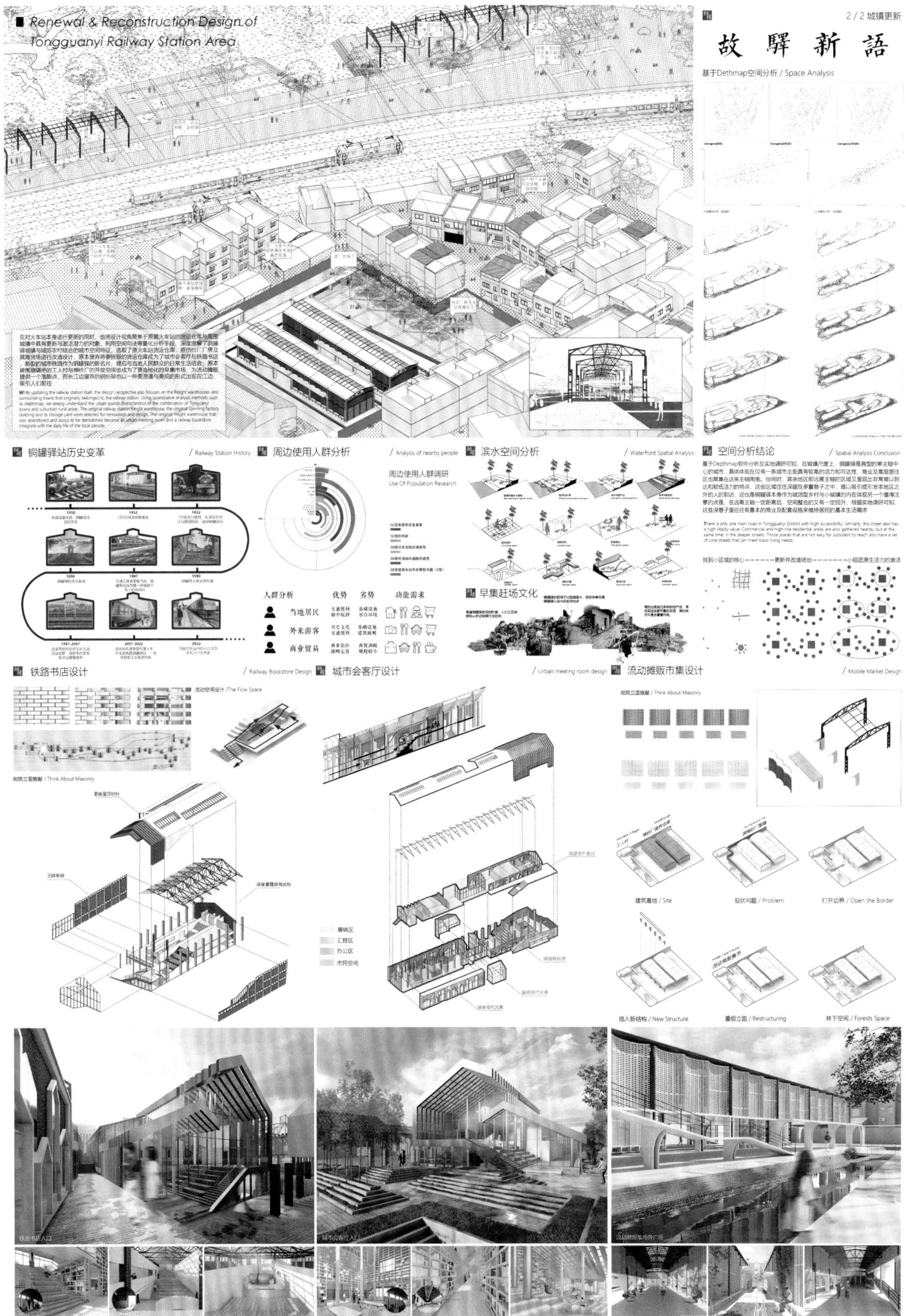
Renewal & Reconstruction Design of Tongguanyi Railway Station Area
2/2 城镇更新
故驛新語
基于Dethmap空间分析 / Space Analysis
铜罐驿站历史变革 / Railway Station History
周边使用人群分析 / Analysis of nearby people
滨水空间分析 / Waterfront Spatial Analysis
空间分析结论 / Spatial Analysis Conclusion
早集赶场文化
人群分析
优势
劣势
功能需求
当地居民
外来游客
商业贸易
铁路书店设计 / Railway Bookstore Design
城市会客厅设计 / Urban meeting room design
流动摊贩市集设计 / Mobile Market Design
建筑基地 / Site
现状问题 / Problem
打开边界 / Open the Border
插入新结构 / New Structure
重组立面 / Restructuring
林下空间 / Forests Space

小酌“夷”情

乡旅融合视角下丝路沿线铁路的复兴——以武夷山市杜坝村为例

The Revival of Railways along the Silk Road from the Perspective of Rural Tourism Integration

参赛学校 University/College
福州大学
Fuzhou University, China

指导教师 Supervisor(s)
武 昕 WU Xin

参赛学生 Participant(s)
潘静涵 PAN Jinghan
毕倩茹 BI Qianru
林晓帆 LIN Xiaofan
林子凌 LIN Ziling

简介 Description

两千多年前，武夷山这片承载着茶马古道历史的沃土，驼铃相闻，茶香氤氲。如今，古老的丝绸之路上建起了现代化的铁路运输网络，加速推进沿线各国经济、文化的合作交融，在新冠肺炎疫情严峻形势下传递了休戚与共的人类命运共同体理念。在茶马古道的独特历史背景下，茶文化成为武夷山人生活不可或缺的一部分，于是，本次的设计我们着眼于该地的历史文化价值，通过对客运站的茶文化植入和数字化系统设计以保留、发扬武夷山独有的茶文化。

More than 2000 years ago, Wuyi Mountain, a fertile land bearing the history of the Ancient Tea Horse Road, was filled with the sound of camel bells and the smell of tea. Today, a modern railway transportation network has been built on the Ancient Silk Road, accelerating the cooperation and integration of the countries along the route in economy and culture and conveying the concept of a community of shared destiny for mankind in a world where the COVID-19 pandemic prevails. With the unique historical background of the Ancient Tea Horse Road, tea culture has become an indispensable part of people living in Wuyi Mountain. Therefore, in this design, we focused on the historical and cultural value of the place. The implanted element of tea and the design of the digital system preserved and promoted the unique tea culture of Wuyi Mountain.

小酌“夷”情 乡旅融合视角下丝路沿线铁路的复兴——以武夷山市杜坝村为例 01

The revival of railways along the Silk Road from the perspective of rural tourism integration

海上丝绸之路沿线城市——福建南平市

铁路线 武夷山市 杜坝村 主要设计区域 Main design area

秦汉时期

海上丝绸之路形成，于三国至隋朝时期发展，繁荣于唐宋时期，转变于明清时期，是已知的最为古老的海上航线。

南平作为海丝真正的“源头”，是闽江之源，海丝之源，多个货物贸易的供应地。

“明永乐年间，郑和七下西洋，携带大量包括武夷茶在内的各种名茶作为礼品赠给各国，打开茶叶之门，外销从此开始。

——《武夷山志·卷七》

1404年

武夷山星村码头 茶叶从此出口转运海外

1880年

武夷山市探索文化创意旅游模式，结合茶文化打造茶博园、风情小镇等一系列文旅项目

2012年

迄今 通铁路/文化交流

用地现状

铁路线(横南线)南北方向经过杜坝村东边，周边现有大量农田和相关旅游基础设施

设计概念

丝路、铁路连通区域边界+茶文化介入+文旅融合实现周边区域振兴

农产品形象提升 传统工业活化 各方需求挖掘 农+文+旅融合

乡旅融合 旅游复兴 铁路更新

茶产业现状

茶产业链解读

现状总结

客运站目标人群

村民：脱贫 茶文化推广 农产品售卖

旅客：购物 休憩 出行

职工：工作 值班 会议

车站内部活动分析

设计说明

两千多年前，武夷山这片承载着茶马古道历史的沃土，驼铃相闻，茶香氤氲。如今，古老的丝绸之路上建起了现代化的铁路运输网络，加速推进沿线各国经济、文化的合作交融，在“新冠”疫情严峻形势下传递了休戚与共的人类命运共同体理念。在茶马古道的独特历史背景下，茶文化成为武夷山人生活不可或缺的一部分，于是，本次的设计我们着眼于该地的历史文化价值，通过对客运站的茶文化置入和数字化系统设计以保留、发扬武夷山独有的茶文化。

小酌"夷"情 乡旅融合视角下丝路沿线铁路的复兴——以武夷山市杜坝村为例 02

The revival of railways along the Silk Road from the perspective of rural tourism integration

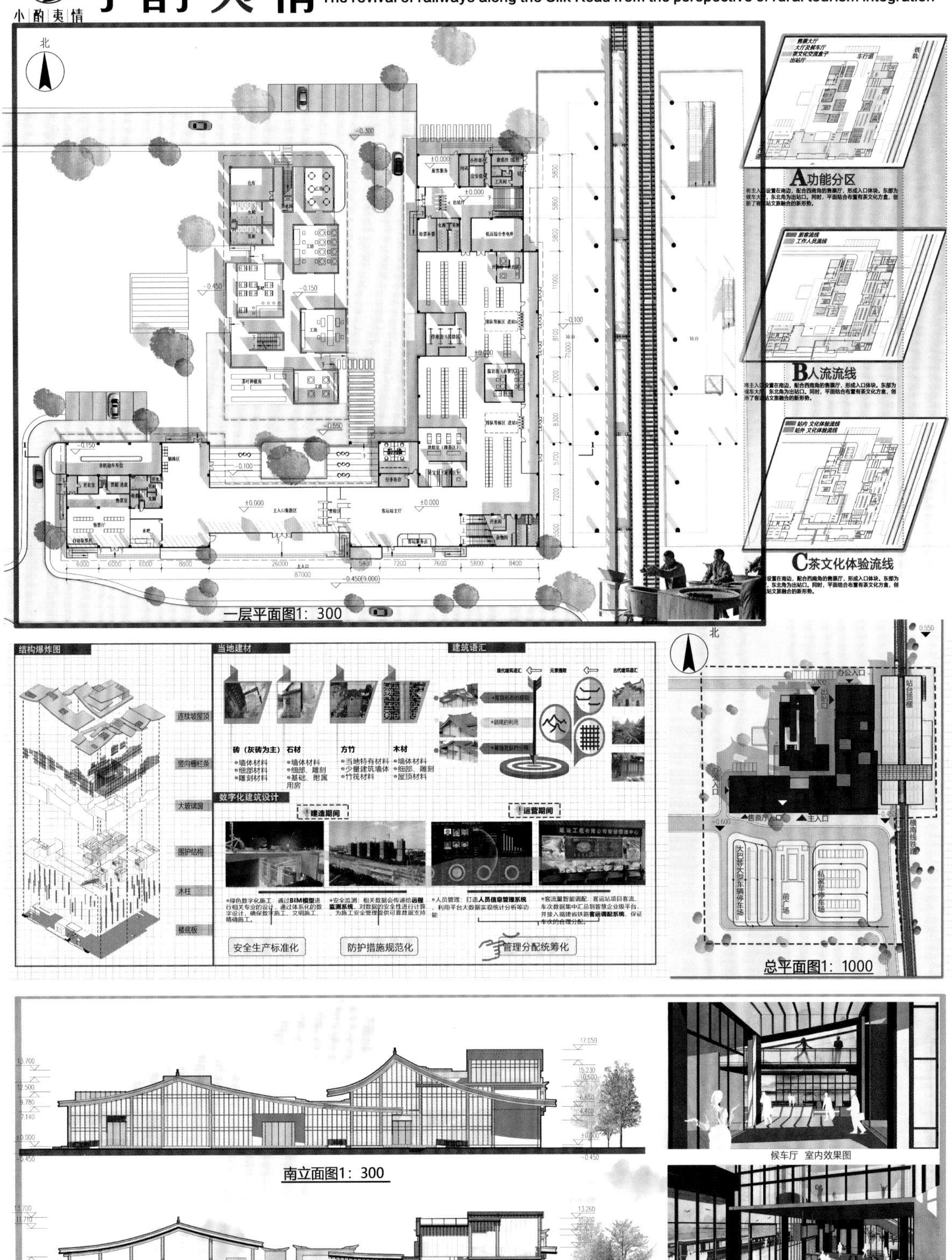

铁路促进发展

Grow From Rail

参赛学校 University/College

俄罗斯莫斯科国立建筑大学

Moscow State University of Civil Engineering, Russia

指导教师 Supervisor(s)

Ekaterina Sergeevna Shafray

参赛学生 Participant(s)

HU Zhiying

MO Siqiao

WANG Ming

DIAO Linfan

简介 Description

“铁路促进发展”项目位于俄罗斯 Nuki 村，这里是东西伯利亚铁路的一个停靠点，村庄以南半公里处就是贝加尔湖高速公路。Nuki 是一个历史悠久的小村庄，有 325 个居民。村庄靠近贝加尔湖，但交通不便限制了当地旅游业的发展。

该建筑项目旨在把 Nuki 村和卡班斯克及贝加尔湖连接起来，将分散的村庄连接成网络，促进 Nuki 村与外界的人口流动。火车站为周边城镇和村庄提供了稳定的交通网络，也支持了旅游业的发展。

火车站由金属和钢筋混凝土建造，建筑两侧均配有通道。车站主要入口上方是一条长人行道，路面铺砖，两侧种树，可作为公共空间，人穿过人行道可直接到达车站。铁路的两条“分支”分别延伸到村庄的两侧，通往更远的地方，这也体现了“铁路促进发展”的理念。

The project “Grow from rail” is in the village Nuki, which is a stopping point of the east Siberian railway and half a kilometre south of the village is the Baikal highway. Nuki is a small historical village with a population of 325. The village is close to Lake Baikal but tourism can not be developed due to traffic inconvenience.

This project intends to link Nuki between blast Kabansk and Baikal, creates connections between these scattered villages and enhance population mobility in Nuki. The railway station allows for a stable connection between surrounding towns and villages, which also supports the development of tourism.

The railway station is made of metal, reinforced concrete and there are pathways on both sides of the building. Above the main entrances of the station is a long sidewalk paved with brick and lined with trees which can be used as a public space that people can come across the station directly. That two “branches” reach both sides of the village and to the further place also demonstrates the concept of “Grow from rail”.

铁路促进发展

GROW FROM RAIL

The project "Grow from rail" is in the village Nuki, which is a stopping point of the east Siberian railway and half a kilometer south of the village is the Baikal highway. This project intends to link Nuki between blast Kabansk and Baikal, create connection between these scattered villages and enhance population mobility in Nuki.

Above the main entrances of the station is a long sidewalk paved with brick and lined with trees which can be use as a public space that people can come across the station directly. Two "branches" reach both sides of the village and to the further place also demonstrates the concept of "Grow from rail".

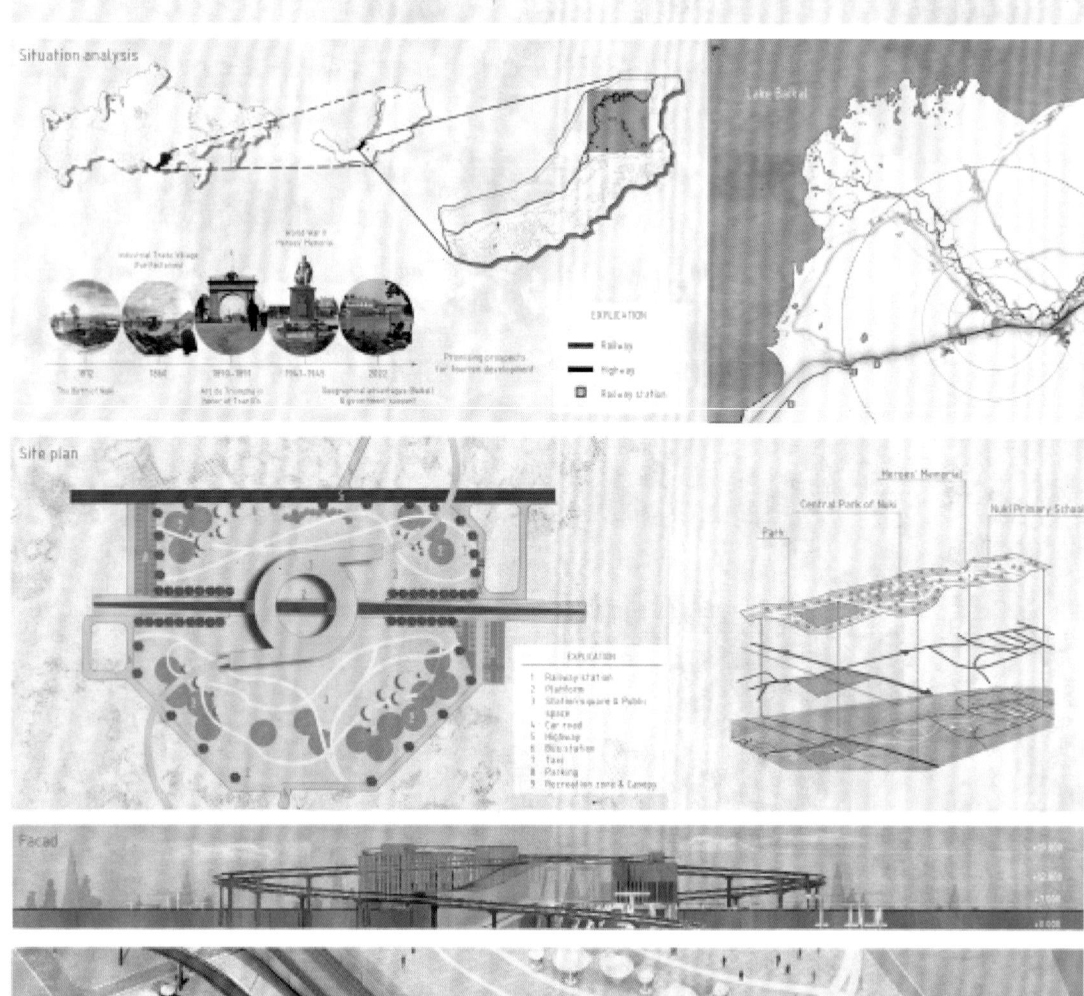

GROW FROM RAIL

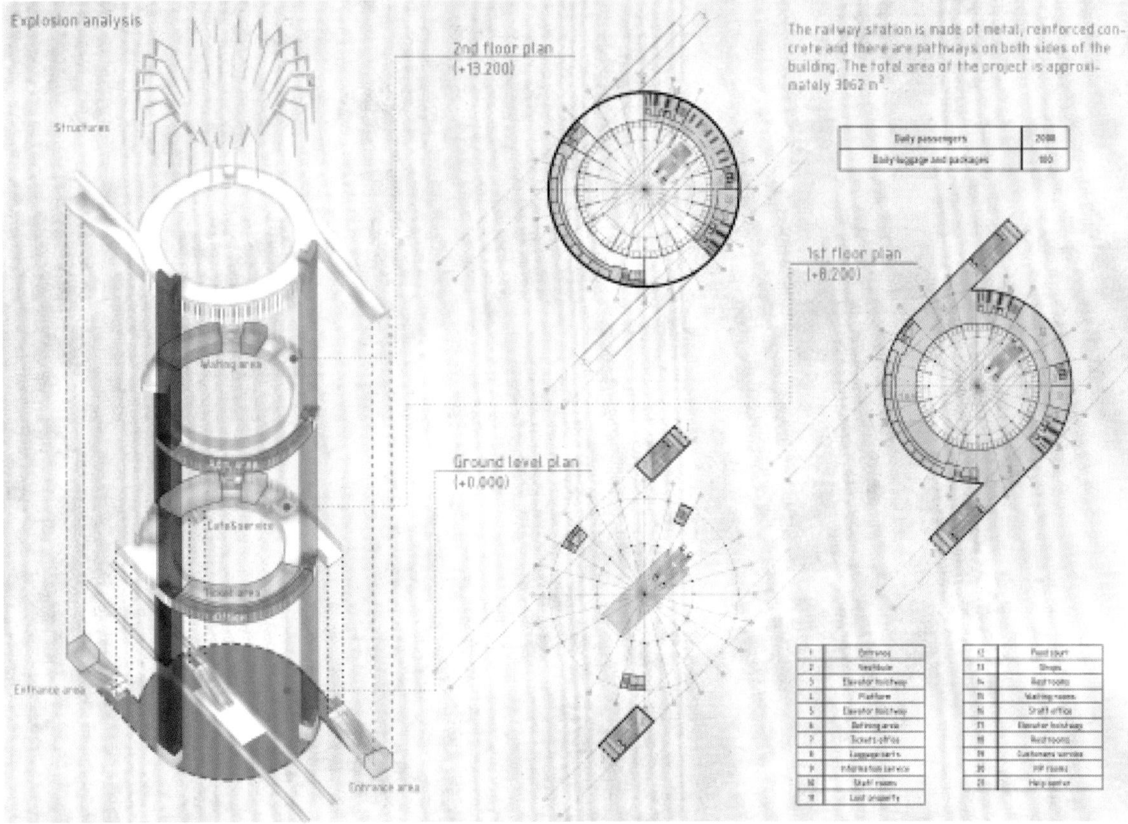

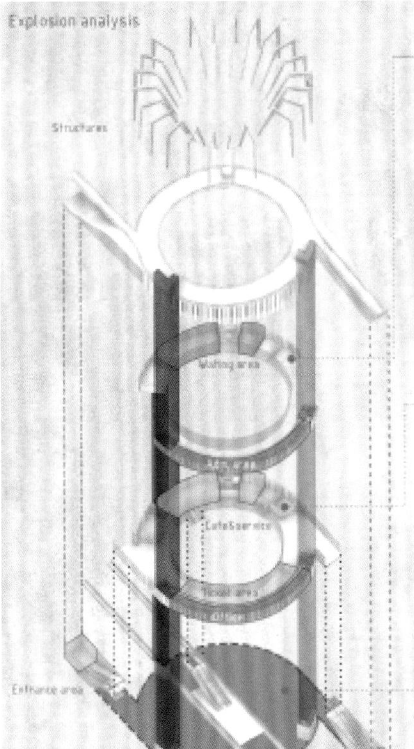
Explosion analysis
Structures
Waiting area
Rest area
Cafe&Service
Ticket area
Office
Entrance area
Entrance area

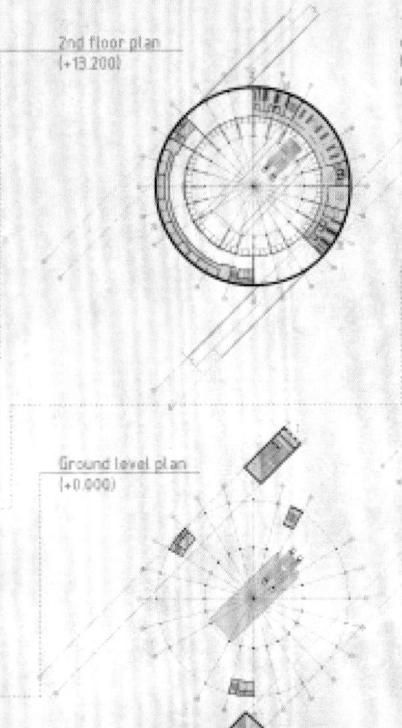
2nd floor plan
(+13.200)
Ground level plan
(+0.000)

The railway station is made of metal, reinforced concrete and there are pathways on both sides of the building. The total area of the project is approximately 3062 m^2.

Daily passengers	2000
Daily luggage and packages	100

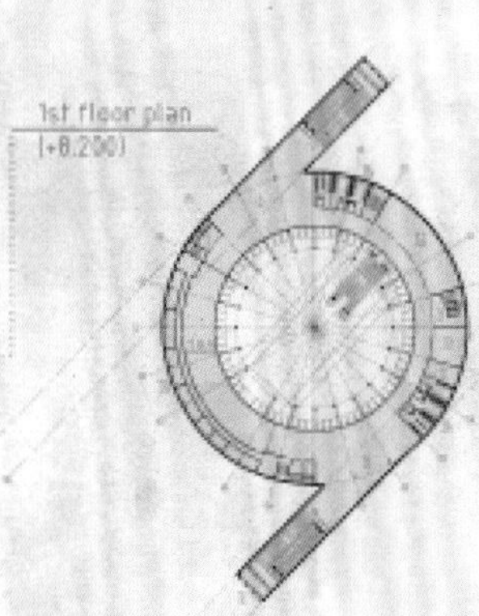
1st floor plan
(+8.200)

1	Entrance
2	Vestibule
3	Elevator hoistway
4	Platform
5	Elevator hoistway
6	Retiring area
7	Tickets office
8	Luggage carts
9	Information service
10	Staff rooms
11	Lost property

12	Food court
13	Shops
14	Restrooms
15	Waiting rooms
16	Staff office
17	Elevator hoistway
18	Restrooms
19	Customers service
20	VIP rooms
21	Help center

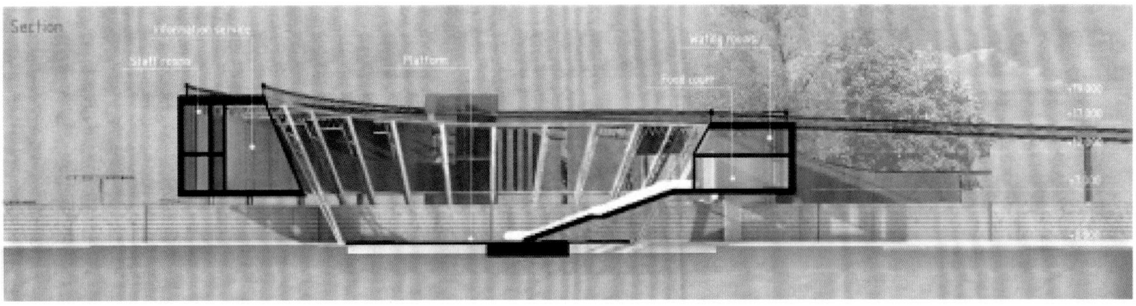
Section
Information service
Staff rooms
Platform
Waiting rooms
Food court

达坂城火车站建筑概念项目

Architectural Conceptual Project of Rail Road Station in Dabancheng

参赛学校 University/College

亚美尼亚国立建筑大学

National University of Architecture and Construction of Armenia

指导教师 Supervisor(s)

Madlena Igitkhanyan

Tatevik Nersesyan

参赛学生 Participant(s)

Hayk Zirakyan

Davit Manukyan

Ararat Midoyan

Aida Ghazaryan

Lilit Poghosyan

简介 Description

该建筑项目的核心是找到最具潜力的突破点，并以此构建整个设计方案。

因此，建筑理念的关键在于，把火车站的建设融入周边具体环境，设想人们会专门来此消磨时间，而并非仅考虑车站空间。

基于此，车站建筑将形成一个新型的公共区域，具有独特的、多功能的活动空间。

上述区域的建筑都基于智能解决方案和当代设计。

基于此，车站和综合建筑楼构造独特，而非普通车站建筑，构成了和以往截然不同的新型公共环境。其名为 DAB 园区，内涵为数字（Digital）、先进（Advanced）、美丽（Beauty），为可持续智能公共区域。

通过这种总体设计，我们希望在达坂城中建立这种大型建筑，吸引人们关注达坂城区日益发展的城镇中新的文化理念和文化创造。

车站为该区域主要建筑，建筑可供多方面观察，角度不同，景色也各不相同，这既丰富了建筑形态，也体现了建筑理念中的哲学。

The main core of the current project was to find the most potential assignment and construct the whole suggestion accordingly.

Due to that, the main key point of the concept is that the building of the station is not considered separately, but as part of the surrounding special environment, where people come especially and spend time.

It is a new public area with its own unique and multifunctional activities.

The formation of the architecture of the mentioned area is based on smart solutions and contemporary design.

From that point of view, the station and general complex building doesn't look like a common station building but a unique structure, part of an absolutely different and new public environment. It calls DAB (Digital, Advanced, Beauty) Park which is a sustainable and smart public area.

Through this general approach, adding such a large-scale building in the city of Dabancheng, the design advice is to attract attention to the new philosophy and creation of the culture in the growing existing town of Dabancheng.

The architectural design of the station, as the dominance of the area, presents itself multilaterally when it's observed from different angles, it looks different, which enriches the building and reflects the philosophy of the concept.

达坂城火车站建筑概念项目

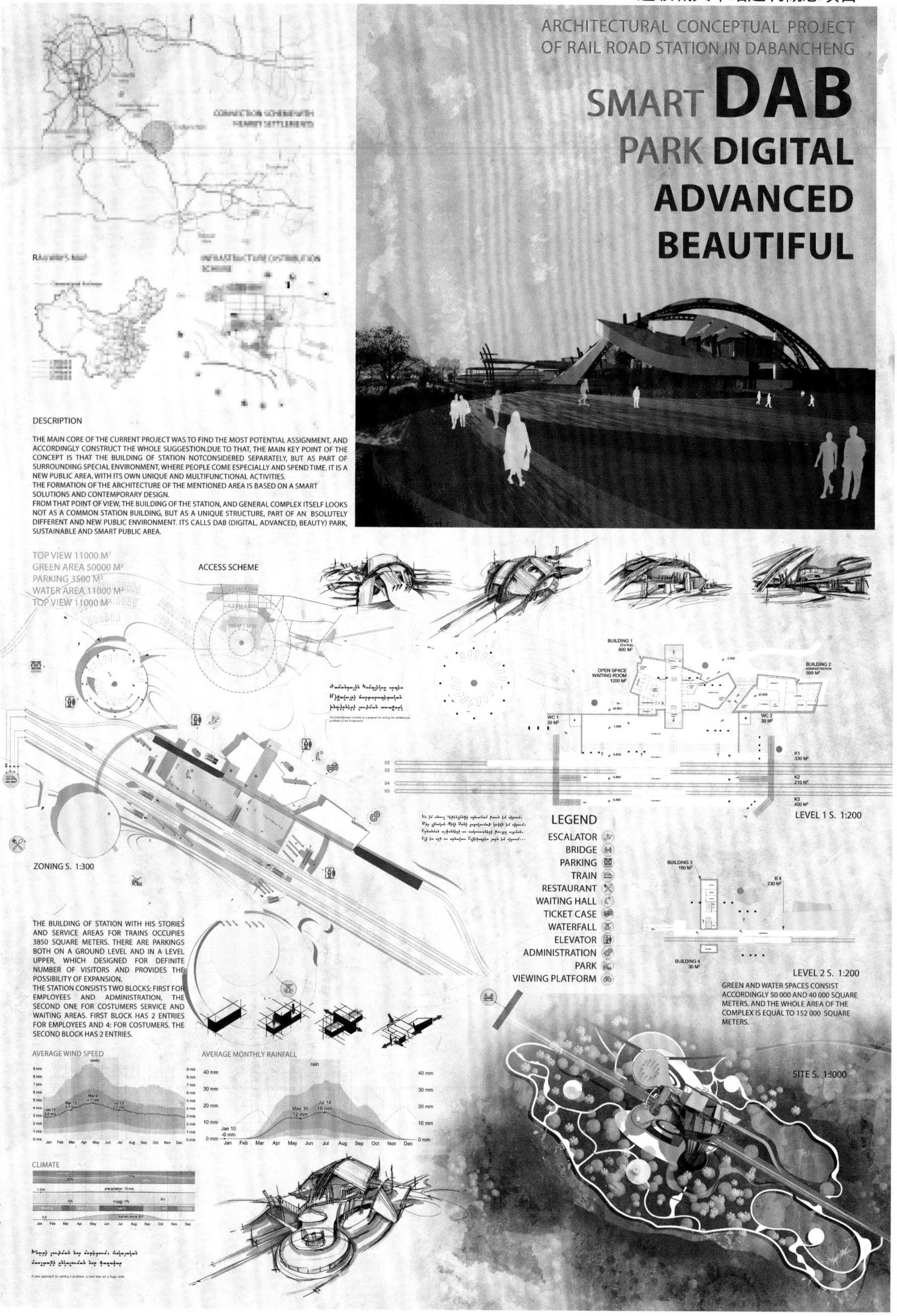

ARCHITECTURAL CONCEPTUAL PROJECT OF RAIL ROAD STATION IN DABANCHENG

SMART DAB PARK DIGITAL ADVANCED BEAUTIFUL

CONNECTION SCHEME WITH NEARBY SETTLEMENTS

RAILWAYS MAP

INFRASTRUCTURE DISTRIBUTION SCHEME

DESCRIPTION

THE MAIN CORE OF THE CURRENT PROJECT WAS TO FIND THE MOST POTENTIAL ASSIGNMENT, AND ACCORDINGLY CONSTRUCT THE WHOLE SUGGESTION.DUE TO THAT, THE MAIN KEY POINT OF THE CONCEPT IS THAT THE BUILDING OF STATION NOTCONSIDERED SEPARATELY, BUT AS PART OF SURROUNDING SPECIAL ENVIRONMENT, WHERE PEOPLE COME ESPECIALLY AND SPEND TIME. IT IS A NEW PUBLIC AREA, WITH ITS OWN UNIQUE AND MULTIFUNCTIONAL ACTIVITIES.

THE FORMATION OF THE ARCHITECTURE OF THE MENTIONED AREA IS BASED ON A SMART SOLUTIONS AND CONTEMPORARY DESIGN.

FROM THAT POINT OF VIEW, THE BUILDING OF THE STATION, AND GENERAL COMPLEX ITSELF LOOKS NOT AS A COMMON STATION BUILDING, BUT AS A UNIQUE STRUCTURE, PART OF AN BSOLUTELY DIFFERENT AND NEW PUBLIC ENVIRONMENT. ITS CALLS DAB (DIGITAL, ADVANCED, BEAUTY) PARK, SUSTAINABLE AND SMART PUBLIC AREA.

TOP VIEW 11000 M^2
GREEN AREA 50000 M^2
PARKING 3500 M^2
WATER AREA 11000 M^2
TOP VIEW 11000 M^2

ACCESS SCHEME

BUILDING 1 STATION 900 M^2
OPEN SPACE WAITING ROOM 1200 M^2
BUILDING 2 ADMINISTRATION 500 M^2
WC 1 30 M^2
WC 2 30 M^2
K1 330 M^2
K2 210 M^2
K3 400 M^2

LEVEL 1 S. 1:200

ZONING S. 1:300

LEGEND
ESCALATOR
BRIDGE
PARKING
TRAIN
RESTAURANT
WAITING HALL
TICKET CASE
WATERFALL
ELEVATOR
ADMINISTRATION
PARK
VIEWING PLATFORM

BUILDING 3 180 M^2
K 4 230 M^2
BUILDING 4 30 M^2

LEVEL 2 S. 1:200

GREEN AND WATER SPACES CONSIST ACCORDINGLY 50 000 AND 40 000 SQUARE METERS. AND THE WHOLE AREA OF THE COMPLEX IS EQUAL TO 152 000 SQUARE METERS.

THE BUILDING OF STATION WITH HIS STORIES AND SERVICE AREAS FOR TRAINS OCCUPIES 3850 SQUARE METERS. THERE ARE PARKINGS BOTH ON A GROUND LEVEL AND IN A LEVEL UPPER, WHICH DESIGNED FOR DEFINITE NUMBER OF VISITORS AND PROVIDES THE POSSIBILITY OF EXPANSION.

THE STATION CONSISTS TWO BLOCKS: FIRST FOR EMPLOYEES AND ADMINISTRATION, THE SECOND ONE FOR COSTUMERS SERVICE AND WAITING AREAS. FIRST BLOCK HAS 2 ENTRIES FOR EMPLOYEES AND 4: FOR COSTUMERS. THE SECOND BLOCK HAS 2 ENTRIES.

AVERAGE WIND SPEED

AVERAGE MONTHLY RAINFALL

SITE S. 1:1000

CLIMATE

THROUGH THIS GENERAL APPROACH, ADDING SUCH A LARGE SCALE BUILDING IN THE CITY DABANCHENG, THE DESIGN ADVICE IS TO ATTRACT ATTENTION TO THE NEW PHILOSOPHY AND CREATING OF THE CULTURE IN EXISTING GROWING CITY OF DABANCHENG. THE ARCHITECTURAL DESIGN OF THE STATION, AS DOMINANCE OF THE AREA, PRESENTS ITSELF MULTILATERALLY, WHEN FROM DIFFERENT ANGLES IT LOOKS IFFERENTLY, WHICH ENRICH THE BUILDING AND REFLECTS THE PHILOSOPHY OF THE CONCEPT.

铁路之滇

Yunnan Railway

参赛学校 University/College
华北水利水电大学
North China University of Water Resources and Electric Power, China

指导教师 Supervisor(s)
张忆萌 ZHANG Yimeng
卢玫珺 LU Meijun

参赛学生 Participant(s)
卫茹冰 WEI Rubing
何安琪 HE Anqi
曾令俭 ZENG Lingjian
孙佳攀 SUN Jiapan
范冰冰 FAN Bingbing

简介 Description

本项目位于中华人民共和国云南省德宏傣族景颇族自治州芒市，与梁河县、陇川县隔龙川江相望，南与缅甸联邦共和国交界。如何通过此铁路站点的设立，带动当地经济发展是我们设计中的重点，同时方案以“连贯东西，归于市井”为主题，打造现时代下基于立体交通构建体系的芒市铁路片区。

方案以火车站为核心，引入站城融合的思想理念，融入浮桥立体交通的设计，结合云南传统民居建筑元素、丰富的夹层配置、耦合植入不同的绿色功能等设计思想，唤醒场地的二次生命。满足当代对于交通枢纽的崭新需求，致力于打造人、自然、城市三位一体的和谐茁壮发展空间，使其成为极具代表性的火车站片区。

The site of the project is located in Mangshi City, Dehong Dai and Jingpo Autonomous Prefecture, Yunnan Province in China. Across the Longchuan River are Lianghe County and Longchuan County, while its neighbor in the south is the Republic of the Union of Myanmar. The focus of our design was to drive local economic development by setting up this railway station. With the theme of "connecting the East and the West, back to the marketplace", this design aims at shaping a railway district of Mangshi City based on a three-dimensional transportation construction system of our era.

With the railway station at its core, the design adopts the concept of station-city integration. A floating bridge was designed for three-dimensional transport. The elements in the traditional residential house in Yunnan are extracted, and together with the design of multiple layers and various green functions, the whole site is to experience rejuvenation. To meet the fresh needs of transportation in today's era, this design is expected to breed a three-pronged development space for people, nature and the city and to make this region a representative railway district.

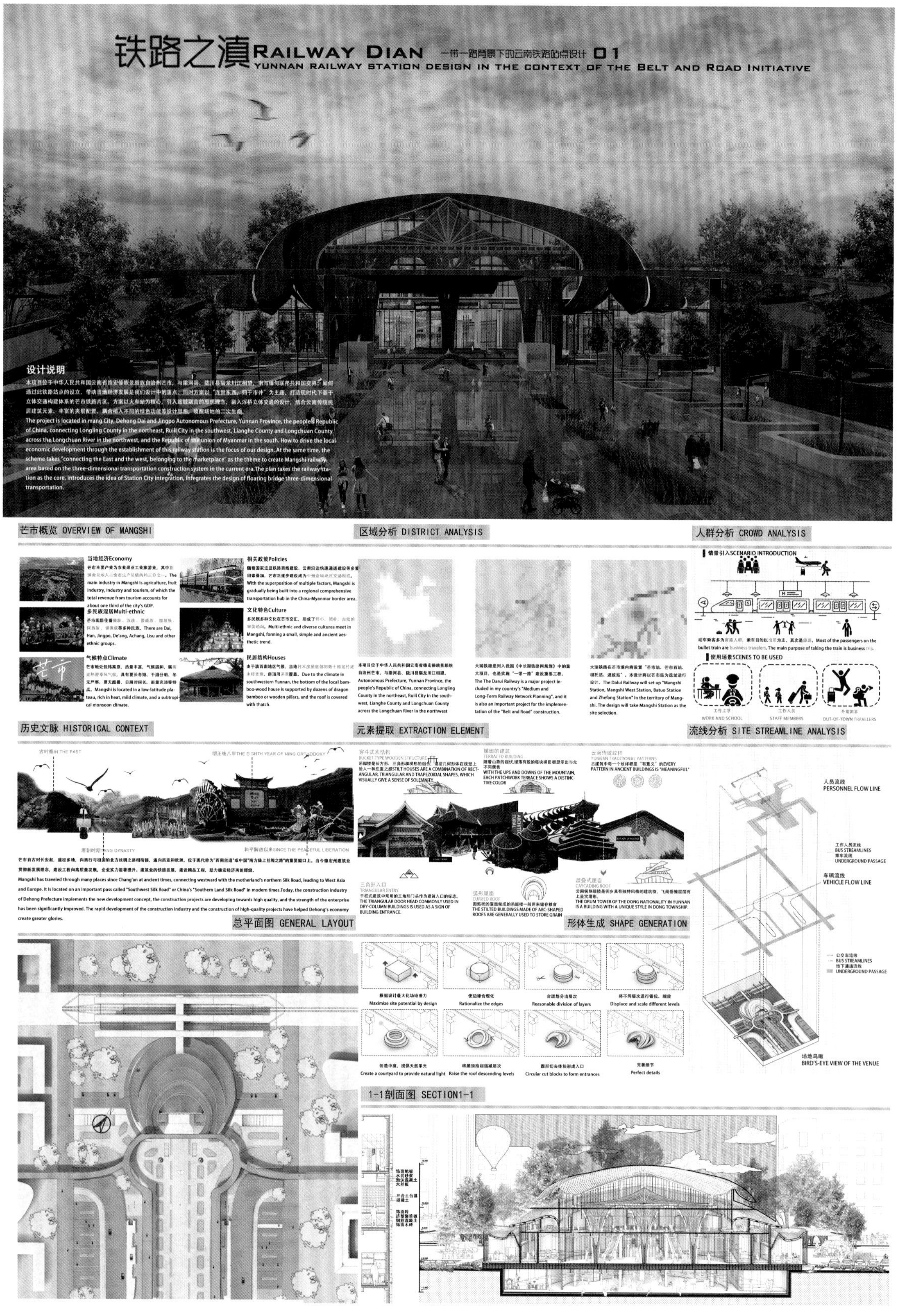

铁路之滇 RAILWAY DIAN

一带一路背景下的云南铁路站点设计 01

YUNNAN RAILWAY STATION DESIGN IN THE CONTEXT OF THE BELT AND ROAD INITIATIVE

设计说明

本项目位于中华人民共和国云南省德宏傣族景颇族自治州芒市，与梁河县、陇川县隔龙川江相望，南与缅甸联邦共和国交界。如何通过此铁路站点的设立，带动当地经济发展是我们设计中的重点。同时方案以"连贯东西，归于市井"为主题，打造现时代下基于立体交通构建体系的芒市铁路片区。方案以火车站为核心，引入站城融合的思想理念，融入浮桥立体交通的设计，结合云南传统民居建筑元素、丰富的夹层配置、耦合植入不同的绿色功能等设计思想，唤醒场地的二次生命。

The project is located in mang City, Dehong Dai and Jingpo Autonomous Prefecture, Yunnan Province, the people's Republic of China, connecting Longling County in the northeast, Ruili City in the southwest, Lianghe County and Longchuan County across the Longchuan River in the northwest, and the Republic of the union of Myanmar in the south. How to drive the local economic development through the establishment of this railway station is the focus of our design. At the same time, the scheme takes "connecting the East and the west, belonging to the marketplace" as the theme to create Mangshi railway area based on the three-dimensional transportation construction system in the current era.The plan takes the railway station as the core, introduces the idea of Station City integration, integrates the design of floating bridge three-dimensional transportation.

芒市概览 OVERVIEW OF MANGSHI

当地经济Economy

芒市主要产业为农业果业工业旅游业，其中旅游业总收入占全市生产总值的约三分之一。The main industry in Mangshi is agriculture, fruit industry, industry and tourism, of which the total revenue from tourism accounts for about one third of the city's GDP.

多民族混居Multi-ethnic

芒市聚居住着傣族、汉族、景颇族、德昂族、阿昌族、傈僳族等多种民族。There are Dai, Han, Jingpo, De'ang, Achang, Lisu and other ethnic groups.

气候特点Climate

芒市地处低纬高原，热量丰富，气候温和，属南亚热带季风气候，具有夏长冬短、干湿分明、冬无严寒、夏无酷暑、日照时间长、雨量充沛等特点。Mangshi is located in a low-latitude plateau, rich in heat, mild climate, and a subtropical monsoon climate.

相关政策Policies

随着国家泛亚铁路西线建设、云南沿边快速通道建设等多重因素叠加，芒市正逐步建设成为中缅边境地区交通枢纽。With the superposition of multiple factors, Mangshi is gradually being built into a regional comprehensive transportation hub in the China-Myanmar border area.

文化特色Culture

多民族多种文化在芒市交汇，形成了纤小、简朴、古拙的审美趋向。Multi-ethnic and diverse cultures meet in Mangshi, forming a small, simple and ancient aesthetic trend.

民居结构Houses

由于滇西南地区气候，当地竹木房屋底部用数十根龙竹或木柱支撑，房顶用茅草覆盖。Due to the climate in southwestern Yunnan, the bottom of the local bamboo-wood house is supported by dozens of dragon bamboo or wooden pillars, and the roof is covered with thatch.

区域分析 DISTRICT ANALYSIS

本项目位于中华人民共和国云南省德宏傣族景颇族自治州芒市……Autonomous Prefecture, Yunnan Province, the people's Republic of China, connecting Longling County in the northeast, Ruili City in the southwest, Lianghe County and Longchuan County across the Longchuan River in the northwest

The Darui Railway is a major project included in my country's "Medium and Long-Term Railway Network Planning", and it is also an important project for the implementation of the "Belt and Road" construction.

The Dalui Railway will set up "Mangshi Station, Mangshi West Station, Batuo Station and Zhefang Station" in the territory of Mangshi. The design will take Mangshi Station as the site selection.

人群分析 CROWD ANALYSIS

情景引入SCENARIO INTRODUCTION

Most of the passengers on the bullet train are business travellers. The main purpose of taking the train is business trip.

使用场景SCENES TO BE USED

WORK AND SCHOOL　STAFF MEMBERS　OUT-OF-TOWN TRAVELLERS

历史文脉 HISTORICAL CONTEXT

IN THE PAST　明正统八年THE EIGHTH YEAR OF MING ORTHODOXY　MING DYNASTY　SINCE THE PEACEFUL LIBERATION

Mangshi has traveled through many places since Chang'an at ancient times, connecting westward with the motherland's northern Silk Road, leading to West Asia and Europe. It is located on an important pass called "Southwest Silk Road" or China's "Southern Land Silk Road" in modern times.Today, the construction industry of Dehong Prefecture implements the new development concept, the construction projects are developing towards high quality, and the strength of the enterprise has been significantly improved. The rapid development of the construction industry and the construction of high-quality projects have helped Dehong's economy create greater glories.

元素提取 EXTRACTION ELEMENT

穿斗式木结构 BUCKET TYPE WOODEN STRUCTURE

……STILT HOUSES ARE A COMBINATION OF RECTANGULAR, TRIANGULAR AND TRAPEZOIDAL SHAPES, WHICH VISUALLY GIVE A SENSE OF SOLEMNITY.

梯田的建筑 TERRACED BUILDING

WITH THE UPS AND DOWNS OF THE MOUNTAIN, EACH PATCHWORK TERRACE SHOWS A DISTINCTIVE COLOR.

云南传统纹样 YUNNAN TRADITIONAL PATTERNS

……PATTERN IN ANCIENT BUILDINGS IS "MEANINGFUL".

三角形入口 TRIANGULAR ENTRY

THE TRIANGULAR DOOR HEAD COMMONLY USED IN DRY-COLUMN BUILDINGS IS USED AS A SIGN OF BUILDING ENTRANCE.

弧形屋面 CURVED ROOF

THE STILTED BUILDINGS MADE OF ARC-SHAPED ROOFS ARE GENERALLY USED TO STORE GRAIN.

叠落式屋面 CASCADING ROOF

THE DRUM TOWER OF THE DONG NATIONALITY IN YUNNAN IS A BUILDING WITH A UNIQUE STYLE IN DONG TOWNSHIP.

流线分析 SITE STREAMLINE ANALYSIS

人员流线 PERSONNEL FLOW LINE

BUS STREAMLINES　UNDERGROUND PASSAGE

车辆流线 VEHICLE FLOW LINE

BUS STREAMLINES　UNDERGROUND PASSAGE

场地鸟瞰 BIRD'S-EYE VIEW OF THE VENUE

总平面图 GENERAL LAYOUT

形体生成 SHAPE GENERATION

根据设计最大化场地潜力 Maximize site potential by design

使边缘合理化 Rationalize the edges

合理划分出层次 Reasonable division of layers

将不同层次进行错位、缩放 Displace and scale different levels

创造中庭，提供天然采光 Create a courtyard to provide natural light

将屋顶抬起递减层次 Raise the roof descending levels

圆形切去体块形成入口 Circular cut blocks to form entrances

完善细节 Perfect details

1-1剖面图 SECTION1-1

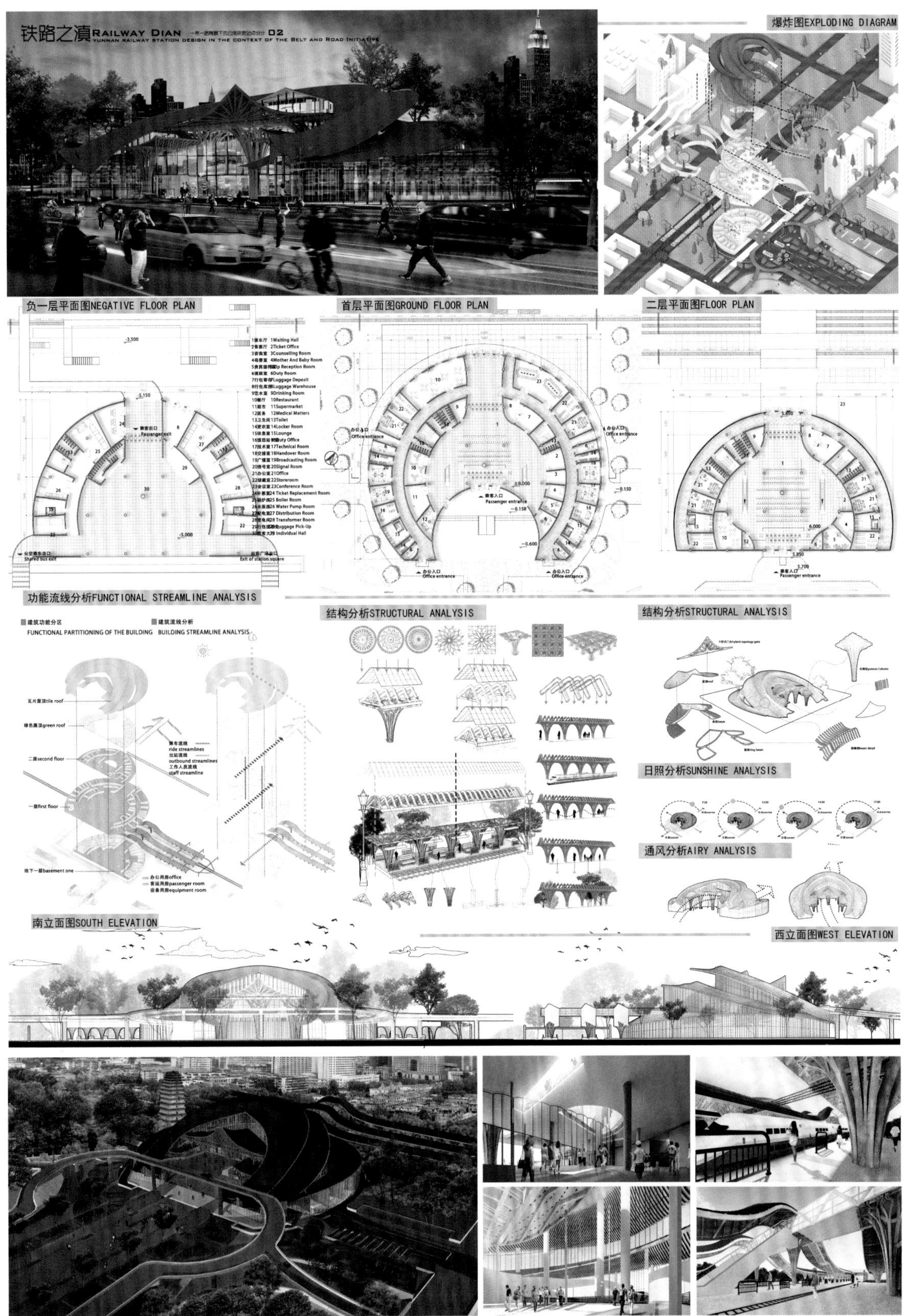
铁路之滇 RAILWAY DIAN
爆炸图EXPLODING DIAGRAM
负一层平面图NEGATIVE FLOOR PLAN
首层平面图GROUND FLOOR PLAN
二层平面图FLOOR PLAN
功能流线分析FUNCTIONAL STREAMLINE ANALYSIS
结构分析STRUCTURAL ANALYSIS
结构分析STRUCTURAL ANALYSIS
日照分析SUNSHINE ANALYSIS
通风分析AIRY ANALYSIS
南立面图SOUTH ELEVATION
西立面图WEST ELEVATION

一带共联，一站寻乡——茶·集火车站规划与设计

Planning and Design of Tea-Market Railway Station

参赛学校 University/College

上海城建职业学院

Shanghai Urban Construction Vocational College, China

指导教师 Supervisor(s)

张雪松 ZHANG Xuesong

卢　楠 LU Nan

参赛学生 Participant(s)

晋鸣骏 JIN Mingjun

马莉艳 MA Liyan

周嘉慧 ZHOU Jiahui

李袁一博 LI Yuanyibo

王　瑞 WANG Rui

简介 Description

“一带共联，一站寻乡”。“茶·集”英德火车客运站建筑设计选址位于广东省清远市下辖县级市英德。为了促进乡村振兴和发挥火车站的经济带动功能，我们将传统市集与火车站进行整合，将场地市集的展销功能融入客运站建筑功能，形成具有“一带一路”产业特色的“茶·集”复合型客运站设计理念。结合当地生活中“后花园”

式的闲适节奏以及岭南地区特有的气候和水文条件，我们将建筑底层架空引入“园”的景观，同时将竖向的景观“筒体”和交通“筒体”与“伞柱”结构构件结合，升腾起的飘板形成屋顶花园的遮阳体系，“架空展销”+“景观筒体”+“生态伞柱”+“屋顶花园”+“遮阳体系”共同构建了集“茶·集·园·站”多功能于一体的数字化绿色客运站。

The belt connects us all and the railway takes us home. The Tea-Market Station is located in the county-level city of Yingde, the jurisdiction of Qingyuan City in China’s Guangdong province. To advance the revitalization of the rural areas and expand the positive impact of railway stations on economic development, we combined the traditional market with the station by integrating the functions of display and market sales into the original function of the building, which could be summarized as a design concept of “Tea-Market” complex railway station with the characteristics of the “Belt and Road” Initiative. The design also combines the slow tempo described as “wandering in the backyard” in the local people’s daily life and the unique climate and hydrologic conditions with an out-door garden sitting on the ground floor of the building, welded tube structures for vertical view and traffic roads as well as a structural constituent of “umbrella bar” and a shading structure in the form of a roof-top garden carried by afloat boards.“Mid-air display and sales”, “scenery tube structure”, “biological umbrella bar”, “roof-top garden” and “shading system” were all included to build such a green digital railway station with multiple identities of Tea, Market, Garden and Station.

“一带” 共联
——“茶·集”火车站规划与设计
Market-Railway Station Planning and Design

①

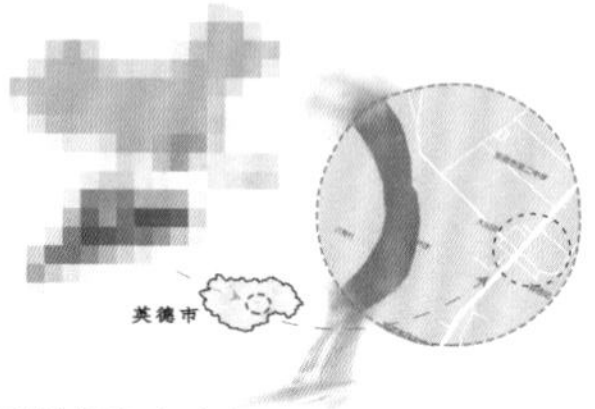

区域分析 Site Analysis

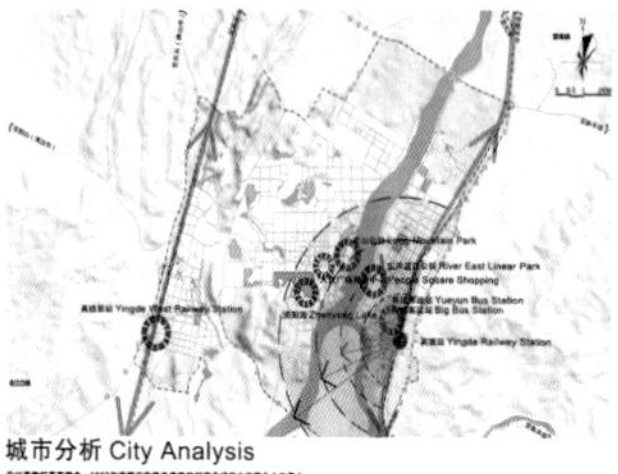

英德市

Yingde is a historical city in the north of Guangdong Province, China. The city is on the Bei River, a tributary of the Pearl River. Administratively, it is part of the Qingyuan prefecture-level city. Yingde is the important transportation hubs for north of Guangdong province and is also the joint connection for the railway network.

城市分析 City Analysis

英德西站 Yingde West Railway Station

It is famous for Yingde hong tea. In 1963 the British royal family popularized Yingde's black tea worldwide after offering the tea to guests at the Queen's Banquet. Yingde's tea history dates back to over 1,200 years ago. It is one of the top three places in the world to grow black tea.

设计说明

“一带共联，一站寻乡”。“茶·集”英德客运站建筑设计选址位于广东省清远市下辖县级市英德。英德站是“八纵八横”铁路建设的节点，与英德市的北江航道、连江航道形成特有的“铁路、水路交通网”。为了促进乡村振兴和发挥火车站的经济带动功能，在本次建筑设计中，我们将传统市集与火车站进行整合，结合英德特有的“中国红茶、绿茶之乡”的茶文化，将场地市集的展销功能融入客运站建筑功能，形成具有“一带一路”产业特色的“茶·集”复合型客运站设计理念。通过客运站建筑在交通网架立起来的一带一路经济网中发挥作用，带动英德当地经济的蓬勃发展。结合当地生活中“后花园”式的闲适节奏、以及岭南地区特有的气候和水文条件，我们将建筑底层架空引入“园”的景观，同时将竖向的景观“筒体”和交通“筒体”与“伞柱”结构构件结合，升腾起的飘板形成屋顶花园的遮阳体系，“架空展销”+“景观筒体”+“生态伞柱”+“屋顶花园”+“遮阳体系”共同构建了具有“茶·集·园·站”多功能一体的数字化绿色客运站。

Design Statement

One-Stop connection and One-stop hometown are slogans of our design. The building located in Yingde city of Guangdong province in China, which is the important transportation hubs for north of Guangdong province and is also the joint connection for the railway network. In the concept design, we develop a idea to integrate local market with traditional railway station to create a new prototype of building. With the local Tea culture, we design a local specialties market including teas in the open space of first floor and create a market-railway station. Combining the green building design principles, we introduce the Chinese Traditional Garden in the roof level to create another gathering space for people relaxing. We also used many green features in building including solar photovoltaic, Self-shading, high performance glass façade, water collection, daylighting and natural ventilation.

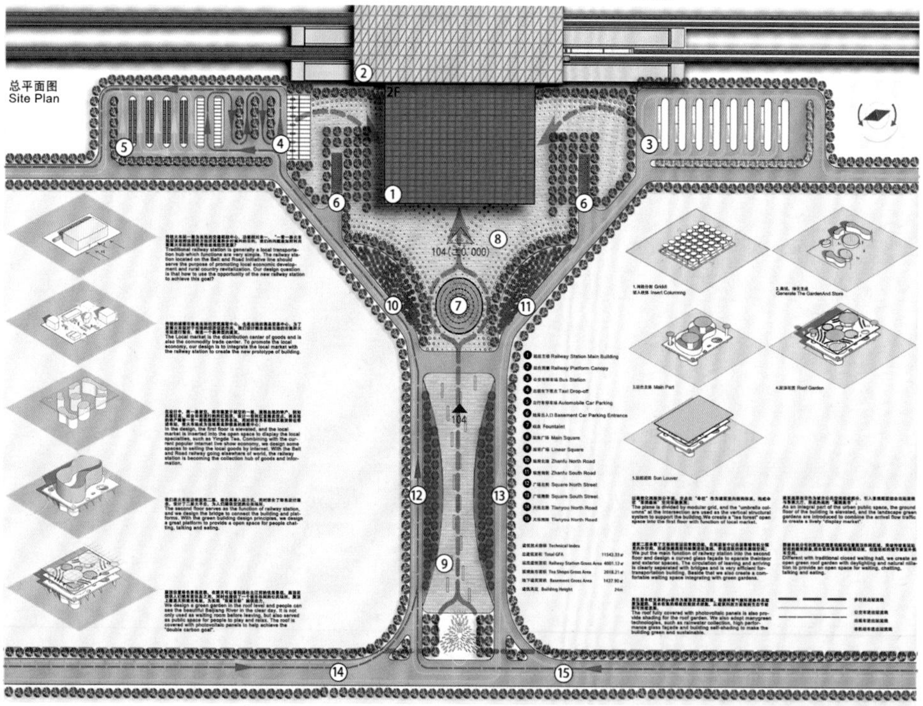

总平面图 Site Plan

1 Railway Station Main Building
2 Railway Platform Canopy
3 Bus Station
4 Taxi Drop-off
5 Automobile Car Parking
6 Basement Car Parking Entrance
7 Fountain
8 Main Square
9 Linear Square
10 Zhanfu North Road
11 Zhanfu South Road
12 Square North Street
13 Square South Street
14 Tianyou North Road
15 Tianyou North Road

Technical Index	
Total GFA	11542.33㎡
Railway Station Gross Area	4001.13㎡
Tea Shops Gross Area	2018.21㎡
Basement Gross Area	1427.90㎡
Building Height	24m

场地现状

场地绿化

场地交通

项目区块

站房现状

站前广场

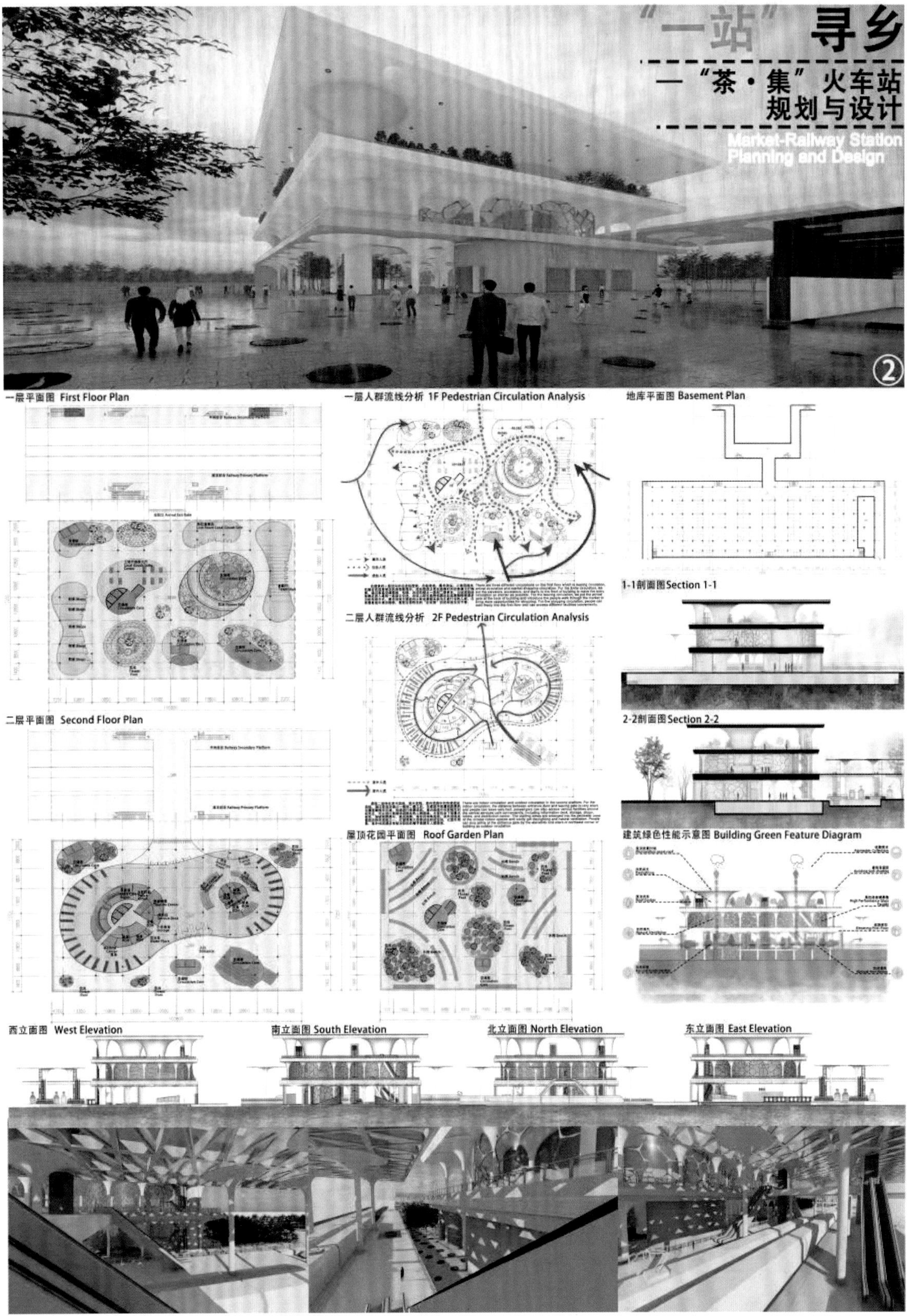
"一站"寻乡
—"茶·集"火车站
规划与设计
Market-Railway Station
Planning and Design
②
一层平面图 First Floor Plan
一层人群流线分析 1F Pedestrian Circulation Analysis
地库平面图 Basement Plan
1-1剖面图 Section 1-1
二层人群流线分析 2F Pedestrian Circulation Analysis
2-2剖面图 Section 2-2
二层平面图 Second Floor Plan
屋顶花园平面图 Roof Garden Plan
建筑绿色性能示意图 Building Green Feature Diagram
西立面图 West Elevation
南立面图 South Elevation
北立面图 North Elevation
东立面图 East Elevation

辞·昔——宁波市江北区慈城火车站设计

Meander of Wave—Design of the Cicheng Railway Station in Jiangbei District

参赛学校 University/College
东南大学
Southeast University, China

指导教师 Supervisor(s)
胡碧琳 HU Bilin
周 霖 ZHOU Lin

参赛学生 Participant(s)
韩 融 HAN Rong
花全均 HUA Quanjun
龚文晨 GONG Wenchen
胡 炯 HU Jiong
冯 薇 FENG Wei

简介 Description

每个城市都有自己的古镇，不论大小、兴衰，它们都是城市历史的缩影和文化的沉淀。慈城位于甬江入海口，所以从唐代中国开始管理海上丝绸之路之时，这一代就是非常著名的港口，是海上丝绸之路的一个重要枢纽。新站房融合老站，延续了慈城的线条之美，波澜起伏的黑瓦屋顶既是江南典型的建筑风格，也是对海洋文化的融合。与错落有致的古建筑一样，庭院的星罗棋布，更是站房不可缺少的灵动风景。小桥流水，一半烟火，一半清欢。

Every city has its ancient towns. Big or small, rising or declining, they are the microcosm of the city's history and the refinement of its culture. The city of Cicheng is located at the estuary of the Yongjiang River in Southeast China, so it has been a very famous port since the Tang Dynasty (618—907) when China started to manage the Maritime Silk Road and it was an important hub along the route. The new station building incorporates the old station to continue the beauty of the lines of Cicheng. The undulating black tiled roof is typical in the architectural style of Jiangnan, and is also a fusion of maritime culture. Like the old buildings scattered unevenly in the region, the dotted courtyards are an animated landscape that is indispensable to the station building. Pleasure combined with hustle and bustle derived from a view that is as simple as a small bridge over the flowing stream.

辞·昔

——宁波市江北区慈城火车站设计

Wave and meander——Design of Cicheng Railway Station in Jiangbei District

每个城市都有自己的古镇，不论大小、兴衰，它们都是城市历史的缩影和文化的沉淀。慈城位于甬江入海口，所以从唐代开始，中国开始管理海上丝绸之路的时候，这一代就是非常著名的港口，是海上丝绸之路的一个重要枢纽。新站房融合老站，延续了慈城的线条之美，波澜起伏的黑瓦屋顶既是江南典型的建筑风格，也是对海洋文化的融合。与错落有致的古建筑一样，庭院的星罗棋布，更是站房不可缺少的灵动风景。小桥流水，一半烟火、一半清欢。

Every city has its own ancient town, regardless of its size, rise or fall, they are a microcosm of the city's history and precipitation of its culture. Cicheng is located at the estuary of the Yong River, so it was a very famous port from the Tang Dynasty, when China started to manage the Maritime Silk Road, and was an important hub of the Maritime Silk Road in this generation. The new station house incorporates the old station and continues the beauty of the lines of Cicheng, with undulating black tiled roofs that typical of the architectural style of Jiangnan and the fusion of maritime culture as well. Like the old buildings in picturesque disorder, the courtyard is a dotted, animated landscape that is indispensable to the station house. Pleasure combined with hustle and bustle derived from a small bridge over the flowing stream.

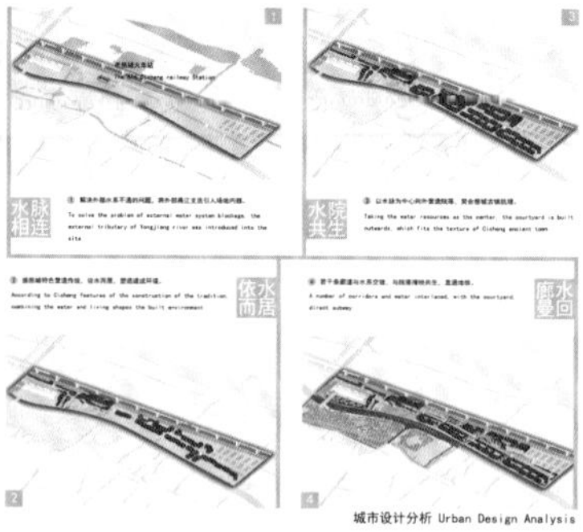

城市设计分析 Urban Design Analysis

建筑密度参数分析 Building density parameter analysis

规划结构图 Planning structure

总平面图 General Layout Plan

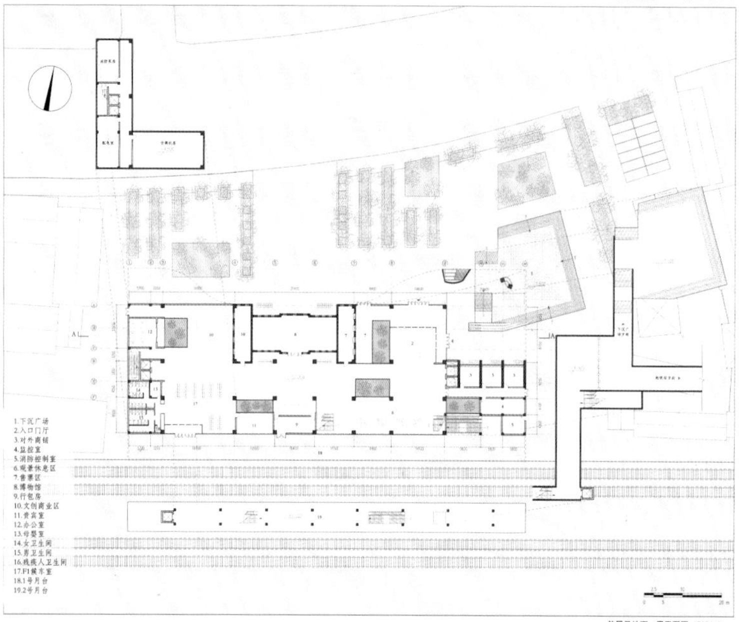

1.下沉广场
2.入口门厅
3.对外商铺
4.监控室
5.消防控制室
6.观景休息区
7.售票区
8.博物馆
9.行包房
10.文创商业区
11.贵宾室
12.办公室
13.母婴室
14.女卫生间
15.男卫生间
16.残疾人卫生间
17.F1候车室
18.1号月台
19.2号月台

首层及地下一层平面图 1F&B1 Plan

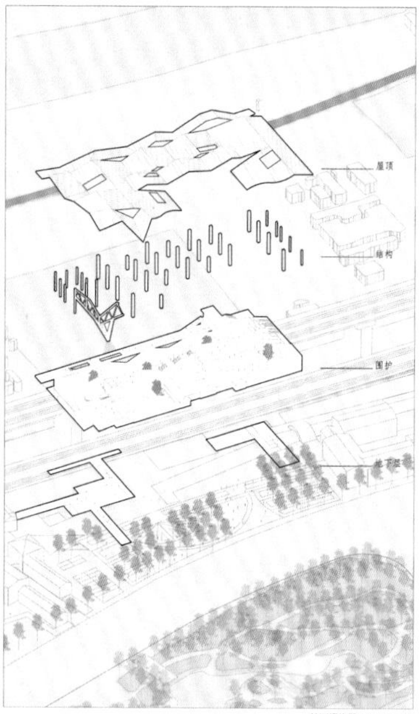

轴测爆炸图 Axonometric

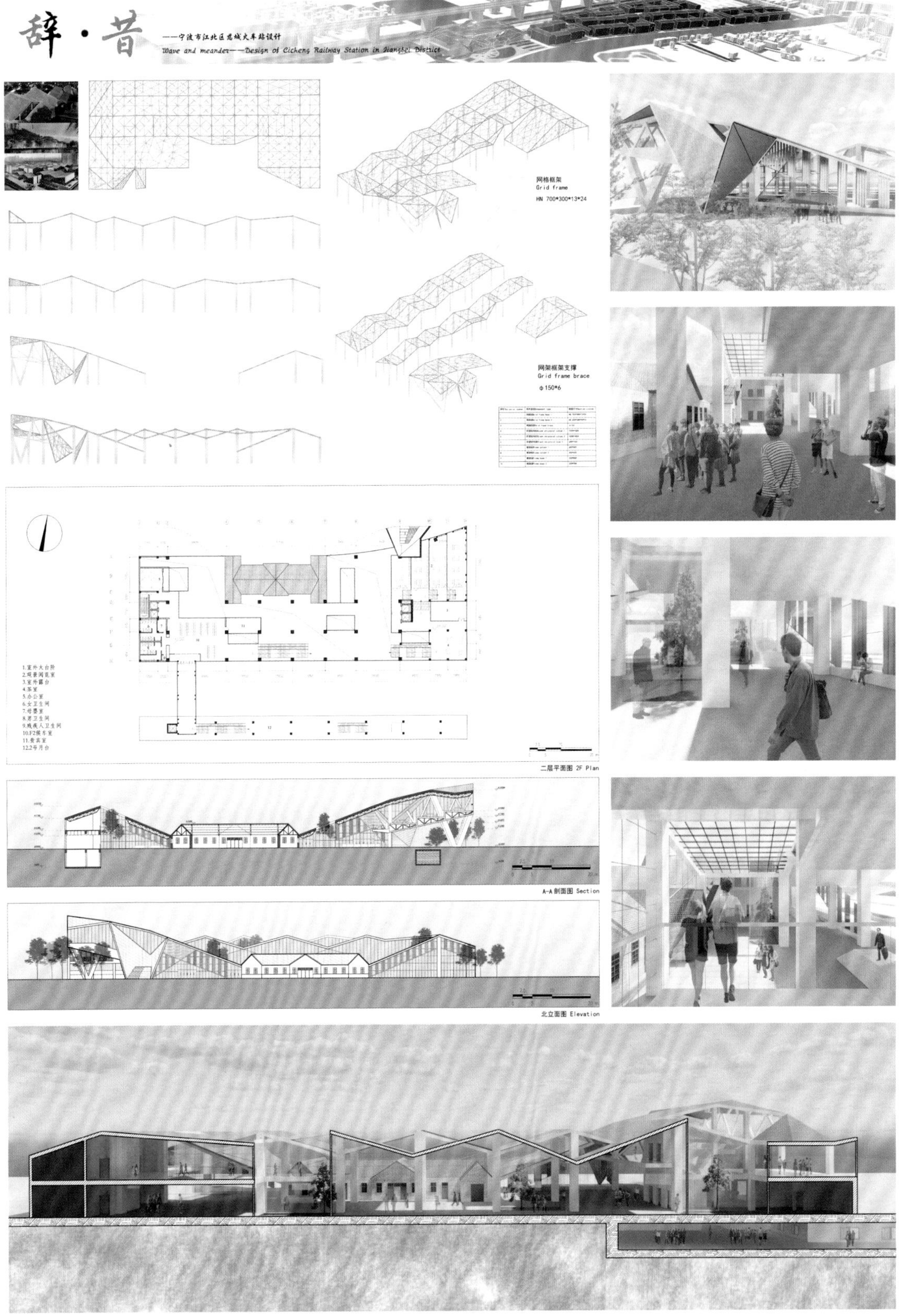
辞·昔
——宁波市江北区慈城火车站设计
Wave and meander——Design of Cicheng Railway Station in Jiangbei District
网格框架
Grid frame
HN 700*300*13*24
网架框架支撑
Grid frame brace
φ150*6
1.室外大台阶
2.观景阅览室
3.室外露台
4.茶室
5.办公室
6.女卫生间
7.母婴室
8.男卫生间
9.残疾人卫生间
10.F2候车室
11.贵宾室
12.2号月台
二层平面图 2F Plan
A-A剖面图 Section
北立面图 Elevation

实现钢管再利用的模块化车站设计

Modular Station with Reuse of Steel Pipes

参赛学校 University/College
英国东伦敦大学
University of East London, the UK

指导教师 Supervisor(s)
Fulvio Wirz

参赛学生 Participant(s)
Hussain Sadliwala
Mohan Dungrani
Alina Klimenteva

简介 Description

英国诺森伯兰郡铁路线重新投入使用，旨在为当地居民和企业提供经过优化的新型交通网络，从而刺激和支持诺森伯兰郡及周边地区的经济增长、重建和社区发展。我们对阿兴顿火车站的设计方案采用了模块化结构，实现了钢管的再利用。重复使用钢管和预制策略可以减少施工期间的碳排放。弯管、接头、立面元素和隔断的预制可以在场外完成，节约能源、水资源同时节省施工时间。我们计划通过车站的设计，整合和振兴城镇的社会和文化活动。因此，我们设计将车站与现有的战争纪念馆连接起来。在建立一个强大社区的同时，我们还计划修建一个毗邻车站大楼的广场，配备餐厅、咖啡馆和商店等设施，满足当地的现实需求，同时有助于吸引游客。

The reopening of the Northumberland line aims to stimulate and support economic growth, regeneration, and community development in Northumberland and the surrounding regions by providing new and improved transport links for local people and businesses. Our proposal for the Ashington railway station follows a modular structure with the reuse of steel pipes. The reuse of steel pipes and a pre-fabrication strategy to reduce carbon emissions during construction. The prefabrication of bent pipes, joints, facade elements, and partitions can be done off-site saving energy, water, and time. Our scheme aims to integrate and enhance the social and cultural activities of the town. Our scheme aims to integrate and enhance the social and cultural activities of the town. Our design connects the station with the existing war memorial on-site. While keeping in mind a strong community, we have proposed a plaza adjoining the station building that will accommodate restaurants, cafes, and shops to meet the contemporary needs of the town. It will also help to attract local tourists to the town.

Ashington

MODULAR STEEL STATION

The reopening of the Northumberland line aims to stimulate and support economic growth, regeneration, and community development in Northumberland and the surrounding regions by providing new and improved transport links for local people and businesses. The proposal for the Ashington railway station follows a modular structure with the reuse of steel pipes. The reuse of steel pipes and a pre-fabrication strategy to reduce carbon emissions during construction. The prefabrication of bent pipes, joints, facade elements, and partitions can be done off-site saving energy, water, and time. The scheme aims to integrate and enhance the social and cultural activities of the town. Proposed design connects the station with the existing war memorial on-site. While keeping in mind a strong community, we have proposed a plaza adjoining the station building that will accommodate restaurants, cafes, and shops to meet the contemporary needs of the town. It will also help to attract local tourists to the town.

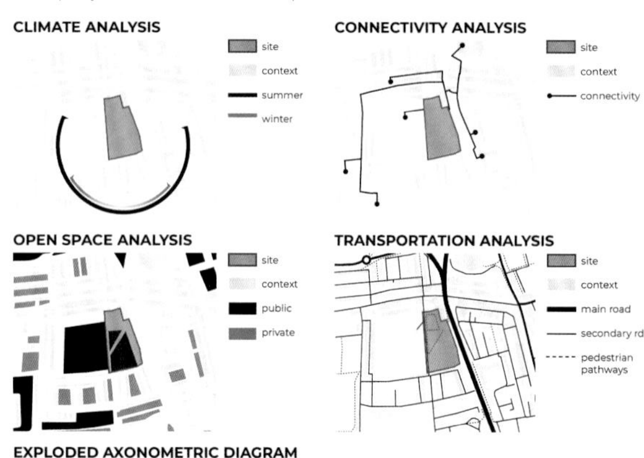

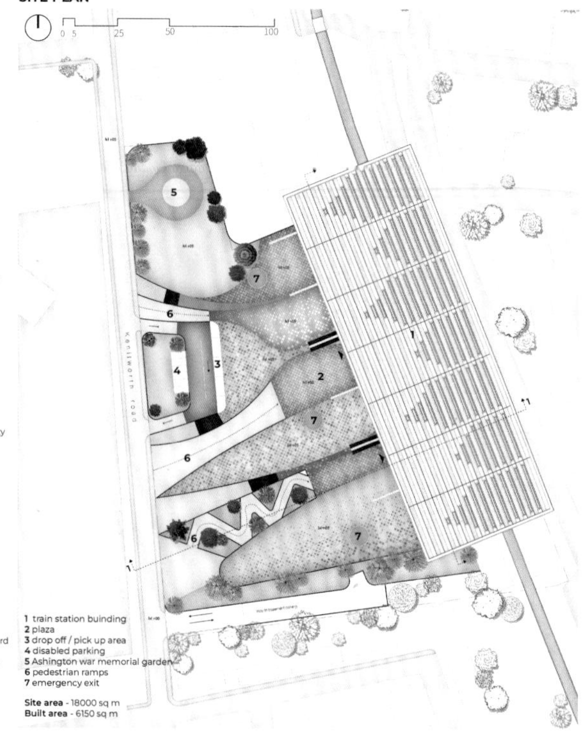

EXPLODED AXONOMETRIC DIAGRAM

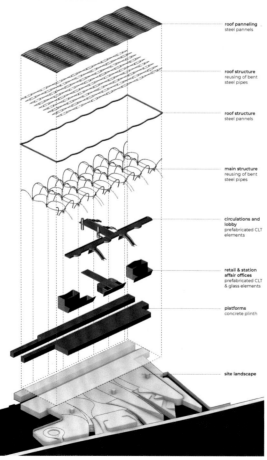

roof panneling
steel pannels

roof structure
reusing of bent steel pipes

roof structure
steel pannels

main structure
reusing of bent steel pipes

circulations and lobby
prefabricated CLT elements

retail & station affair offices
prefabricated CLT & glass elements

platforms
concrete plinth

site landscape

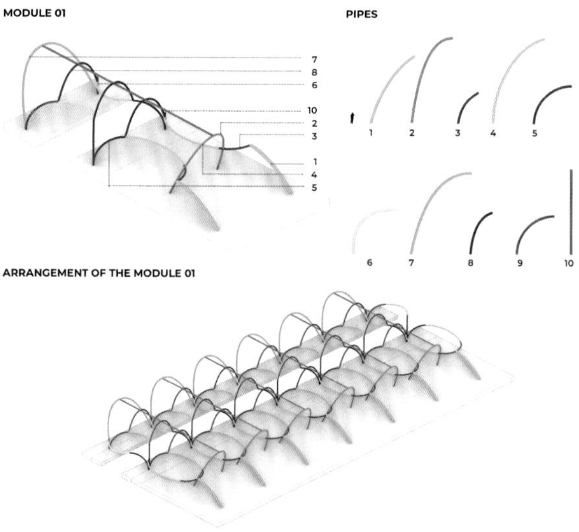

PROTOTYPE OF JOINTS SCALE 1:100

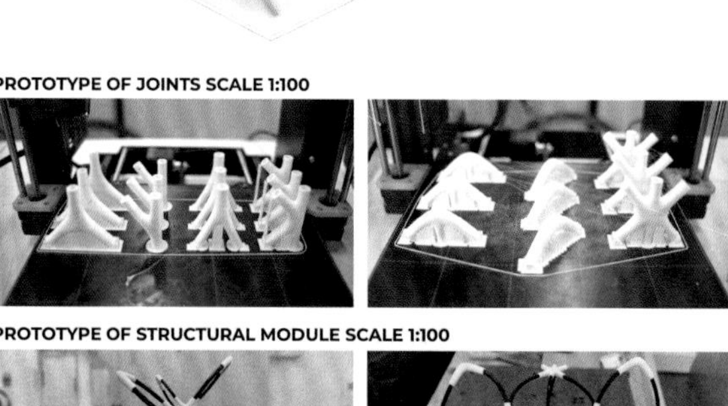

PROTOTYPE OF STRUCTURAL MODULE SCALE 1:100

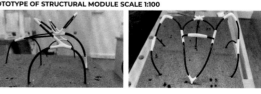

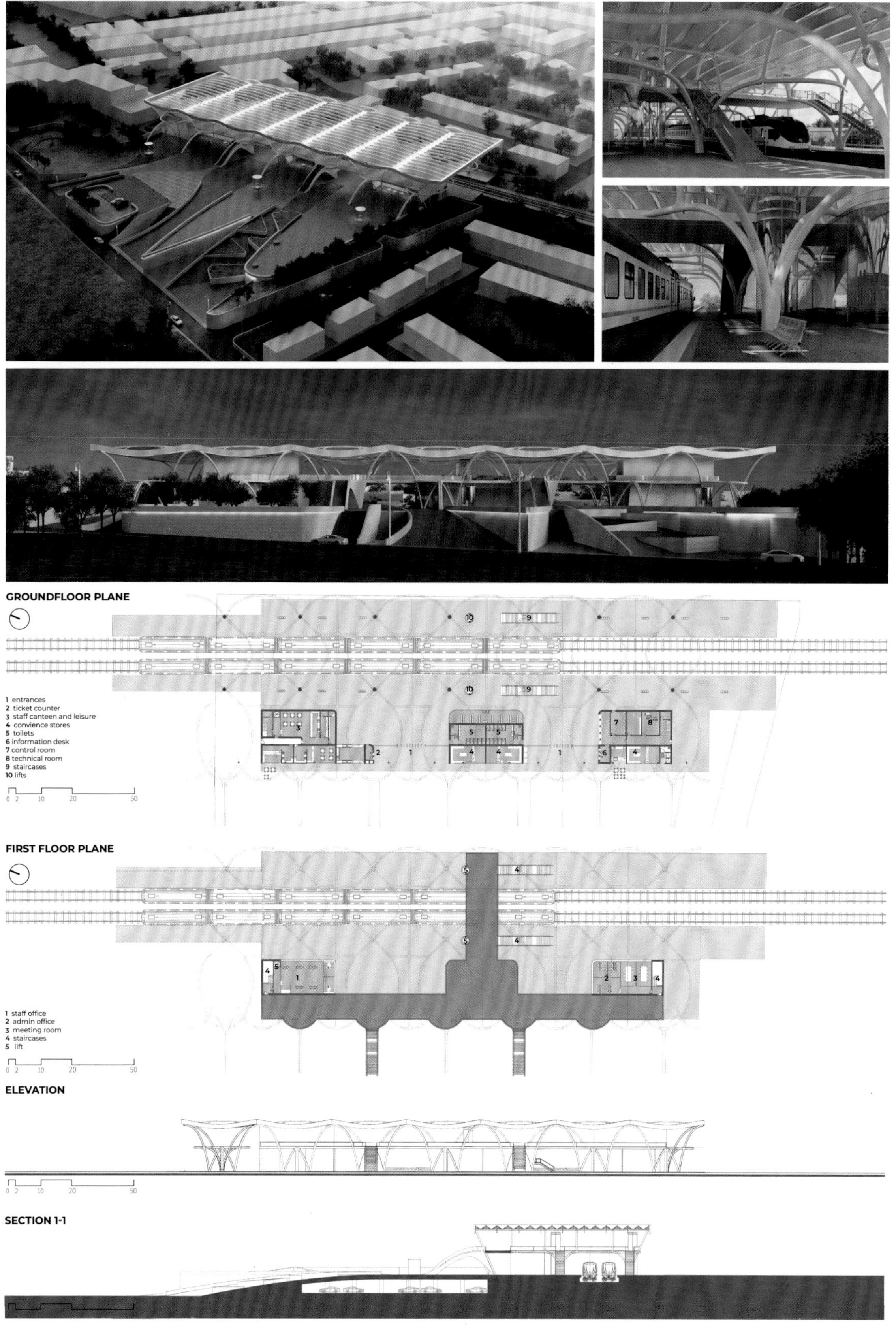
GROUNDFLOOR PLANE
1 entrances
2 ticket counter
3 staff canteen and leisure
4 convience stores
5 toilets
6 information desk
7 control room
8 technical room
9 staircases
10 lifts
0 2 10 20 50
FIRST FLOOR PLANE
1 staff office
2 admin office
3 meeting room
4 staircases
5 lift
0 2 10 20 50
ELEVATION
0 2 10 20 50
SECTION 1-1

三角设计：黏土拱顶和 3D 打印加固的车站设计

Triangulation: Station Design with Reinforced Clay Vaults with 3D Printed Reinforcements

参赛学校 University/College

英国东伦敦大学

University of East London, the UK

指导教师 Supervisor(s)

Fulvio Wirz

参赛学生 Participant(s)

Binal Patel

Martina Cossu

Ubhanayah Pathmanathan

Lokesh Emmidi

简介 Description

三角设计是英格兰南海岸海斯火车站的再开发方案。海斯是英格兰汉普郡南安普敦附近的一个小镇。它位于南安普敦溺湾的岸边，渡轮服务使得身处南安普敦的人们能够到达附近的小型购物区、码头和游艇码头。20 世纪初，海斯还是一个“小渔村”；每小时有一班汽船开往南安普敦，码头尽头有海斯游艇俱乐部的会所。海斯火车站建在海岸路外一条短路尽头的路堤上。车站有一个长 350 英尺、宽 12 英尺的站台。以前的海斯火车站现已改建为河畔遗址中心（Waterside Heritage Centre）。我们提议采用一种创新方案来解决城市设计难题，借助计算机进行设计构思，将几何形状视为灵感来源。多边形由直线构成。三角形是唯一在一侧受到压力时也能保持稳定的多边形，而如果正方形的一边受到压力，就会变成菱形。

Triangulated is a proposal for the redevelopment of a railway station in Hythe, in the South coast of England. Hythe is a town near Southampton, Hampshire, England. It is located by the shore of Southampton Water, and has a ferry service connecting it to Southampton a small shopping area, a pier, and a marina for yachts. At the beginning of the 20th century, Hythe was a “little fishing village”; with an hourly steamboat service to Southampton, and the clubhouse of Hythe Yacht Club was at the end of the pier. Hythe station was built on an embankment at the end of a short approach road off Shore Road. The station had a single platform 350ft long by 12ft wide on the downside of the line. The previous Hythe station is now the Waterside Heritage Centre. Our proposal is an innovative solution to solve the difficulty of urban design by using a computational design strategy and a geometrical shape as the origin pattern. polygon is a shape made from a straight line, and the triangle is the only polygon that will not shift under pressure as, if pressure is applied to one side of a square, it will shift into a rhombus.

三角设计：黏土拱顶和 3D
打印加固的车站设计

Triangulated

Triangulated is a proposal for the redevelopment of a railway station in Hythe, in the South coast of England.

Hythe is a town near Southampton, Hampshire, England. It is located by the shore of Southampton Water, and has a ferry service connecting it to Southampton a small shopping area, a pier, and a marina for yachts.

At the beginning of the 20th century, Hythe was a "little fishing village";with an hourly steamboat service to Southampton, and with the clubhouse of Hythe Yacht Club at the end of the pier.

Hythe station was built on an embankment at the end of a short approach road off Shore Road. The station had a single platform 350ft long by 12ft wide on the downside of the line The previous Hythe station is now the Waterside Heritage Centre.

Our proposal is an innovative solution to solving the difficulty of urban design by using a computational design strategy and a geometrical shape as the origin pattern.
A polygon is a shape made from a straight lines, and the triangle is the only polygon that will not shift under pressure as, if pressure is applied to one side of a square, it will eventually shift into a rhombus.

All triangles are stable.
When a force (the load) is applied to one of the corners of a triangle, it is distributed down each side. The two sides of the triangle are squeezed. Another word for this squeezing is compression. The third side of the triangle is pulled, or stretched sideways- tension.

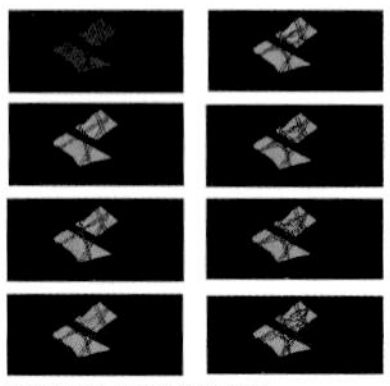

Spatial planning and urban design strategy.
Definition of the urban strategy by analysis of the networks.

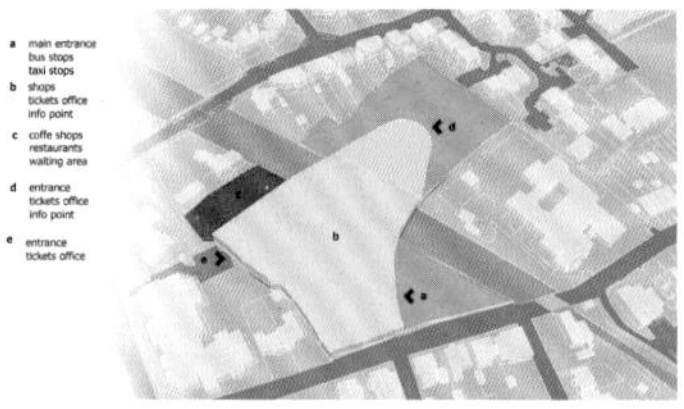

1.Massing diagram.

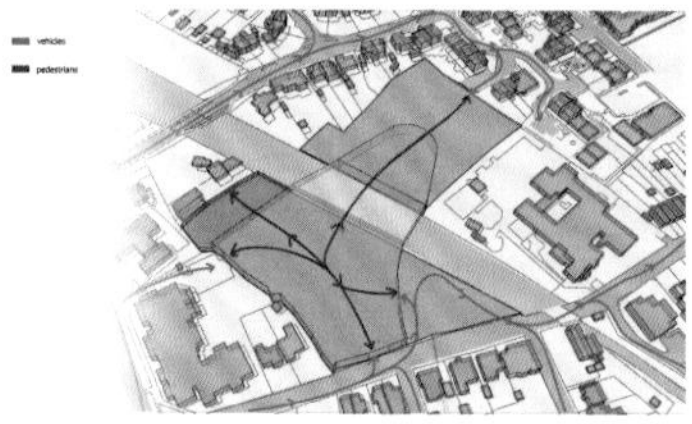

2.Circulation diagram.

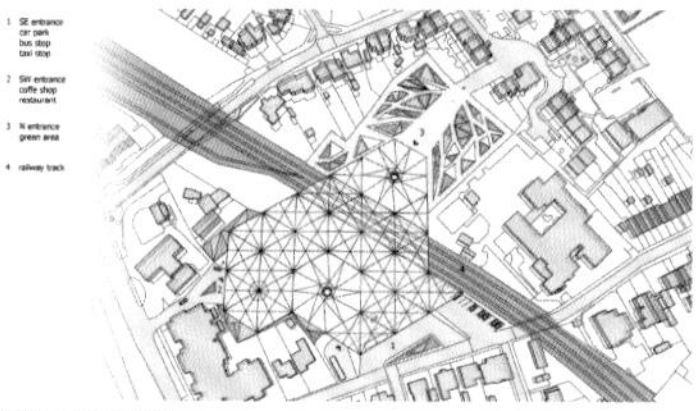

3. Final allocation of functions.

1. Study of pathways.

2. Definition of pathways and urban design.

3.Final result.

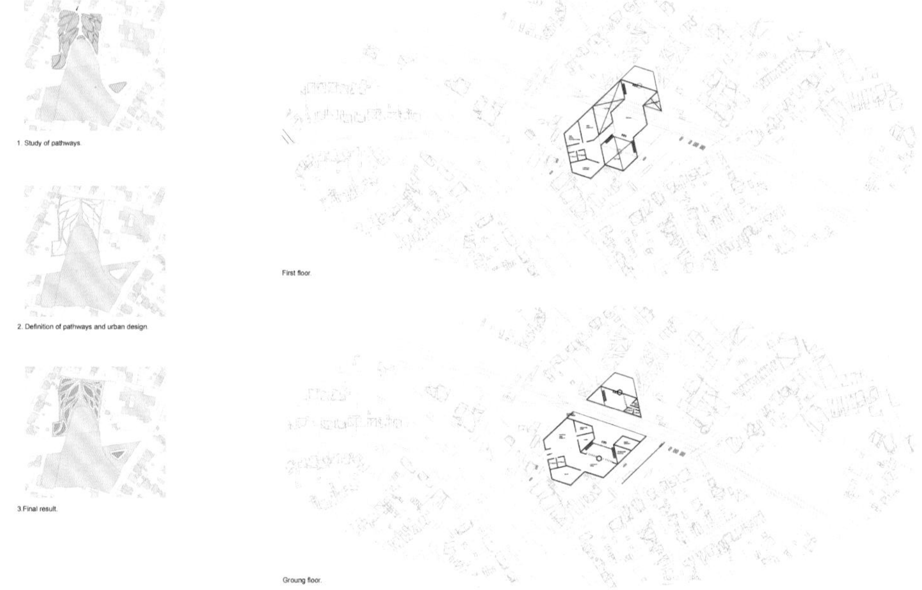

First floor.

Groung floor.

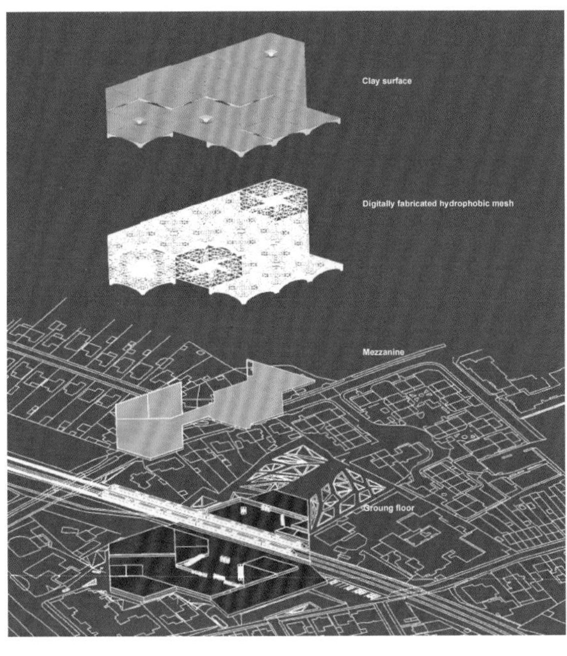

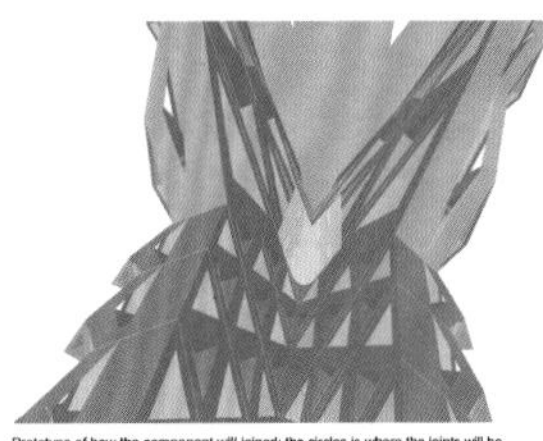

Prototype of how the component will joined; the circles is where the joints will be.

The components parts will be split and printed separately. The connecters of the joint will be printed separately, and the joints of it will be printed attached to the mesh elements. Doing this joinery method allows for low-cost and less wastage of materials.

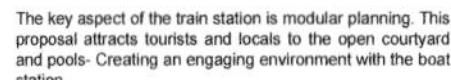

The key aspect of the train station is modular planning. This proposal attracts tourists and locals to the open courtyard and pools- Creating an engaging environment with the boat station.

Using digital fabrication to construct a train station establishes better management in production, speed, accuracy, zero waste (making it eco-friendly), lower cost structures, and the capability to build in any climate.

As clay has a high plasticity level- because of the amount of moisture applied and absorbed, 3D reinforcement with hydrophobic properties for clay application can prevent the moisture level from damaging the structure and decreasing its life span. Using the 3D printing method to create a hydrophobic surface can reduce the need for material coating. The components will be printed in plates and bolted together to spray the clay on top.
The components parts will be split and printed separately. The connecters of the joint will be printed separately, and the joints of it will be printed attached to the mesh elements. Doing this joinery method allows for low-cost and less wastage of materials.

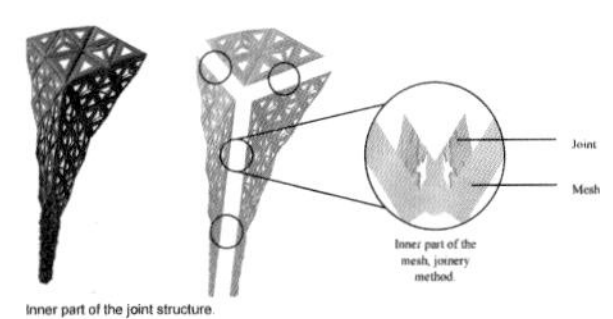

Joint
Mesh
Inner part of the mesh, joinery method.

Inner part of the joint structure.

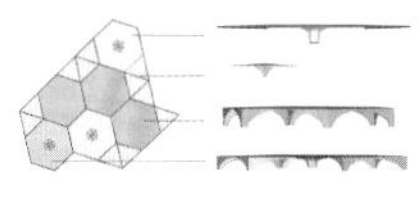

Four components modularity.

1. 3D printed high density mesh.

2. 3D printed high density mesh with clay application.

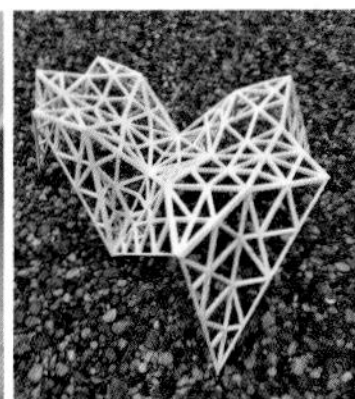

3. 3D printed low density mesh.

4. 3D printed low density mesh with clay application.

5. 3D printed mesh with clay application.

NW side facing elevation.

1. X-ray picture of digitally fabricated hydrophobic mesh reinforcement for clay.

2. Hydrophobic mesh design.

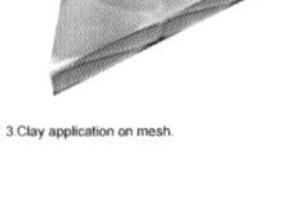

3.Clay application on mesh.

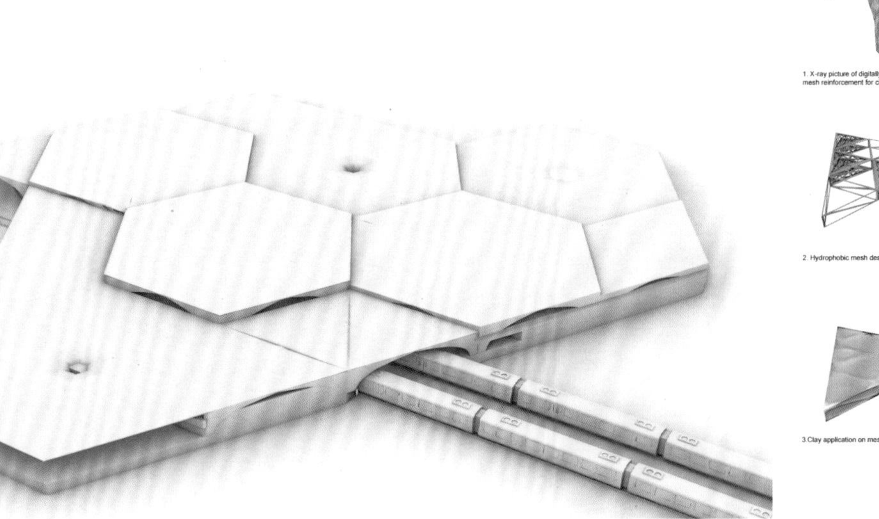

Perspective render.

绵延·织情

Stretched Hills and Weaved Emotions

参赛学校 University/College
安徽建筑大学
Anhui Jianzhu University, China

指导教师 Supervisor(s)
解玉琪 XIE Yuqi
陈萨如拉 CHEN Sarula

参赛学生 Participant(s)
王昊平 WANG Haoping
贺晓萱 HE Xiaoxuan
沈隆怀 SHEN Longhuai
许雨佳 XU Yujia

简介 Description

铁路车站往往承担着人员和物资运输的重要功能。但是铁路却又会因铁轨铺设而造成一系列问题。针对引发的这些问题，我们提出一种新解法，试图给予车站一种多义性空间的可能。而基地选在张掖这座文化灿烂的交通重镇，我们亦是延续上文思路。车站本连接兰新铁路上的城市与村落，而我们的新车站跨铁路，连接铁路两侧居民，以两个方向的“连接”来进一步凸显张掖“交流之城”的名片，然后结合甘肃当地多彩文化，最终打造一个承担着人文重量与人情温度的建筑。

Railway stations usually function for the transportation of people and goods. However, traditional railways may bring about a series of problems due to the laying of tracks. Our solution is to provide a possibility for the space to be of multiple senses. Regarding that, we set the site in Zhangye, a city in northwest China, also a transportation hub with brilliant culture. The station originally connects the cities and villages along the Lanzhou-Xinjiang Railway, while our new station is set across the tracks and connects residents on both sides of them. Such a two-way connection highlights the reputation of Zhangye as a "city of communication". By combining with the colorful culture of the Gansu Province, the ultimate goal of the design is to create a building manifesting profound humanity and a close bond between people.

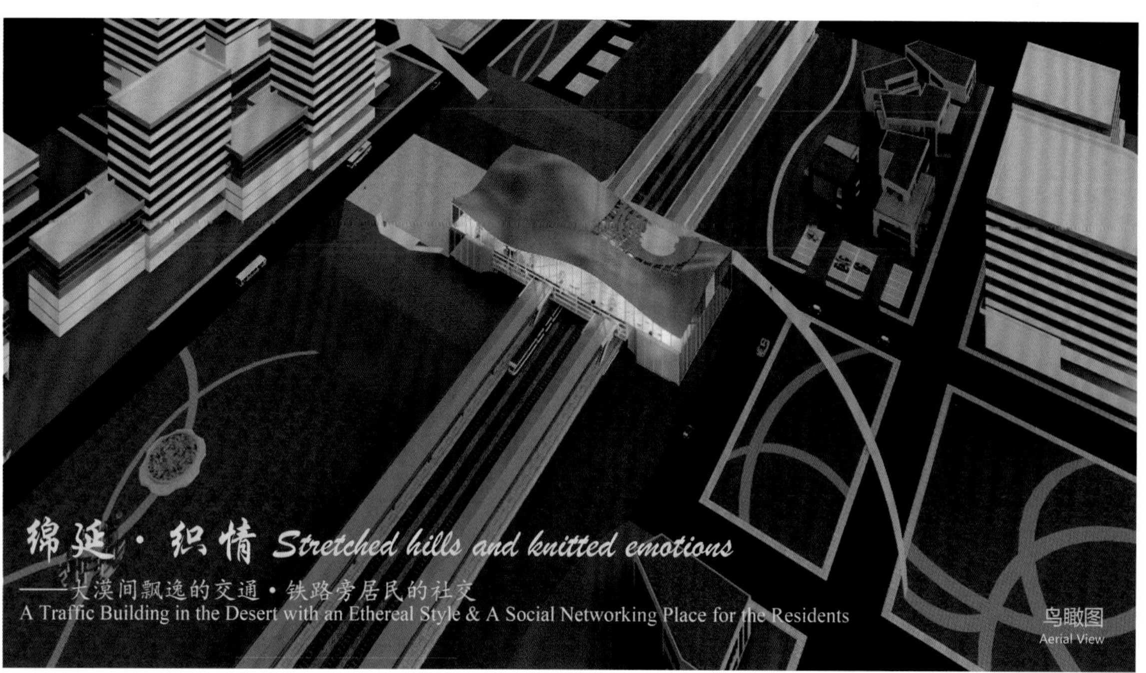

绵延·织情 Stretched hills and knitted emotions
——大漠间飘逸的交通·铁路旁居民的社交
A Traffic Building in the Desert with an Ethereal Style & A Social Networking Place for the Residents
鸟瞰图
Aerial View

区位
Location

中国甘肃 Gansu Province,China
甘肃张掖 Zhangye City,Gansu
张掖甘州 Ganzhou District, Zhangye

位于甘肃省的张掖自古以来就是商贾云集之地与丝绸之路重镇。旧时，张掖“云飞丝路飘花雨，风动驼铃运锦绸”，骆驼商旅来往于茫茫戈壁与绵延大漠之间。新中国成立以来，张掖借着一条兰新铁路迅速发展，再续丝路辉煌。张掖这座城市因“交流”与“贸易”而焕发无限生机。我们的基地位于张掖甘州区新兰铁路沿线的小村落旁。

Zhangye in Gansu Province has been a place of merchants and an important town on the Silk Road since ancient times. In the old days, flowers and rain floating above the road in Zhangye,clouds floating Wind blew the camel bell which accompanied the trade caravan to transport silk. Camel business travelled between the vast Gobi and the vast desert. Since the founding of new China, Zhangye has developed rapidly with the Lanzhou-Xinjiang railway and continued the glory of the Silk Road. Zhangye city is full of vitality because of "exchange" and "trade." Our base is located besides a small village along the Lanzhou-Xinjiang railway.

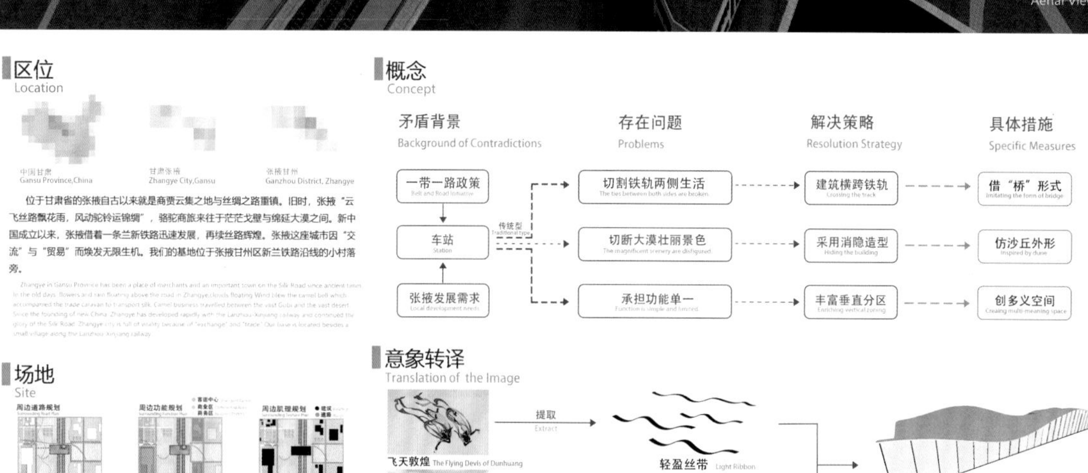

场地
Site

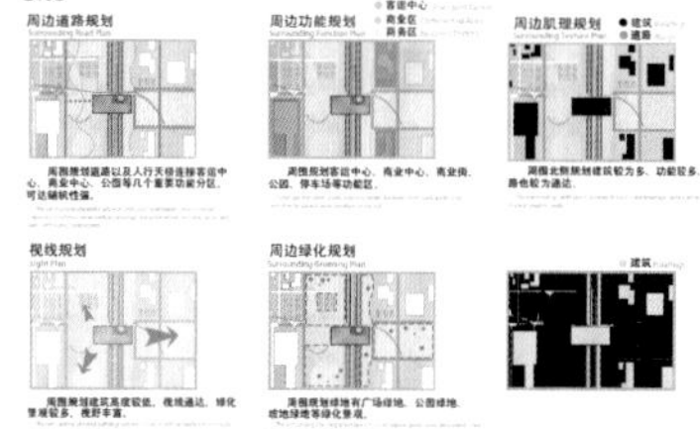

设计说明
Design Description

铁路车站往往承担着人员物资运输的重要功能。但是铁路却又会因铁轨铺设而造成一系列问题。针对引发的这些问题，我们试图给予车站一种多义性空间的可能。

而基地选在张掖，这座文化灿烂的交通重镇，我们亦是延续上文思路。车站连接兰新铁路上的城市与村落，而我们的新车站跨铁路，连接铁路两侧居民，以两个方向的“连接”来进一步凸显张掖“交流之城”的名片，结合甘肃当地多彩文化，最终打造一个承担着人文重量与人情温度的建筑。

Railway stations often undertake the important function of transportation of personnel and materials. However, the traditional railway can cause a series of problems due to the laying of tracks. Aiming at the problems, we propose a new solution, which tries to give the station a possibility of multi-meaning space.

The base is chosen in Zhangye, the traffic town with brilliant culture, which is also the result that we decided to continue the above ideas. The station originally connects the cities and villages along the Lanzhou-Xinjiang Railway, while our new station crosses the tracks and connects residents on both sides of the tracks. The "connection" of the two directions further highlights the name card of Zhangye "City of communication". Meanwhile, we combines the local colorful culture of Gansu and finally creates a building bearing the weight of humanity culture and the temperature between people.

人群需求及应对策略
Population Demand and Corresponding Strategies

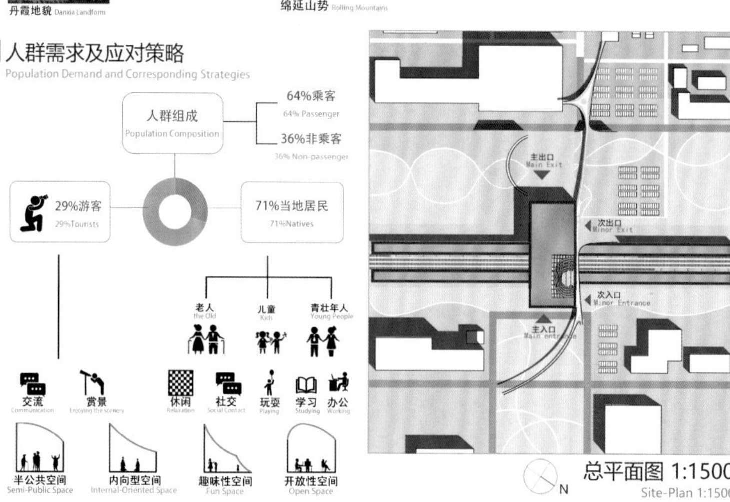

总平面图 1:1500
Site-Plan 1:1500

候车大厅
Waiting Hall

综合活动区
Comprehensive Activity Area

站台
Platform

天桥
Footbridge

观景区
Viewing Area

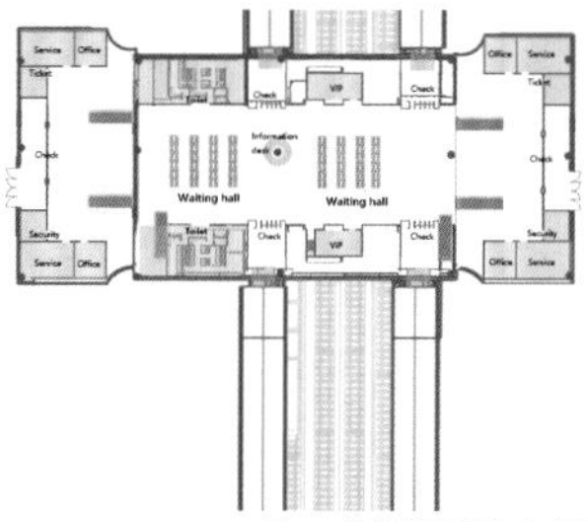

一层平面图 1:800
Ground Floor Plan 1:800

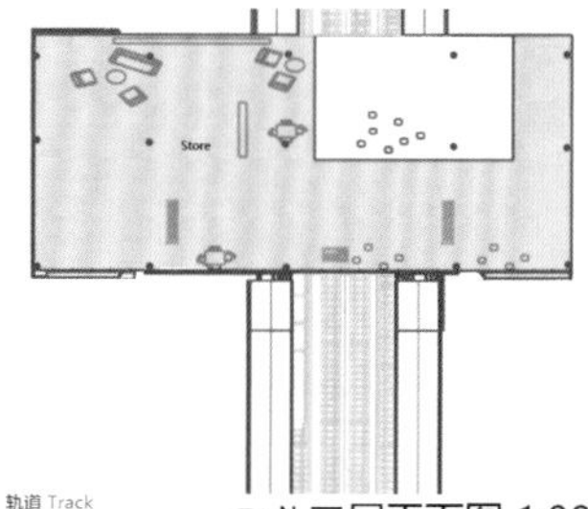

轨道 Track
交通空间 Traffic Space
活动空间 Activity Space

二层平面图 1:800
First Floor Plan 1:800

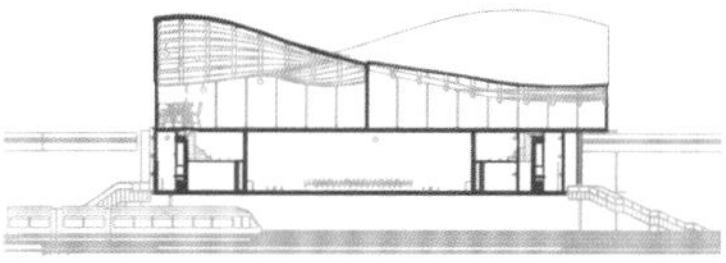

剖面图 1:400
Section Plan 1:400

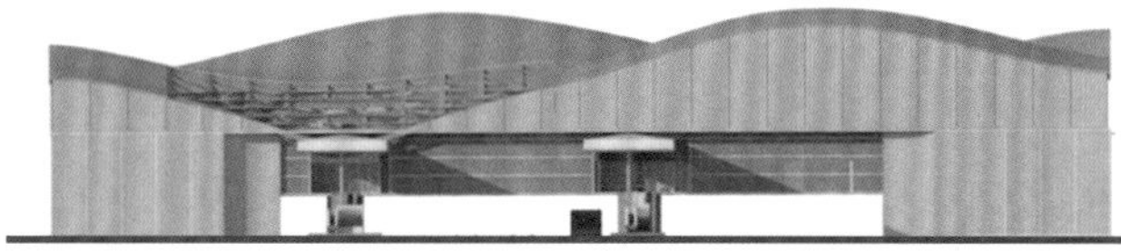

立面图 1:400
Facade Plan 1:400

复合型图书室
Compound Library

以可移动书架提供灵活布局
Provides flexible layout with removable shelves

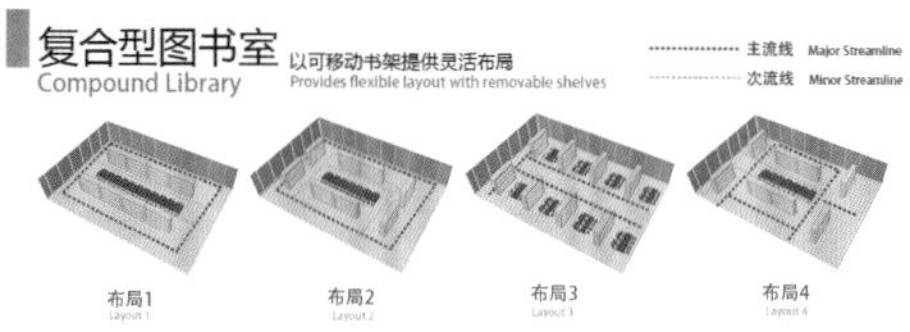

主流线 Major Streamline
次流线 Minor Streamline

布局1 Layout 1
布局2 Layout 2
布局3 Layout 3
布局4 Layout 4

社会影响
Social Impact

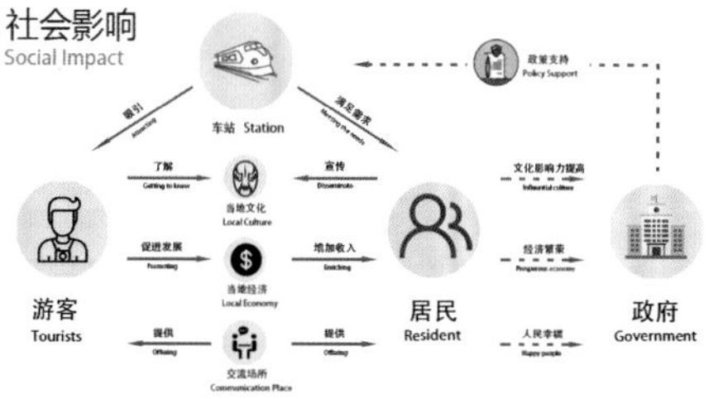

车站 Station
政策支持 Policy Support
吸引 Attracting
满足需求 Meeting the needs
了解 Getting to know
宣传 Disseminate
当地文化 Local Culture
促进发展 Promoting
增加收入 Enriching
当地经济 Local Economy
提供 Offering
交流场所 Communication Place
游客 Tourists
居民 Resident
政府 Government
文化影响力提高 Influential culture
经济繁荣 Prosperous economy
人民幸福 Happy people

爆炸图
Exploded Diagram

铝板屋面
Aluminum Sheeting Roof

弧形变截面箱型构件
Curved variable section box members

木纹铝薄板
Wood grain aluminum sheet

玻璃幕墙及混凝土楼板
Glass Curtain Walls and Concrete Floor

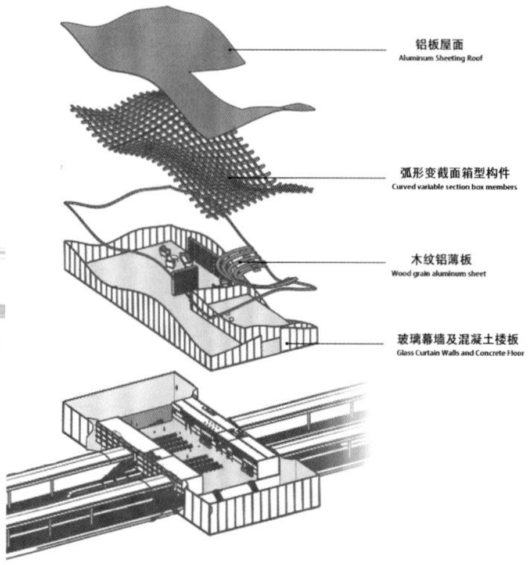

蚕鸣

Silkworm Singing

参赛学校　University/College

安徽科技学院

Anhui Science and Technology University, China

指导教师　Supervisor(s)

宋永伟　SONG Yongwei

刘昕烁　LIU Xinshuo

参赛学生　Participant(s)

杨雅雅　YANG Yaya

赖　澜　LAI Lan

王梓杰　WANG Zijie

肖家乐　XIAO Jiale

许　陈　XU Chen

简介 Description

本设计以风为设计理念切入，追寻着张骞出使西域的那股古风，思考怎么通过火车站的植入为低迷的路网区域带来新的经济提升，最终找到通过打造“建筑代言人”的方法，扩大区域的知名度的同时又增加了客流量，故通过提取“风的形状”展开建筑设计。因风而起，随风而动，伴其蚕鸣，抵达何处？所到之处焕然新生。

Starting from the concept of wind, this design was inspired by the long-passed wind that accompanied the envoy Zhang Qian to the west as the first explorer in the Han Dynasty, the design is to find the right way of elevating the sluggish economy in the region with quite a few previously roads expanded by the building of a new railway station. It is expected that the building speaks for the region to increase its popularity and expand the passenger flow. This is why the design is trying to capture and employ the "shape of the wind", thus the station may look like being blown, breathing in the constantly waving wind with the singing silkworm that has never. Where is the railway leading to? —— wherever there is new hope of life.

抵达之谜

the wind slikworm singing--arrived in mystery

设计说明 Design specification

本设计以风为设计理念切入，追寻着张骞出使西域的那股古风，思考怎么通过火车站的植入为低迷路网区域带来新的经济提升。最终找到通过打造“建筑代言人”的方法，扩大的区域的知名度同时有增加的人流量。故通过提取风的形状，展开建筑设计

The design starts with the design concept of wind, tracing the ancient style of Zhang Qian's mission to the Western Regions, and how to bring new economic improvement to the depressed road network region through the implantation of the railway station. Finally, we found a way to expand the visibility of the area and increase the flow of people by creating an "architectural spokesperson". Therefore, by extracting the shape of the wind, an architectural design

区位介绍 Location introduction

火车站选址位于海北藏族自治州祁连县内，北邻古丝绸之路的河西走廊，故有青海“北大门”之称。呈现了草原广袤、雪山连绵，冰川过境的独特地貌，哈木铁路为其和刚察县两地的交流提供了条件。

The site of the railway station is located in Qilian County, Haibei Tibetan Autonomous Prefecture, adjacent to the Hexi Corridor of the ancient Silk Road to the north, so it is called the "North Gate" of Qinghai. It presents the unique landform of vast grasslands, continuous snow-capped mountains, and glaciers passing through. The Hamu Railway provides conditions for the exchange between it and Gangcha County.

多民族
Multi-ethnic

天境祁连
Heavenly Realm Qilian

流线分析 Streamline analysis

为避免人流交叉，采用两侧分流进出入口，同时站内便捷换乘

多文化
Multicultural

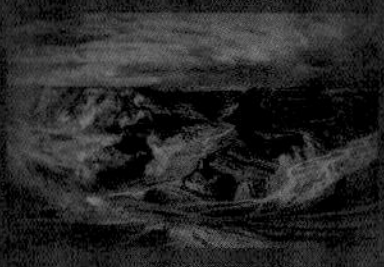

矿产资源
mineral resources

河西走廊
Hexi Corridor

生态保护
Ecological protection

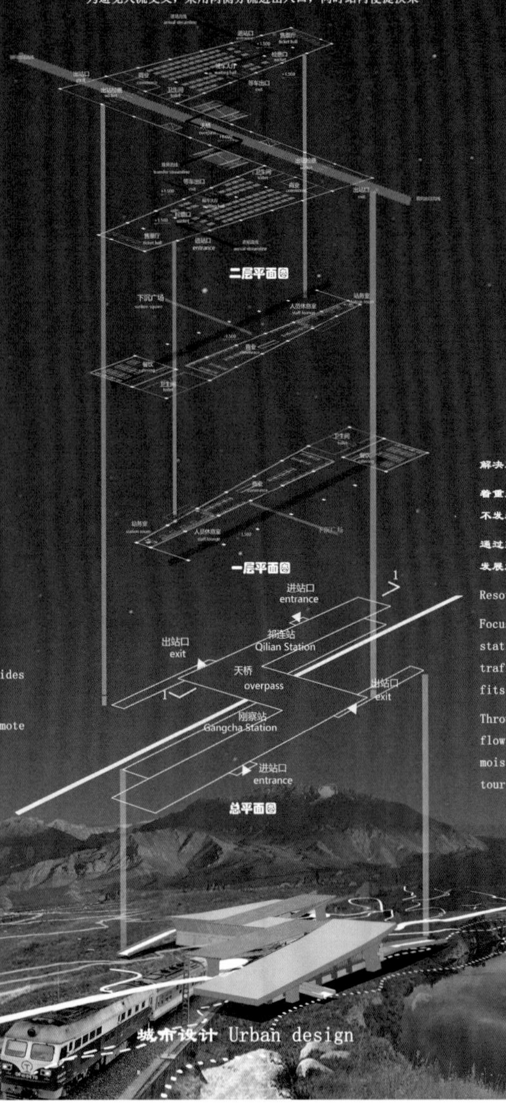

二层平面图

一层平面图

进站口
entrance
出站口
exit
祁连站
Qilian Station
天桥
overpass
出站口
exit
刚察站
Gangcha Station
进站口
entrance

总平面图

问题思考

停用的工业铁路如何新生？

火车除交通联系纽带价值以外的价值探索？

如何利用停靠站带动城市经济发展？

Thinking

How to revitalize obsolete industrial railways?

How to explore the value of train besides the value of transportation link?

How to use the docking station to promote economic development of the city?

解决策略

着重火车站建筑的自身美学价值给偏，远，交通不发达地区带来的效益。

通过建筑吸引人流进而结合周围优渥的地理环境发展旅游等产业。

Resotlution strategy

Focus on the aesthetic value of railway station construction to the partial, far, traffic undeveloped areas to bring benefits

Through the building to attract human flow and then combined with surrounding moist geographical environment to develop tourism and other industries.

祁连

刚察
Gangcha county

城市设计 Urban design

原停用工业铁路
The industrial railway was deactivated

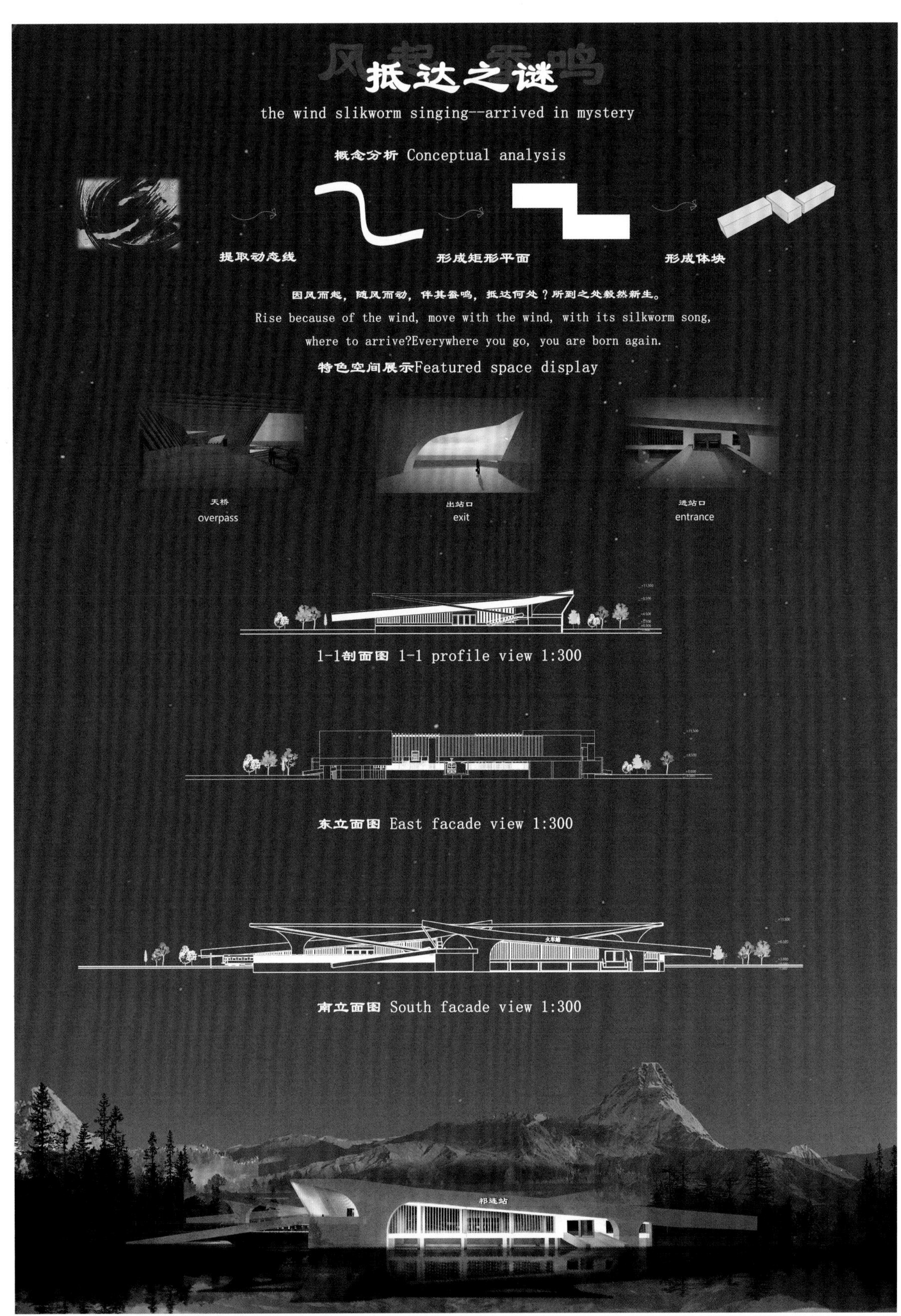
风蚕·蚕鸣
抵达之谜
the wind slikworm singing--arrived in mystery
概念分析 Conceptual analysis
提取动态线
形成矩形平面
形成体块
因风而起，随风而动，伴其蚕鸣，抵达何处？所到之处毅然新生。
Rise because of the wind, move with the wind, with its silkworm song,
where to arrive?Everywhere you go, you are born again.
特色空间展示Featured space display
天桥
overpass
出站口
exit
进站口
entrance
1-1剖面图 1-1 profile view 1:300
东立面图 East facade view 1:300
南立面图 South facade view 1:300

山国里的绵延

Inheritance and Transmission in the Mountain Country

参赛学校　University/College

安徽科技学院

Anhui Science and Technology University, China

指导教师　Supervisor(s)

刘昕烁　LIU Xinshuo

参赛学生　Participant(s)

丁秦杰　DING Qinjie

叶　雯　YE Wen

邓义环　DENG Yihuan

房科臣　FANG Kechen

程　琪　CHENG Qi

简介 Description

基地位于云南省西双版纳傣族自治州景洪市内，场地邻接清泉路。铁路客运站设计充分考虑景洪市旅游发展的战略目标，站房与站前广场设计迎合当地建筑风貌与文化特色，站房总体为三层，根据各功能之间的密切程度进行划分，合理组织进站、出站、办公流线。站房造型提取当地建筑重檐陡坡屋顶、拱形元素，采用红色和橙黄色调，尽显开放与包容，将成为区域新地标。

The site is located in Jinghong City, Xishuangbanna Dai Autonomous Prefecture,

Yunnan Province in China and is adjacent to the Qingquan Road. The design of the railway station takes full consideration of the strategic goal of tourism development in Jinghong City. The station and the square in the front are consistent with the local architectural style and cultural characteristics. The building has three floors, arranged with different functions loosely or closely related to each other. Entering, exiting and working flows are properly designed. The building reflects elements like double eaves, steep inclined roofs and arch typically used in the local traditional architecture. The colors red and orange were employed to express openness and inclusiveness, making the station become a new landmark in the region.

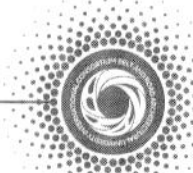

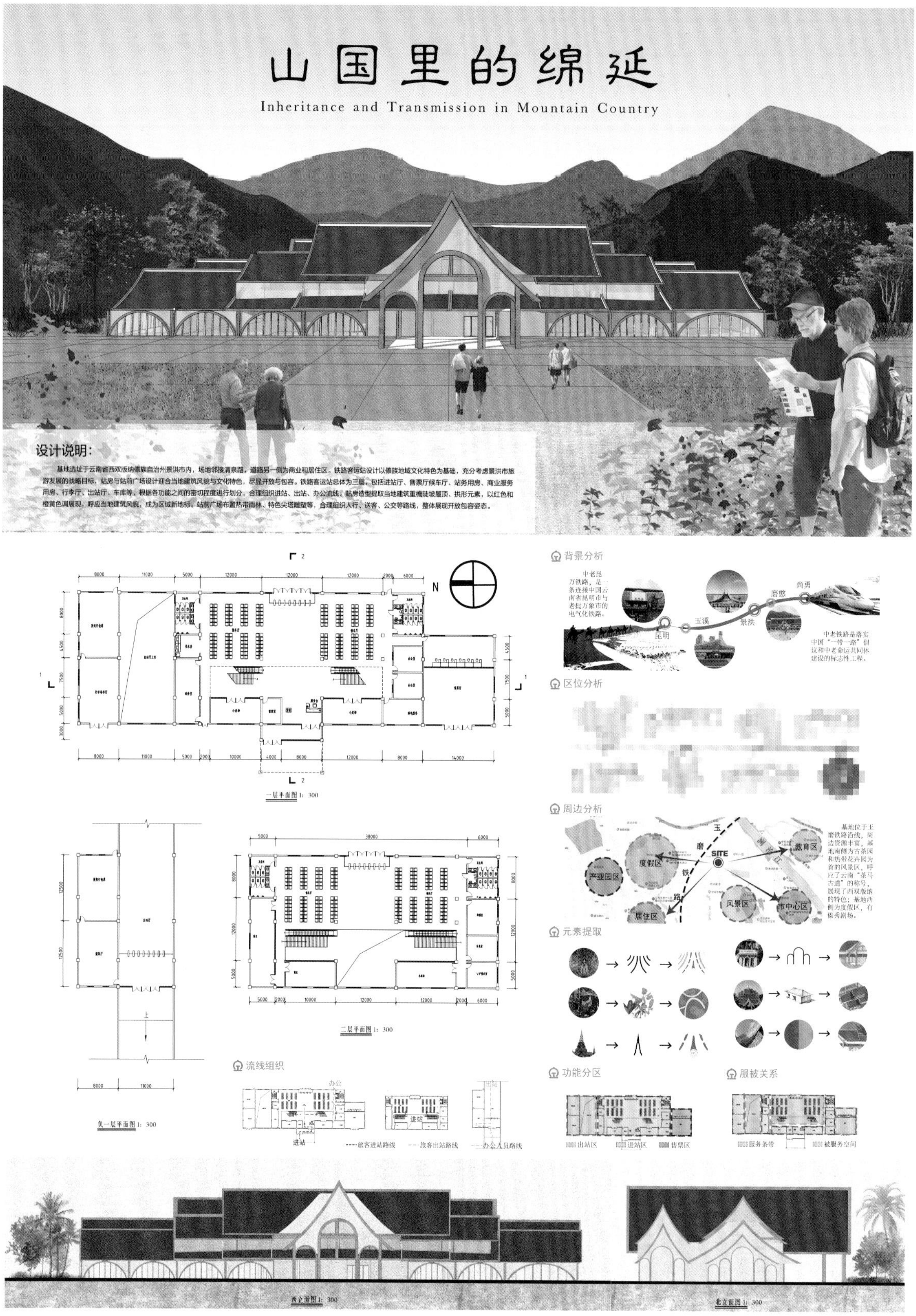

山国里的绵延

Inheritance and Transmission in Mountain Country

设计说明：

基地选址于云南省西双版纳傣族自治州景洪市内，场地邻接清泉路，道路另一侧为商业和居住区。铁路客运站设计以傣族地域文化特色为基础，充分考虑景洪市旅游发展的战略目标，站房与站前广场设计迎合当地建筑风貌与文化特色，尽显开放与包容。铁路客运站总体为三层，包括进站厅、售票厅候车厅、站务用房、商业服务用房、行李厅、出站厅、车库等，根据各功能之间的密切程度进行划分，合理组织进站、出站、办公流线。站房造型提取当地建筑重檐陡坡屋顶、拱形元素，以红色和橙黄色调展现，呼应当地建筑风貌，成为区域新地标。站前广场布置热带雨林、特色尖塔雕塑等，合理组织人行、送客、公交等路线，整体展现开放包容姿态。

一层平面图 1：300

背景分析

中老昆万铁路，是一条连接中国云南省昆明市与老挝万象市的电气化铁路。

昆明　玉溪　景洪　磨憨　尚勇

中老铁路是落实中国“一带一路”倡议和中老命运共同体建设的标志性工程。

区位分析

周边分析

玉磨铁路　SITE　产业园区　度假区　居住区　教育区　风景区　市中心区

基地位于玉磨铁路沿线，周边资源丰富，基地南侧为古茶园和热带花卉园为首的风景区，呼应了云南“茶马古道”的称号，展现了西双版纳的特色；基地西侧为度假区，有傣秀剧场。

元素提取

负一层平面图 1：300

二层平面图 1：300

流线组织

办公　进站　出站

旅客进站路线　旅客出站路线　办公人员路线

功能分区

出站区　进站区　售票区

服被关系

服务条带　被服务空间

西立面图 1：300　　北立面图 1：300

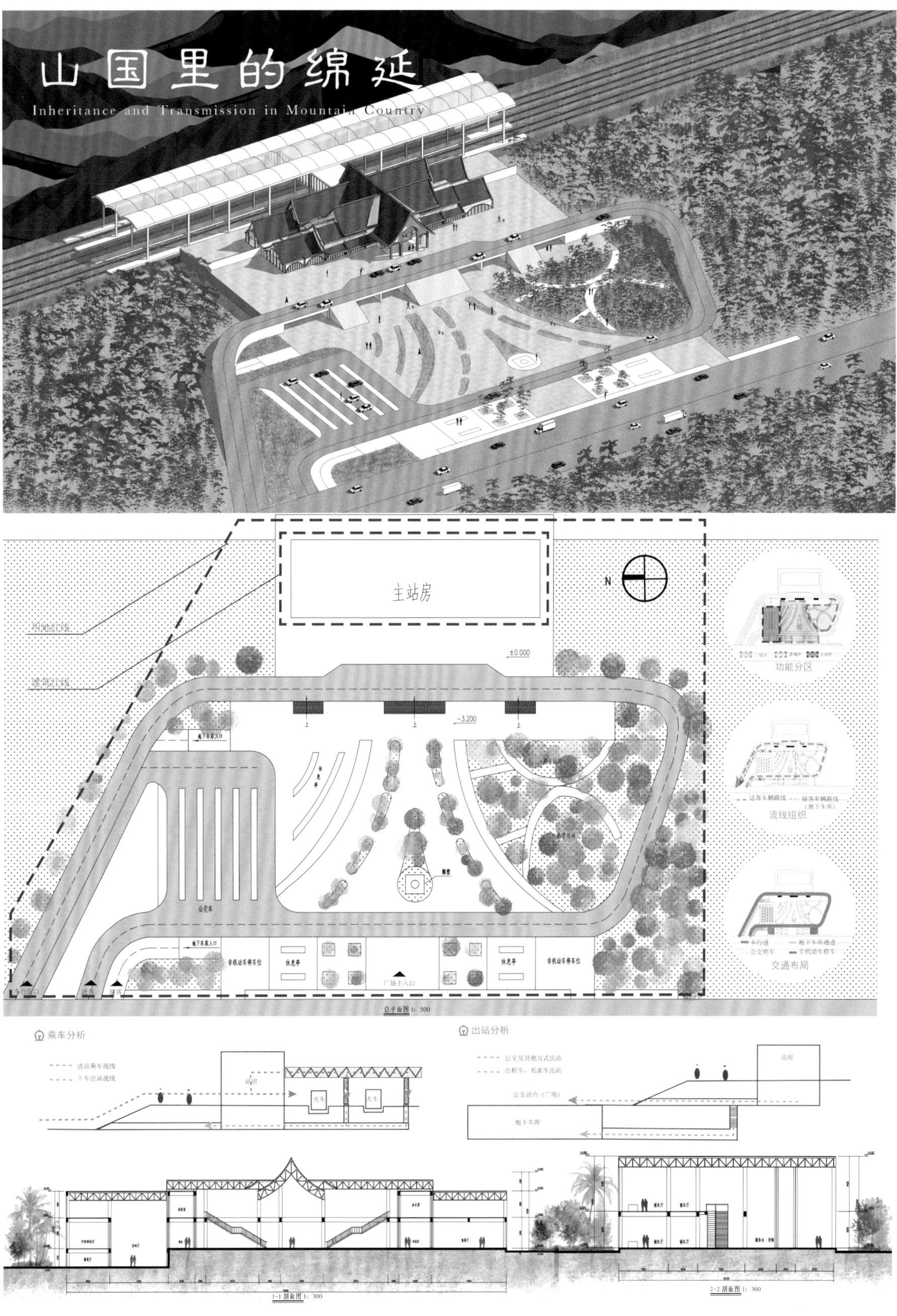
山国里的绵延
Inheritance and Transmission in Mountain Country
主站房
±0.000
-3.200
公交车
非机动车停车位
休息亭
广场主入口
功能分区
流线组织
交通布局
总平面图 1：500
乘车分析
进站乘车流线
下车出站流线
出站分析
公交及其他方式出站
出租车、私家车出站
公交站台（广场）
地下车库
站房
1-1 剖面图 1：300
2-2 剖面图 1：300

古瓷新生

Ancient Porcelain Repair

参赛学校 University/College
安徽理工大学
Anhui University of Science and Technology, China

指导教师 Supervisor(s)
黄云峰 HUANG Yunfeng

参赛学生 Participant(s)
黄逸凡 HUANG Yifan
沈丰乐 SHEN Fengle
高　琪 GAO Qi
江　珊 JIANG Shan
张涵博 ZHANG Hanbo

简介 Description

千年丝路，阵阵驼铃声中，远方的启明星尚未落下，一件件精美的瓷器飘向远方。

千年变换，驼铃声熄，瓷器历经千年已然只剩碎片。时空轮转，远方的驼铃声响，天上的明星再度升起，新时代丝绸之路悄然而生，向着启明星始的方向，伴来这千年的低吟，携着新时代的新使命蜿蜒而生。串起破碎的千年碎片，在这“新”的战场上，架起属于新时代的新桥梁。

In the past millennium, on the Silk Road, spells of camel bells kept ringing in the desert, the Venus high above in the sky had not yet fallen. Exquisite porcelain was carried to distant places one piece after another.

A thousand years have passed and camel bells have vanished, the porcelains only remain in fragments. But now, the old days are back and the dynamic Silk Road is awakened. Again, we can hear the camel bells in the distance and see Venus rising. The Silk Road of the modern era has emerged quietly. Facing the direction of the rising star, accompanied by the voice telling the past, today's Silk Road has resurrected with its new mission: piecing together the remains and serving as the new bridge across this revitalized playing field.

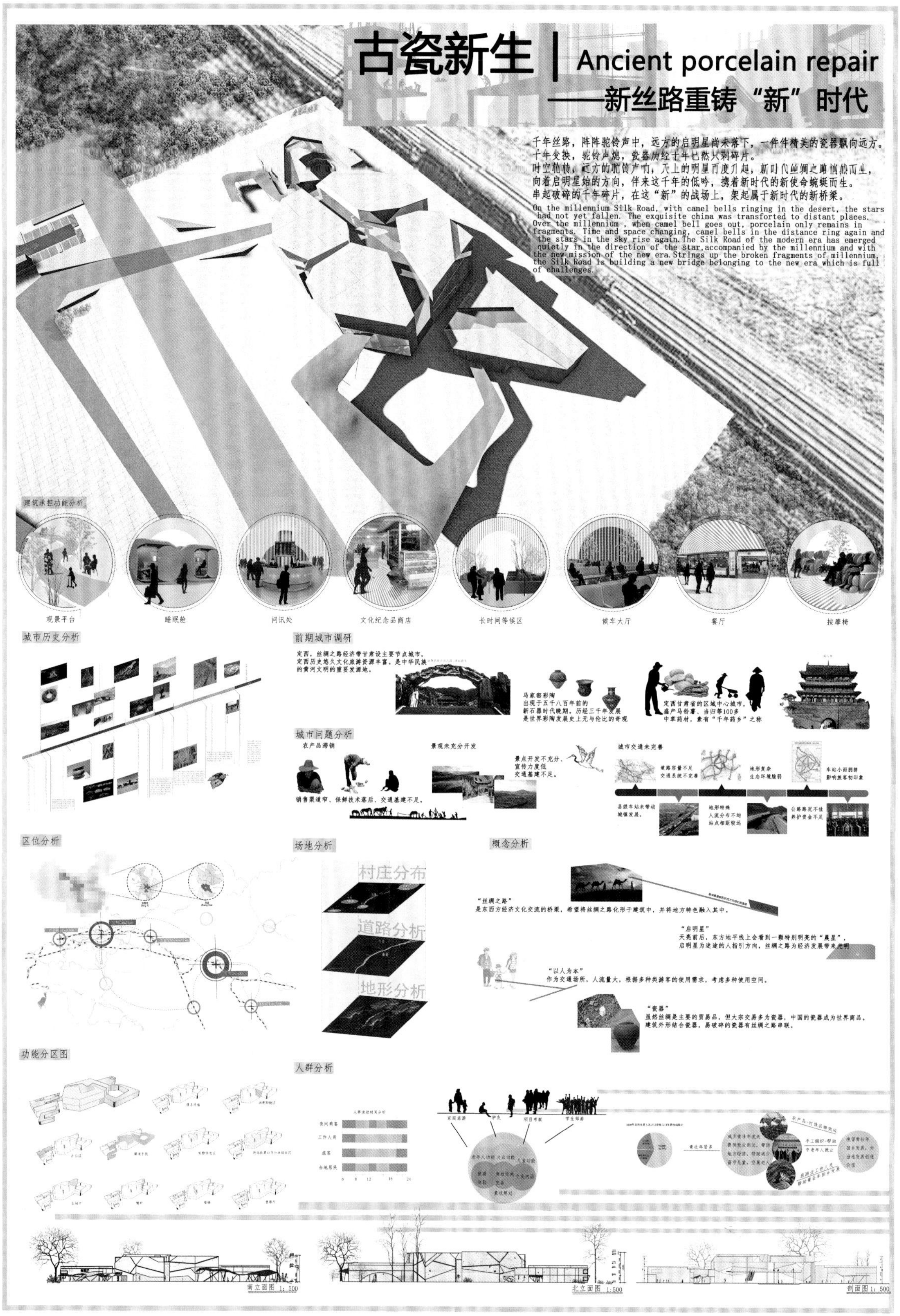

古瓷新生 | Ancient porcelain repair

——新丝路重铸“新”时代

千年丝路，阵阵驼铃声中，远方的启明星尚未落下，一件件精美的瓷器飘向远方。
千年变换，驼铃声熄，瓷器历经千年已然只剩碎片。
时空轮转，远方的驼铃声响，天上的明星再度升起，新时代丝绸之路悄然而生，
向着启明星始的方向，伴来这千年的低吟，携着新时代的新使命蜿蜒而生。
串起破碎的千年碎片，在这“新”的战场上，架起属于新时代的新桥梁。

On the millennium Silk Road, with camel bells ringing in the desert, the stars had not yet fallen. The exquisite china was transforted to distant places. Over the millennium , when camel bell goes out, porcelain only remains in fragments. Time and space changing, camel bells in the distance ring again and the stars in the sky rise again. The Silk Road of the modern era has emerged quietly in the direction of the star, accompanied by the millennium and with the new mission of the new era. Strings up the broken fragments of millennium, the Silk Road is building a new bridge belonging to the new era which is full of challenges.

建筑承担功能分析

观景平台　睡眠舱　问讯处　文化纪念品商店　长时间等候区　候车大厅　餐厅　按摩椅

城市历史分析

前期城市调研

定西，丝绸之路经济带甘肃设主要节点城市，定西历史悠久文化旅游资源丰富。是中华民族的黄河文明的重要发源地。

马家窑彩陶
出现于五千八百年前的新石器时代晚期。历经三千年发展是世界彩陶发展史上无与伦比的奇观

定西甘肃省的区域中心城市，盛产马铃薯、当归等100多中草药材，素有“千年药乡”之称

城市问题分析

农产品滞销
销售渠道窄、保鲜技术落后、交通基建不足。

景观未充分开发
景点开发不充分、宣传力度低交通基建不足。

城市交通未完善
道路容量不足 交通系统不完善
地形复杂 生态环境脆弱
车站小而拥挤 影响旅客初印象
县级车站未带动城镇发展。
地形特殊 人流分布不均 站点相距较远
公路路况不佳 养护资金不足

区位分析

场地分析

村庄分布
道路分析
地形分析

概念分析

“丝绸之路”
是东西方经济文化交流的桥梁，希望将丝绸之路化形于建筑中，并将地方特色融入其中。

“启明星”
天亮前后，东方地平线上会看到一颗特别明亮的“晨星”，启明星为迷途的人指引方向，丝绸之路为经济发展带来光明

“以人为本”
作为交通场所，人流量大，根据多种类游客的使用需求，考虑多种使用空间。

“瓷器”
虽然丝绸是主要的贸易品，但大宗交易多为瓷器，中国的瓷器成为世界商品。建筑外形结合瓷器，易破碎的瓷器有丝绸之路串联。

功能分区图

人群分析

家庭旅游　驴友　项目考察　学生郊游

青壮年居多

南立面图 1:500　北立面图 1:500　剖面图 1:500

眷故

Family

参赛学校 University/College
长春建筑学院
Changchun University of Architecture and Civil Engineering, China

指导教师 Supervisor(s)
滕佳佳 TENG Jiajia
李　冲 LI Chong

参赛学生 Participant(s)
姬孟汐 JI Mengxi
王泽会 WANG Zehui
陈兴华 CHEN Xinghua
杨　航 YANG Hang
王志勇 WANG Zhiyong

简介 Description

本案基地位于甘肃省临夏回族自治州北部永靖县内。基地形状近似三角形；东北侧背靠青山，接铁路专线和大庄路；西南侧面朝黄河，接滨河北路；西侧有一座公园，接折达公路；东侧靠近居住区。

基于以上条件，建筑设计以回应场地，激活当地为理念，去描写作为古新丝绸之路节点上的永靖，应该如何定位破局，积极去融入新的发展机会。

建筑定位为火车站，主要功能包括接发旅客服务、展览空间、商业三部分以及配套设施。

The site of this design is located in Yongjing County, north of Linxia Hui Autonomous Prefecture in Gansu Province in China. The shape of the site is an approximate triangle. Sitting on the northeast side is the Qingshan Mountain, connected to the railway line and Dazhuang Road; on the southwest side is the Yellow River, joint with the Binhe North Road; on the west side is a park right by the Zheda Highway; its east side is close to the residential area.

Based on the above conditions, the design of the building reflects the principle of interacting with the surroundings and activating the area. The aim of building the architecture is to mirror the Yongjing County standing on the conjunction of the ancient and modern Silk Rood and how it seeks to make breakthroughs in today's new role by actively participating in the local development.

The building is a railway station with passenger services, exhibition space, business as main functions and relevant supporting facilities.

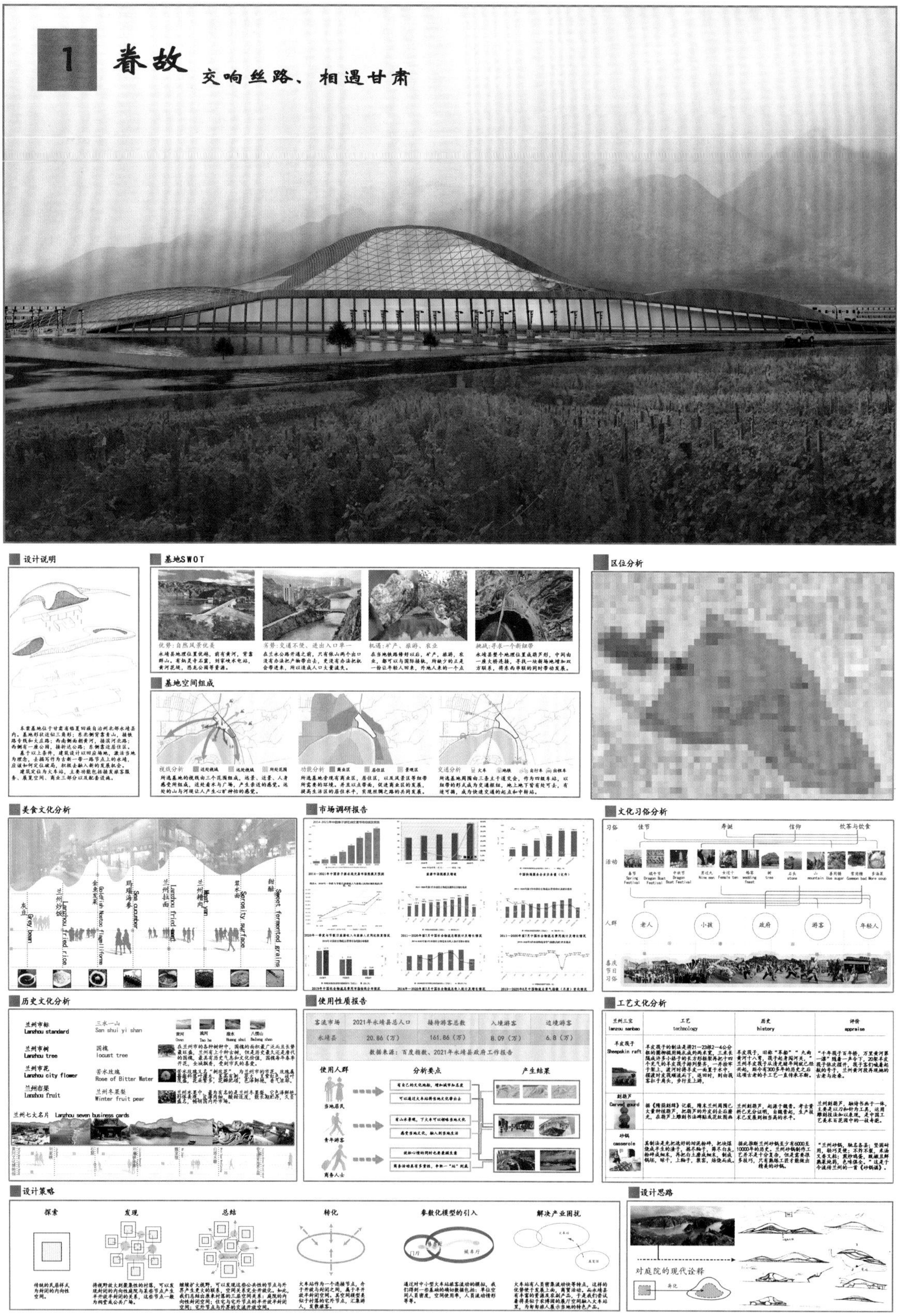

1 眷故 交响丝路、相遇甘肃

设计说明

基地SWOT

区位分析

基地空间组成

美食文化分析

市场调研报告

文化习俗分析

历史文化分析

使用性质报告

客流市场	2021年永靖县总人口	接待游客总数	入境游客	过境游客
永靖县	20.86（万）	161.86（万）	8.09（万）	6.8（万）

数据来源：百度指数、2021年永靖县政府工作报告

工艺文化分析

设计策略

探索 发现 总结 转化 参数化模型的引入 解决产业困扰

设计思路

对庭院的现代诠释

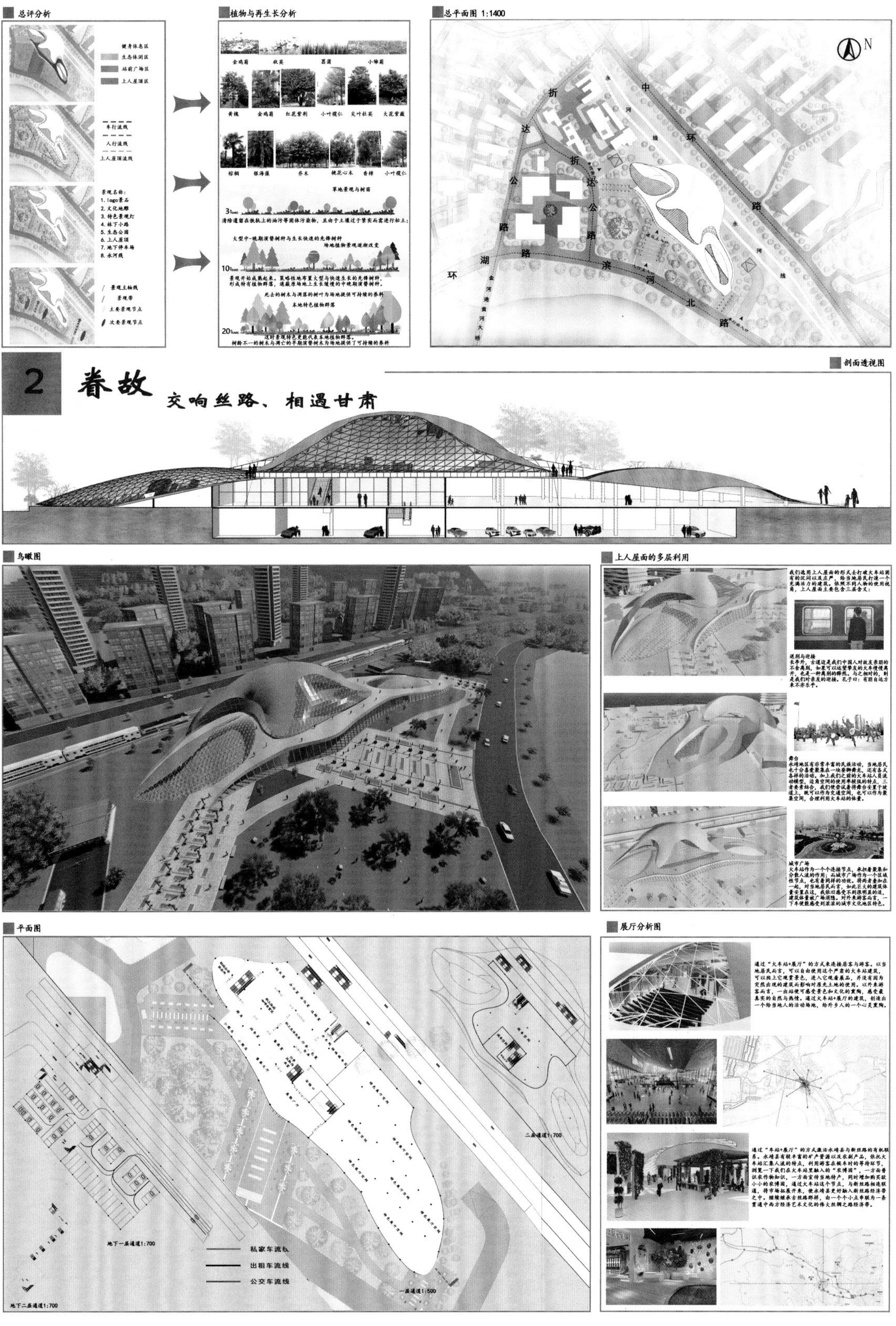
总评分析
植物与再生长分析
总平面图 1:1400
2
眷故
交响丝路、相遇甘肃
剖面透视图
鸟瞰图
上人屋面的多层利用
平面图
展厅分析图
私家车流线
出租车流线
公交车流线

回忆

Memories

参赛学校 University/College

广东建设职业技术学院

Guangdong Construction Polytechnic, China

指导教师 Supervisor(s)

张文新 ZHANG Wenxin

参赛学生 Participant(s)

唐超君 TANG Chaojun

彭丝杨 PENG Siyang

陈鹳锨 CHEN Guanxian

谭 骏 TAN Jun

梁子健 LIANG Zijian

简介 Description

建筑面积：783 平方米

建筑风格：复古

本项目体现的是远离家乡长期在外工作的人们的思乡之情。因此本次设计是根据建筑物周边的使用性质、所处环境和相应标准，运用建筑美学原理，创造功能合理、舒适优美、满足人们物质和精神生活需要的四等火车站。这一空间环境既具有使用价值，满足相应的功能要求，同时也反映了历史文脉、建筑风格、环境气氛等精神因素。

Building area: 783m²

Architecture style: Vintage

This project represents the long-held homesickness of people who work far away from home. Based on its location, surroundings and relevant standards, the designed four-class station, with aesthetics in architecture applied, has reasonable functions, a neat environment and comfortable facilities, and can satisfy people's physical and mental needs. The spatial environment of the station is of practical use and can fully fulfill its function. Moreover, it is of spiritual value by reflecting the history, the architectural style and the environmental atmosphere.

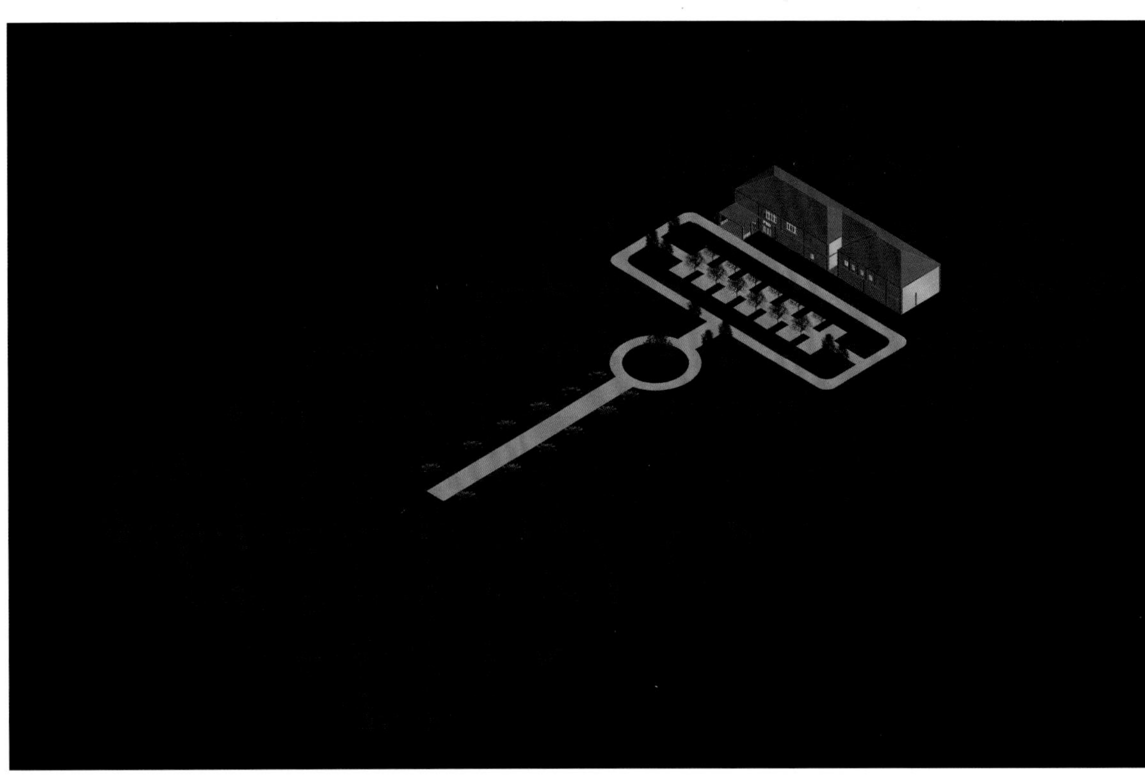

铁路客运站 + 集市综合体

The "Railway Station + Market" Complex

参赛学校 University/College

河北建筑工程学院

Hebei University of Architecture, China

指导教师 Supervisor(s)

郭晓君 GUO Xiaojun

谷文静 GU Wenjing

参赛学生 Participant(s)

楚越敏 CHU Yuemin

张　灏 ZHANG Hao

张懂懂 ZHANG Dongdong

张可心 ZHANG Kexin

张　瑜 ZHANG Yu

简介 Description

该设计从地域性出发，坚持以人为本理念。功能与空间上，不拘泥传统，探索敦煌车站的体验式空间，将集市与车站相结合。以历史中胡商贸易为原型，车站两侧设模块化坊市，传承敦煌文化的同时，弥补城市功能空间，连通城市两侧。打破了固有印象，内外交融，营造出公共交往的文化氛围，响应“一带一路”建设。建筑形态上，对应敦煌和莫高窟“始于汉魏，盛于隋唐”的历史特征，融合石窟艺术的神韵，体现在大屋顶、拱形立面、地形回应等元素中。

Starting from the uniqueness of this region, the design upholds a people-oriented principle. In our effort to explore the experiential space of the Dunhuang station in northwest China, we combined the market and the station itself, instead of restricting the design of functions and space in traditional practices. Taking the ancient market for business with Hu (western) merchants as the prototype, shops and markets divided into modules are to be set on both sides of the station. In this way, traditional Dunhuang culture is preserved, urban space for different functions is more complete and both sides of the city are connected. The design is trying to update people's stereotypes of the city, integrate the space within and outside the city, create an atmosphere of massive communication and respond positively to the building of the "Belt and Road". The station itself reflects the historical features of the Dunhuang City and the Mogao Caves described as "bred in the Han and Wei Dynasties and prosperous in the Sui and Tang Dynasties". The charm of grottoes could be found in the huge roof, arch facades, topographic response, etc.

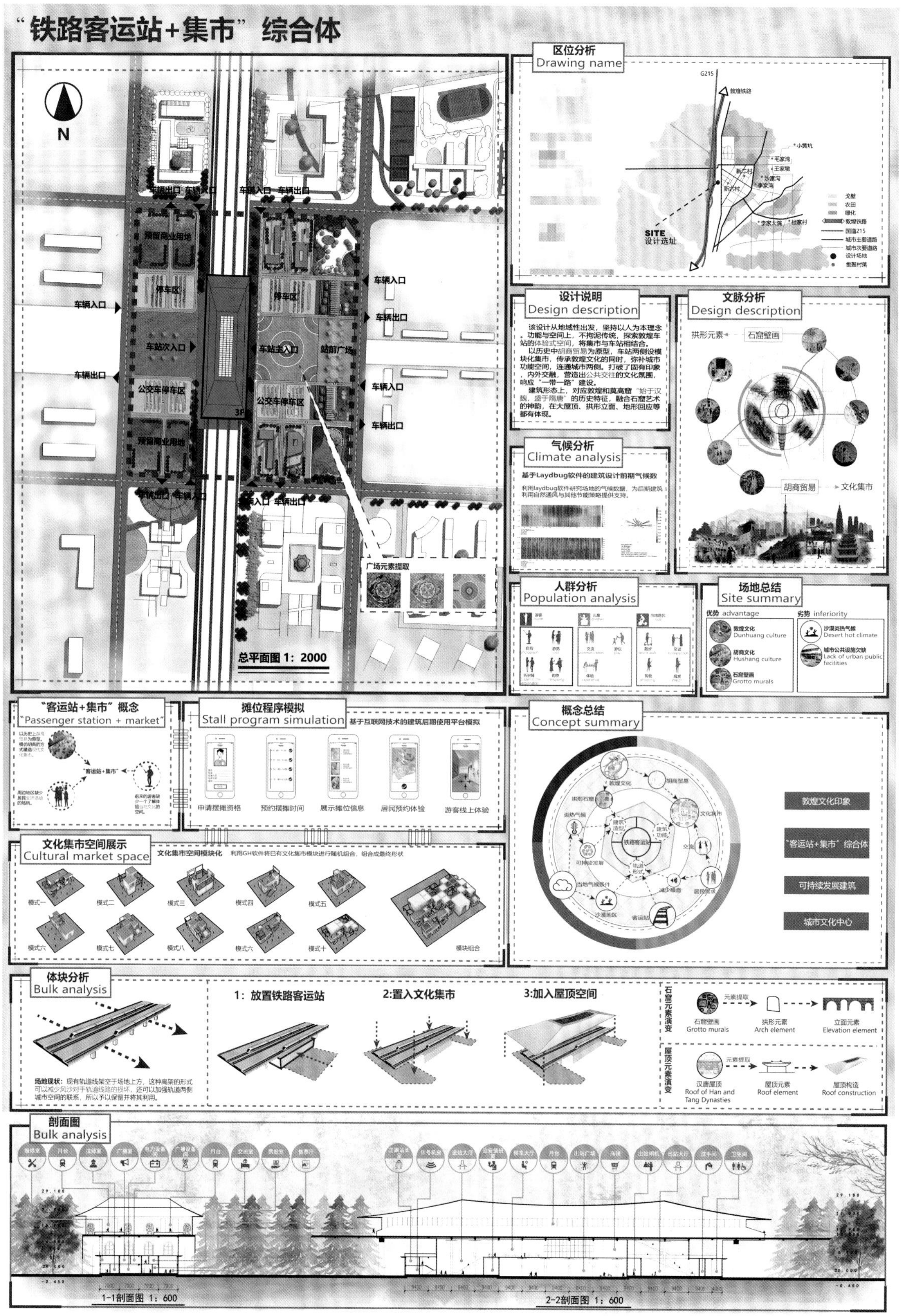

"铁路客运站+集市"综合体

区位分析 Drawing name

设计说明 Design description

该设计从地域性出发，坚持以人为本理念。功能与空间上，不拘泥传统，探索敦煌车站的体验式空间，将集市与车站相结合。

以历史中胡商贸易为原型，车站两侧设模块化集市，传承敦煌文化的同时，弥补城市功能空间，连通城市两侧，打破了固有印象，内外交融，营造出公共交往的文化氛围，响应"一带一路"建设。

建筑形态上，对应敦煌和莫高窟"始于汉魏，盛于隋唐"的历史特征，融合石窟艺术的神韵，在大屋顶、拱形立面、地形回应等都有体现。

文脉分析 Design description

气候分析 Climate analysis

基于Laydbug软件的建筑设计前期气候数

利用laydbug软件研究场地的气候数据，为后期建筑利用自然通风与其他节能策略提供支持。

人群分析 Population analysis

场地总结 Site summary

优势 advantage

- 敦煌文化 Dunhuang culture
- 胡商文化 Hushang culture
- 石窟壁画 Grotto murals

劣势 inferiority

- 沙漠炎热气候 Desert hot climate
- 城市公共设施欠缺 Lack of urban public facilities

总平面图 1：2000

"客运站+集市"概念 "Passenger station + market"

摊位程序模拟 Stall program simulation

基于互联网技术的建筑后期使用平台模拟

申请摆摊资格　预约摆摊时间　展示摊位信息　居民预约体验　游客线上体验

概念总结 Concept summary

敦煌文化印象

"客运站+集市"综合体

可持续发展建筑

城市文化中心

文化集市空间展示 Cultural market space

文化集市空间模块化　利用GH软件将已有文化集市模块进行随机组合，组合成最终形状

体块分析 Bulk analysis

1：放置铁路客运站　2:置入文化集市　3:加入屋顶空间

场地现状：现有轨道线架空于场地上方，这种高架的形式可以减少风沙对于轨道线路的损坏，还可以加强轨道两侧城市空间的联系，所以予以保留并将其利用。

石窟元素演变：石窟壁画 Grotto murals → 拱形元素 Arch element → 立面元素 Elevation element

屋顶元素演变：汉唐屋顶 Roof of Han and Tang Dynasties → 屋顶元素 Roof element → 屋顶构造 Roof construction

剖面图 Bulk analysis

1-1剖面图 1：600

2-2剖面图 1：600

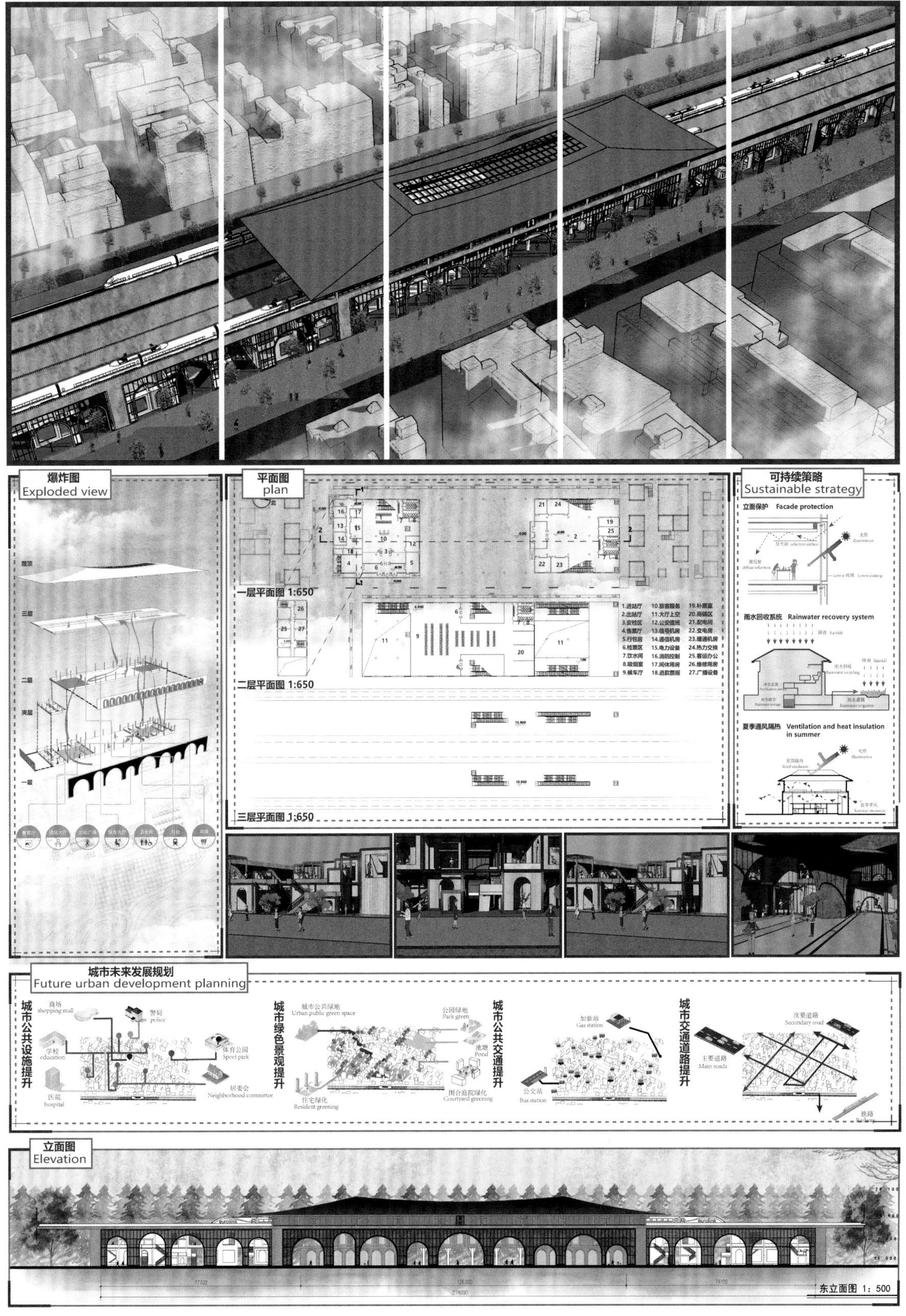
爆炸图
Exploded view
平面图
plan
一层平面图 1:650
二层平面图 1:650
三层平面图 1:650
可持续策略
Sustainable strategy
立面保护 Facade protection
雨水回收系统 Rainwater recovery system
夏季通风隔热 Ventilation and heat insulation in summer
城市未来发展规划
Future urban development planning
城市公共设施提升
城市绿色景观提升
城市公共交通提升
城市交通道路提升
立面图
Elevation
东立面图 1：500

思接千载，视通万里

Thinking for Thousands of Years, Seeing for Thousands of Miles

参赛学校 University/College

河南城建学院

Henan University of Urban Construction, China

指导教师 Supervisor(s)

申红田 SHEN Hongtian

殷许鹏 YIN Xupeng

参赛学生 Participant(s)

郑　璐 ZHENG Lu

沈欣悦 SHEN Xinyue

李琛博 LI Chenbo

任　帅 REN Shuai

鲁轩宇 LU Xuanyu

简介 Description

该项目场地位于河南省平顶山市郏县冢头镇陈寨村，郏县是千年古城，历史可追溯至西汉，郏县古建筑拥有中国北方民居的特点，立柱、斗拱、梁架的组合构筑了这里的建筑文化语言，我们提取了传统建筑的元素，采用与传统建筑语言相呼应、相统一的设计手法，重视以人为核心的场景体验的，集交通、娱乐、购物为一体的火车站设计。

用新手法、新材料承接千年的文化，用新意境、新思想视通万里江山。

This project is located in the Chenzhai village, Zhongtou Town, Jiaxian County, Pingdingshan City, Henan Province in China.

Traced back to the Western Han Dynasty (202 BC—8 AD), the Jiaxian county has a history of thousands of years. The ancient architecture in the Jiaxian county reflects the characteristics of the residential houses in northern China. The combination of columns, Dougong (cap block) and beam frames consists of the style of the architectural culture here. We extracted the elements of traditional architecture and adopted the design method that echoes and complies with the traditional architectural language to make a railway station featuring people-centered scenario experience and integrates the functions of transportation, entertainment and shopping. We inherit the culture of thousands of years with new techniques and materials and look beyond thousands of miles of rivers and mountains with a new artistic conception and new ideas.

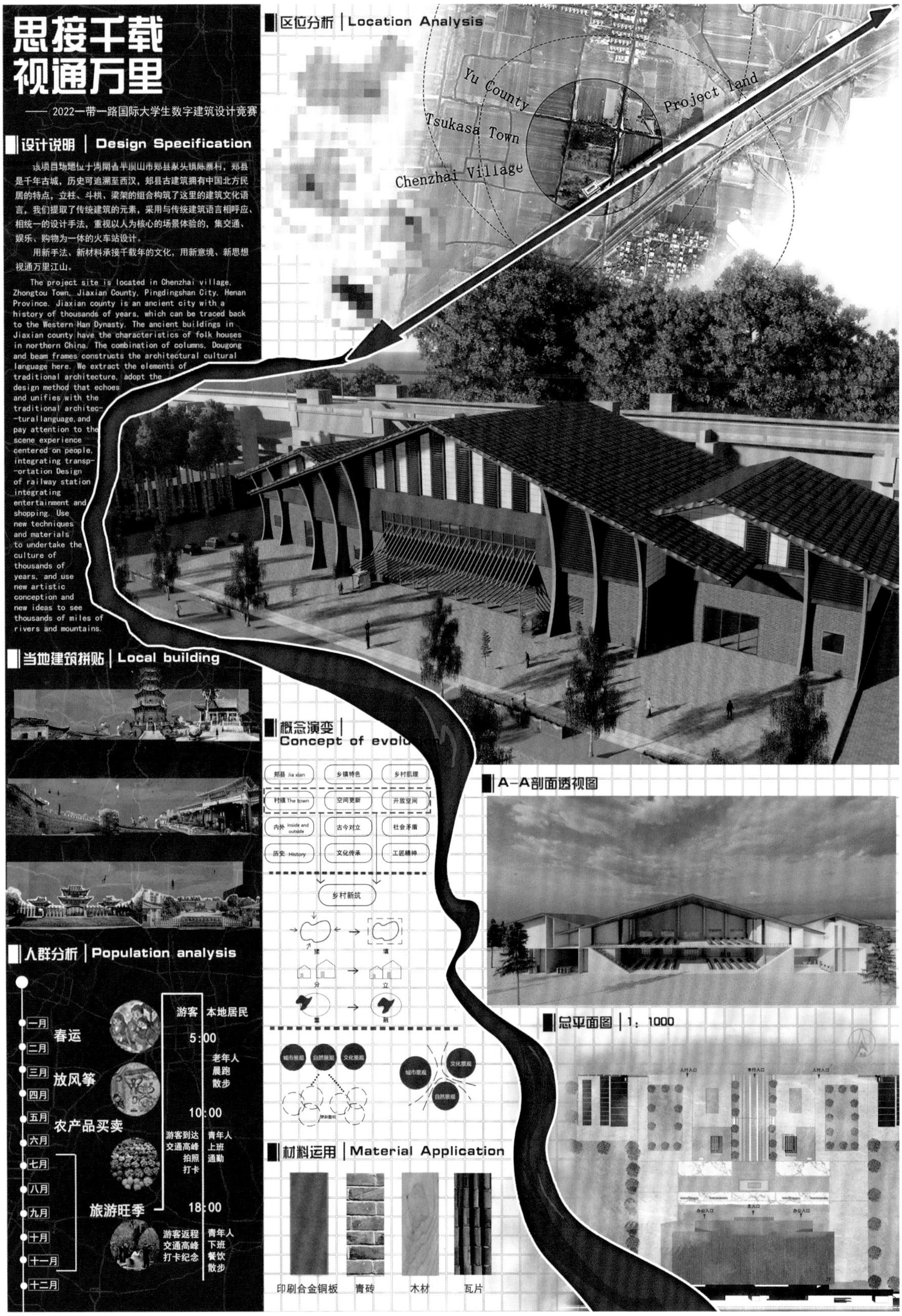

思接千载 视通万里

—— 2022一带一路国际大学生数字建筑设计竞赛

设计说明 | Design Specification

该项目场地位于河南省平顶山市郏县冢头镇陈寨村，郏县是千年古城，历史可追溯至西汉，郏县古建筑拥有中国北方民居的特点，立柱、斗拱、梁架的组合构筑了这里的建筑文化语言，我们提取了传统建筑的元素，采用与传统建筑语言相呼应、相统一的设计手法，重视以人为核心的场景体验的，集交通、娱乐、购物为一体的火车站设计。

用新手法、新材料承接千载年的文化，用新意境、新思想视通万里江山。

The project site is located in Chenzhai village, Zhongtou Town, Jiaxian County, Pingdingshan City, Henan Province. Jiaxian county is an ancient city with a history of thousands of years, which can be traced back to the Western Han Dynasty. The ancient buildings in Jiaxian county have the characteristics of folk houses in northern China. The combination of columns, Dougong and beam frames constructs the architectural cultural language here. We extract the elements of traditional architecture, adopt the design method that echoes and unifies with the traditional architec-tural language, and pay attention to the scene experience centered on people, integrating transp-ortation Design of railway station integrating entertainment and shopping. Use new techniques and materials to undertake the culture of thousands of years, and use new artistic conception and new ideas to see thousands of miles of rivers and mountains.

区位分析 | Location Analysis

Yu County
Tsukasa Town
Chenzhai Village
Project land

当地建筑拼贴 | Local building

人群分析 | Population analysis

一月
春运
二月
三月
放风筝
四月
五月
农产品买卖
六月
七月
八月
九月
旅游旺季
十月
十一月
十二月

游客	本地居民
5:00	
	老年人 晨跑 散步
10:00	
游客到达 交通高峰 拍照 打卡	青年人 上班 通勤
18:00	
游客返程 交通高峰 打卡纪念	青年人 下班 餐饮 散步

概念演变 | Concept of evolution

郏县 Jia xian
乡镇特色
乡村肌理
村镇 The town
空间更新
开放空间
内外 inside and outside
古今对立
社会矛盾
历史 History
文化传承
工匠精神
乡村新筑

揉
填
分
立

材料运用 | Material Application

印刷合金铜板
青砖
木材
瓦片

A-A剖面透视图

总平面图 | 1：1000

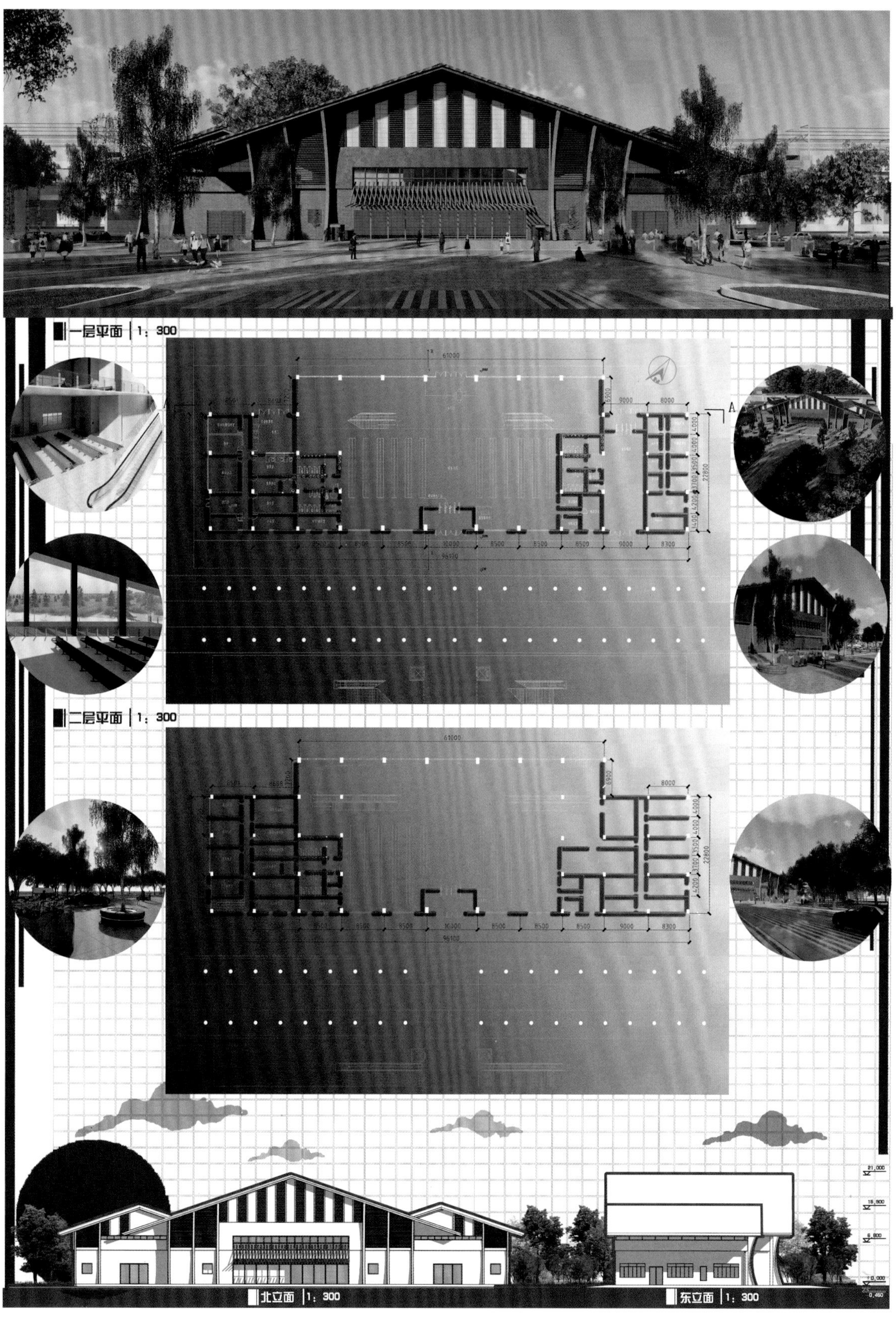
一层平面 1：300
二层平面 1：300
北立面 1：300
东立面 1：300

江苏省镇江市宝盖山铁路遗址保护及更新设计研究

Study on the Protection and Renewal Design of Baogaishan Railway Site in Zhenjiang City, Jiangsu Province

参赛学校 University/College
江苏大学
Jiangsu University, China

指导教师 Supervisor(s)
李 晓 LI Xiao

参赛学生 Participant(s)
朱彦榕 ZHU Yanrong
韩 旭 HAN Xu
印 璐 YIN Lu
黄 杨 HUANG Yang
赵心怡 ZHAO Xinyi

简介 Description

基于升级利用废弃铁路遗址文化资源，探究现存老旧工业遗址及绿色林地基底的生态价值，对废弃铁路遗址进行活化与更新研究和设计。文章选取镇江宝盖山铁路遗址为研究对象，采用 GIS、RS 技术与景观规划设计结合的方法，对镇江宝盖山铁路遗址沿线的植被及绿地景观状况进行深入研究与设计。提出合理利用植被资源，分层协调资源分配，健全绿地生态景观科学管理等重塑生态景观的设计策略。

The design aims to upgrade and utilize the cultural resources of the abandoned railway sites and explore the ecological value of the existing old industrial sites and green woodland basement. The desolated railway sites were studied and designed for cremation and renewal. By adopting the approach of combining technologies including GIS and RS with the planning and designing of the views, we conducted an in-depth study and design for the plantation and greenland along the Baogaishan railway site in Zhenjiang. Based on that, we put forward the design strategies of reshaping the ecological landscape, including using vegetation resources rationally, coordinating a hierarchical allocation of resources and improving the scientific management of a green ecological landscape.

镇江宝盖山铁路遗址保护及更新设计 RESEARCH ON PROTECTION AND RENEWAL DESIGN OF ZHENJIANG BAOGAISHAN RAILWAY SITE 01

区位分析Location analysis

是北至宝盖路，南至中山西路，西至京畿路，东至太古山路，主要的研究对象为该范围内沪宁铁路线沿线的遗产点

It extends to Baogai road in the north, Zhongshan West Road in the south, Jingji road in the West and taigushan road in the East It is the heritage sitealong the Shanghai Nanjing railway line within the scope

历史沿革development of history

1905 1906 1937 1977 2004 2014 2022

气候分析 Climate analysis

江苏省位于亚洲大陆东岸中纬度地带，属东亚季风气候区，处在亚热带和暖温带的气候过渡地带。江苏省地势平坦，介于30°46′～35°07′N之间，116°22′～121°55′E之间，一般以淮河、苏北灌溉总渠一线为界，以北地区属暖温带湿润、半湿润季风气候；以南地区属亚热带湿润季风气候。

Jiangsu Province is located in the mid-latitude zone on the east coast of the Asian continent, belonging to the East Asian monsoon climate zone, in the subtropical and warm temperate climate transition zone.The terrain of Jiangsu Province is flat, between 30 ° 46 ′ ~ 35 ° 07 ′ N, between 116 ° 22 ′ ~ 121 ° 55 ′ E, generally bounded by the Huai River and the main irrigation canal in northern Jiangsu, and the northern area is warmT he temperate humid and semi-humid monsoon climate;

平面图 Floor plan

设计思路Design ideas

场地策略Venue strategy

科普教育 亲子活动 休闲娱乐 领略风光 缅怀历史

草地模式 Grass mode

穿过花海 穿过隧道 穿过梯田 穿过密林 穿过疏林

场地分析 Site analysis

植物设计Plant design

站台花园

儿童游憩区

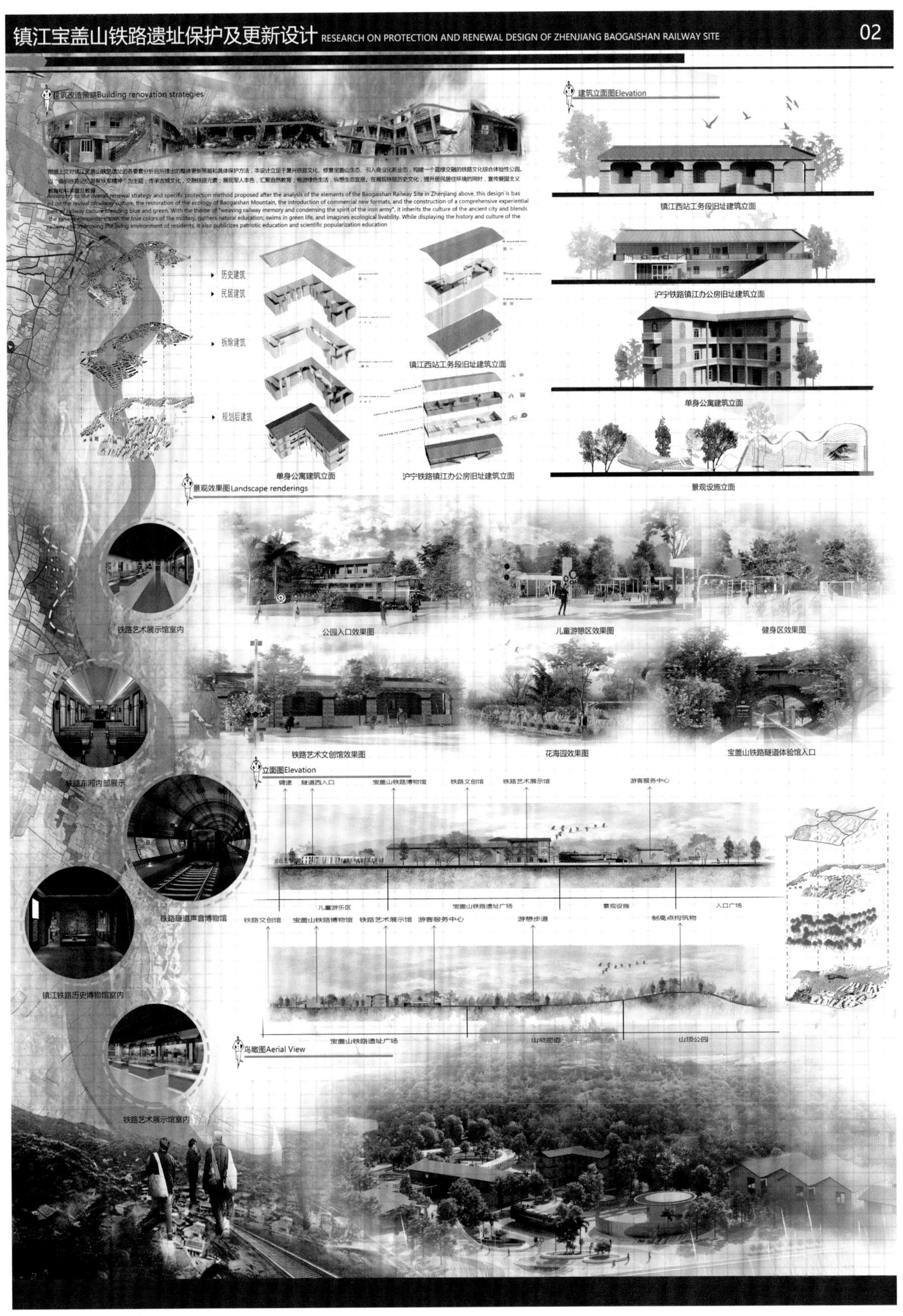
镇江宝盖山铁路遗址保护及更新设计 RESEARCH ON PROTECTION AND RENEWAL DESIGN OF ZHENJIANG BAOGAISHAN RAILWAY SITE
02
建筑改造策略Building renovation strategies
历史建筑
民居建筑
拆除建筑
规划后建筑
单身公寓建筑立面
沪宁铁路镇江办公房旧址建筑立面
镇江西站工务段旧址建筑立面
建筑立面图Elevation
单身公寓建筑立面
景观设施立面
景观效果图Landscape renderings
公园入口效果图
儿童游憩区效果图
健身区效果图
铁路艺术文创馆效果图
花海园效果图
宝盖山铁路隧道体验馆入口
铁路艺术展示馆室内
铁路车厢内部展示
铁路隧道声音博物馆
镇江铁路历史博物馆室内
立面图Elevation
宝盖山铁路遗址广场
山顶公园
鸟瞰图Aerial View

畔上凌华

Ice Flower by the Lake

参赛学校	University/College
	吉林建筑大学
	Jilin Jianzhu University, China
指导教师	Supervisor(s)
	宋义坤　SONG Yikun
	林　铓　LIN Mang
参赛学生	Participant(s)
	李　润　LI Run
	左一轩　ZUO Yixuan
	郭　璐　GUO Lu
	范馨穗　FAN Xinsui
	王炳璇　WANG Bingxuan

简介 Description

火车问世后，一个地方的火车站会成为这个地方兴衰的见证者，车站陪伴着无数人度过无数个日夜，就像长岭湖边的冰凌子一样，看着人们来来往往，看哈尔滨在“一带一路”的道路上阔步前行。这朵冰凌子坐落于哈尔滨西郊长岭湖湖畔的长岭湖村，旨在为这块蓬勃发展的土地注入新的力量，促进当地经济文化多元化的发展。

After the advent of the train, the local railway station will become a witness to the rise and fall of the place. Stations accompanied numerous people through countless days and nights, just like the ice flower by the Changling Lake that watches people come and go and the Harbin City in northeast China strides forward on the "Belt and Road". Located in Changling Lake Village on the western outskirts of Harbin, this "ice flower" was designed to inject new impetus into this booming land and promote the development of local economic and cultural diversification.

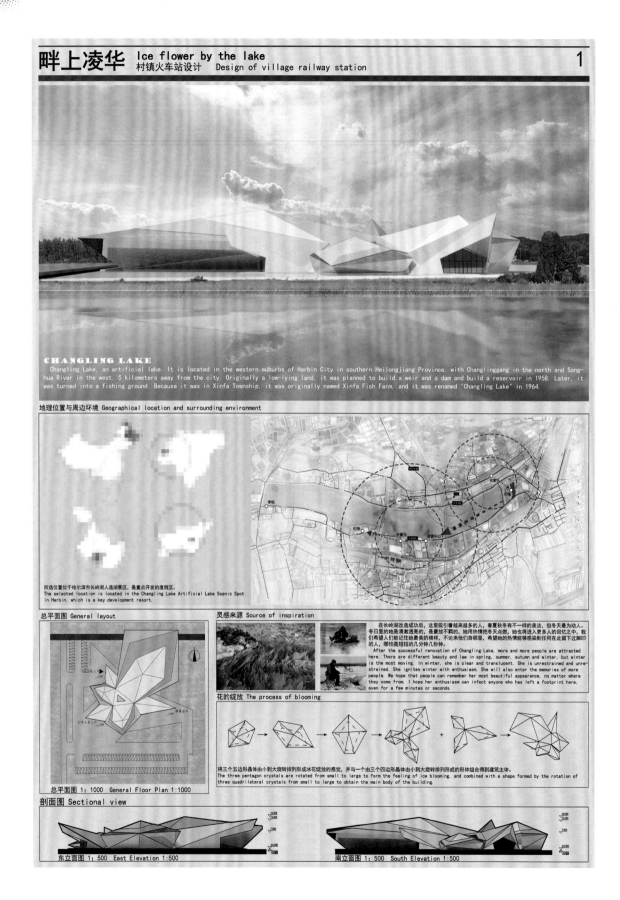

畔上凌华 Ice flower by the lake 1

村镇火车站设计 Design of village railway station

CHANGLING LAKE

Changling Lake, an artificial lake. It is located in the western suburbs of Harbin City in southern Heilongjiang Province, with Changlinggang in the north and Songhua River in the west. 5 kilometers away from the city. Originally a low-lying land, it was planned to build a weir and a dam and build a reservoir in 1958. Later, it was turned into a fishing ground. Because it was in Xinfa Township, it was originally named Xinfa Fish Farm, and it was renamed "Changling Lake" in 1964.

地理位置与周边环境 Geographical location and surrounding environment

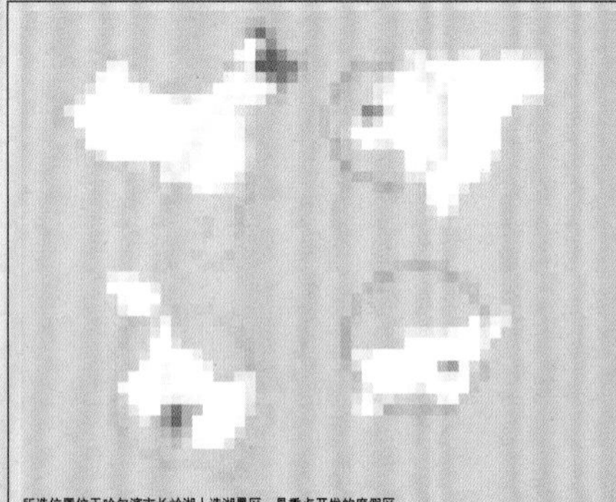

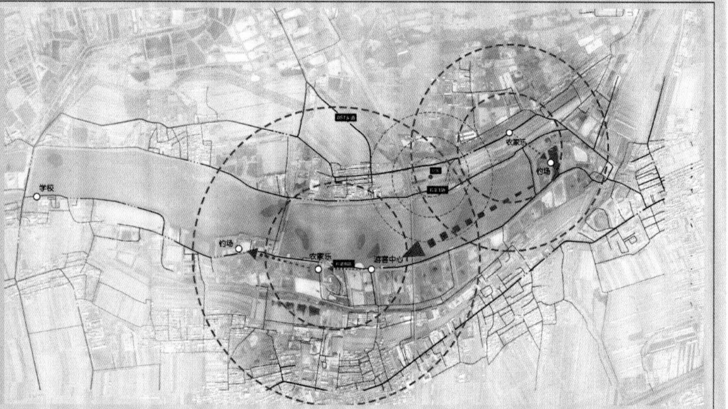

所选位置位于哈尔滨市长岭湖人造湖景区，是重点开发的度假区。
The selected location is located in the Changling Lake Artificial Lake Scenic Spot in Harbin, which is a key development resort.

总平面图 General layout

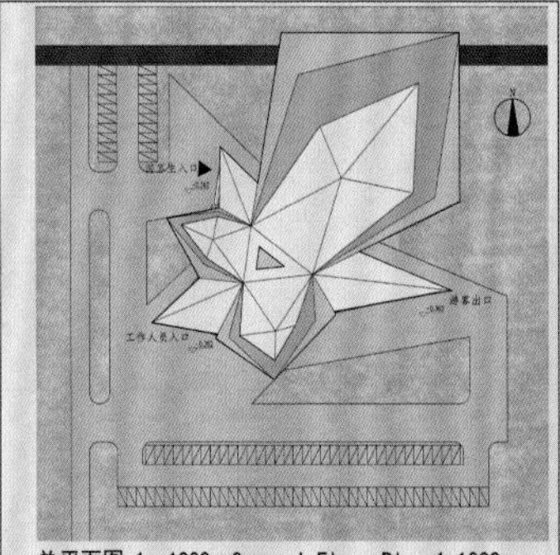

总平面图 1：1000 General Floor Plan 1:1000

灵感来源 Source of inspiration

在长岭湖改造成功后，这里吸引着越来越多的人，春夏秋冬有不一样的美法，但冬天最为动人。冬日里的她是清澈透亮的，是豪放不羁的，她用热情把冬天点燃，她也将进入更多人的回忆之中，我们希望人们能记住她最美的模样，不论来他们自哪里，希望她的热情能够感染到任何在此留下过脚印的人，哪怕是短短的几分钟几秒钟。

After the successful renovation of Changling Lake, more and more people are attracted here. There are different beauty and law in spring, summer, autumn and winter, but winter is the most moving. In winter, she is clear and translucent. She is unrestrained and unrestrained. She ignites winter with enthusiasm. She will also enter the memories of more people. We hope that people can remember her most beautiful appearance, no matter where they come from, I hope her enthusiasm can infect anyone who has left a footprint here, even for a few minutes or seconds.

花的绽放 The process of blooming

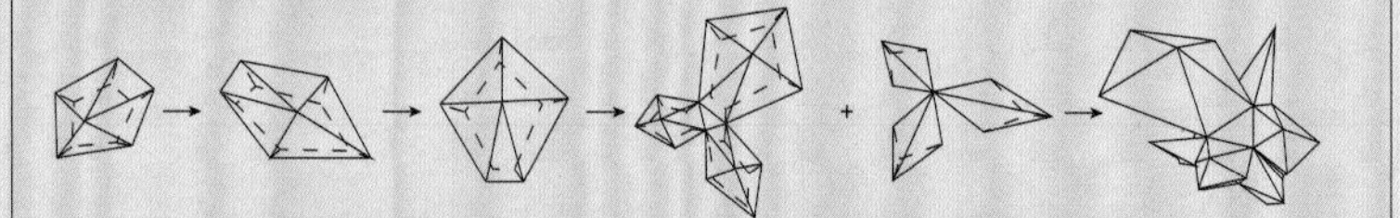

将三个五边形晶体由小到大旋转排列形成冰花绽放的感觉，并与一个由三个四边形晶体由小到大旋转排列形成的形体组合得到建筑主体。
The three pentagon crystals are rotated from small to large to form the feeling of ice blooming, and combined with a shape formed by the rotation of three quadrilateral crystals from small to large to obtain the main body of the building.

剖面图 Sectional view

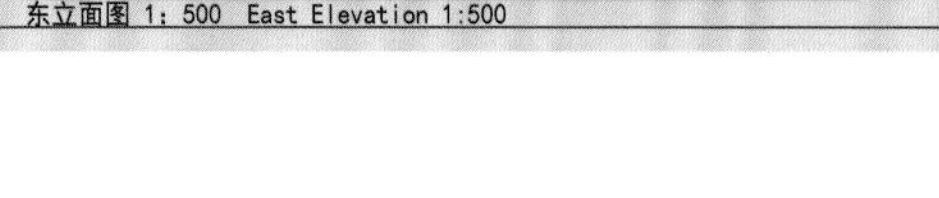

东立面图 1：500 East Elevation 1:500

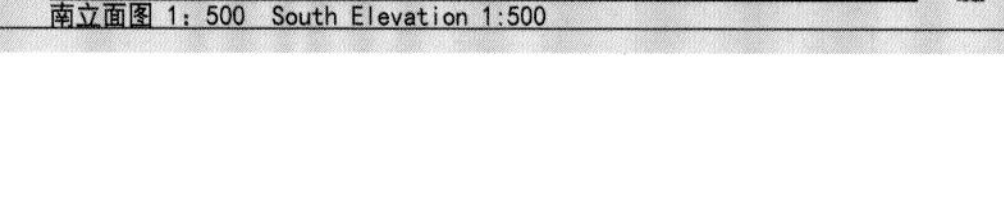

南立面图 1：500 South Elevation 1:500

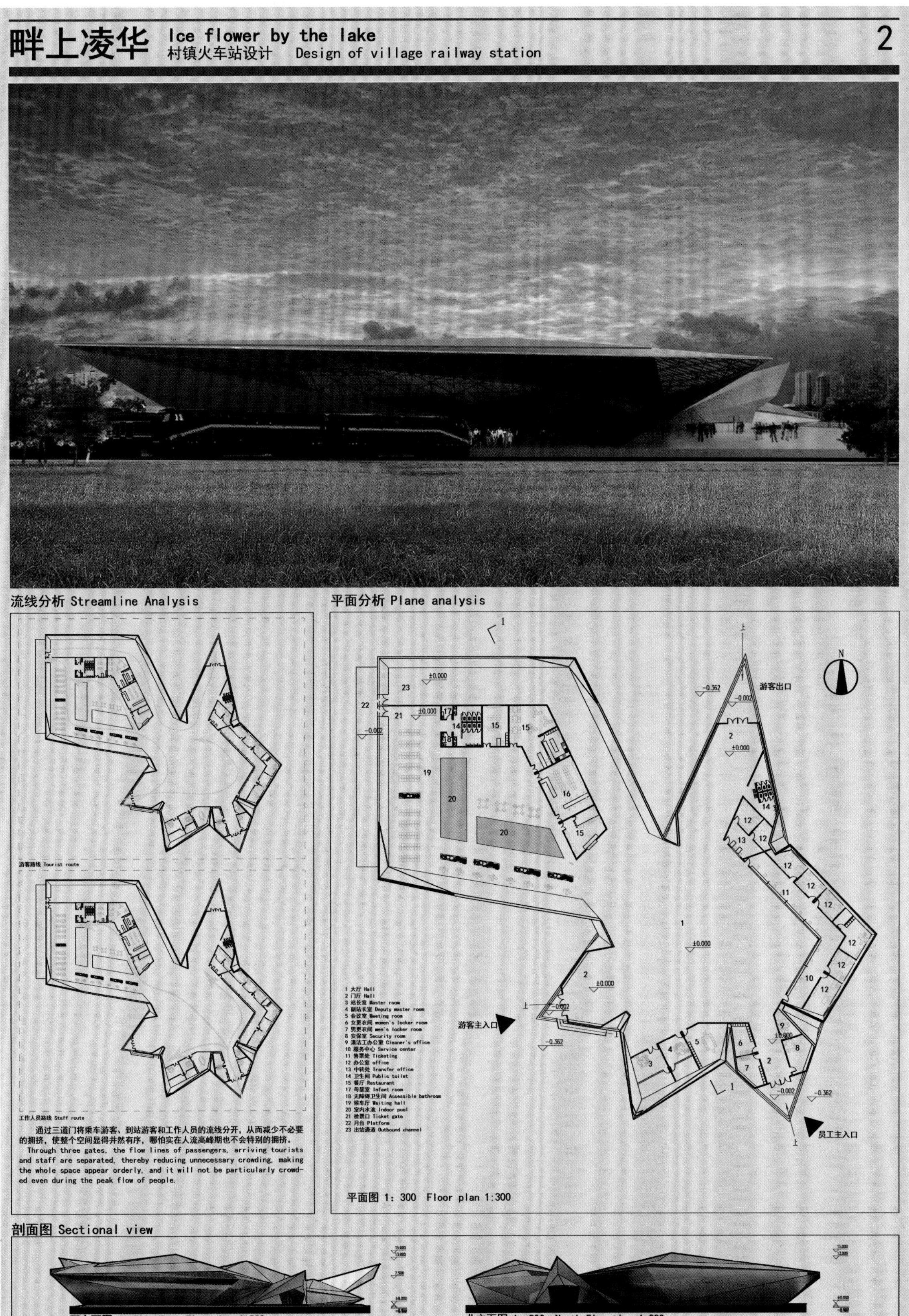
畔上凌华 Ice flower by the lake
村镇火车站设计 Design of village railway station
2
流线分析 Streamline Analysis
游客路线 Tourist route
工作人员路线 Staff route
通过三道门将乘车游客、到站游客和工作人员的流线分开，从而减少不必要的拥挤，使整个空间显得井然有序，哪怕实在人流高峰期也不会特别的拥挤。
Through three gates, the flow lines of passengers, arriving tourists and staff are separated, thereby reducing unnecessary crowding, making the whole space appear orderly, and it will not be particularly crowded even during the peak flow of people.
平面分析 Plane analysis
游客出口
游客主入口
员工主入口
1 大厅 Hall
2 门厅 Hall
3 站长室 Master room
4 副站长室 Deputy master room
5 会议室 Meeting room
6 女更衣间 women's locker room
7 男更衣间 men's locker room
8 安保室 Security room
9 清洁工办公室 Cleaner's office
10 服务中心 Service center
11 售票处 Ticketing
12 办公室 office
13 中转处 Transfer office
14 卫生间 Public toilet
15 餐厅 Restaurant
17 母婴室 Infant room
18 无障碍卫生间 Accessible bathroom
19 候车厅 Waiting hall
20 室内水池 Indoor pool
21 检票口 Ticket gate
22 月台 Platform
23 出站通道 Outbound channel
平面图 1: 300 Floor plan 1:300
剖面图 Sectional view
西立面图 1: 500 West Elevation 1:500
北立面图 1: 500 North Elevation 1:500

离·合

Separation & Retogether

参赛学校 University/College

吉林建筑大学

Jilin Jianzhu University, China

指导教师 Supervisor(s)

宋义坤 SONG Yikun

金 莹 JIN Ying

参赛学生 Participant(s)

陈嘉怡 CHEN Jiayi

陈佳宁 CHEN Jianing

李卓十 LI Zhuoshi

汪学林 WANG Xuelin

刘佳钰 LIU Jiayu

简介 Description

铁路，见证了多少来来往往；车站，承载着多少分分合合。身处车站中的，可能是求学的游子；也可能是背井离乡打拼生存，再也不会回到故土的人。大庆，曾因石油产业兴盛一时，但随着石油资源的日渐枯竭，大批青年人踏上了归家的列车。在“一带一路”的背景下，大庆的转型升级已被积极践行。因此，此次建筑设计选取大庆市向荣村，旨在发挥车站作用之余带动周边地区经济发展，为转型尽一份力。

The railway witnessed numerous trips; the station represents countless separations and reunions. Those who are on the stop may be students going to schools or people leaving for other cities to work and will never come back. Daqing, a city in northeast China, was once prosperous because of its oil industry. However, as oil gradually runs up, a large number of young people working here went back home. With the "Belt and Road" Initiative, Daqing's transformation and update have been ongoing. Therefore, we chose the Xiangrong Village in Daqing in this architectural design to drive the economic development of the surrounding areas with this station and contribute to Daqing's transformation on top of relying on the basic functions of the station.

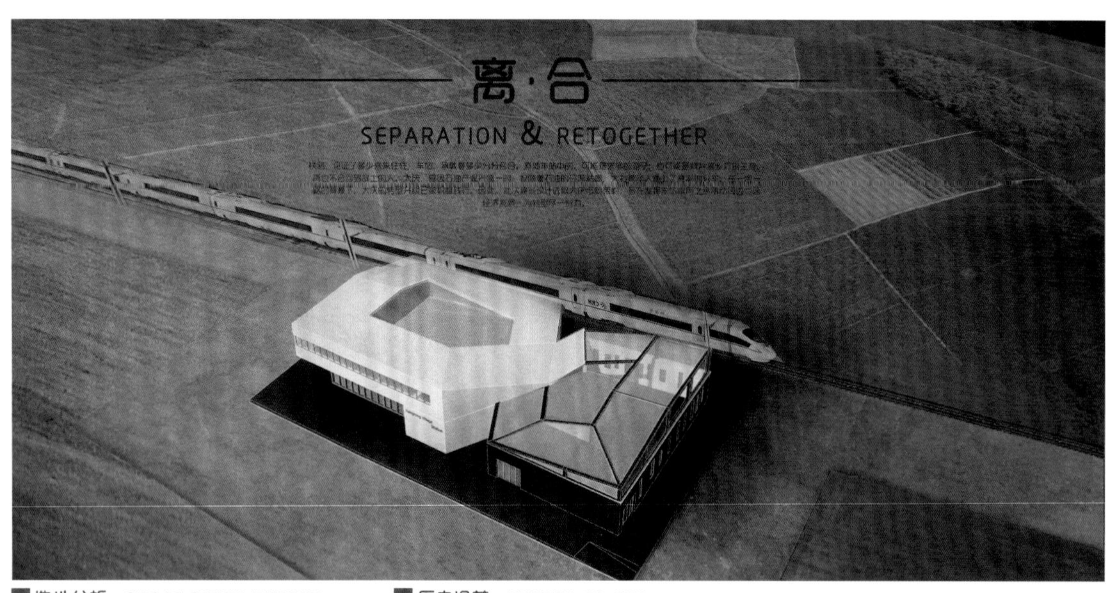

选址分析 SITE SELECTION ANALYSIS

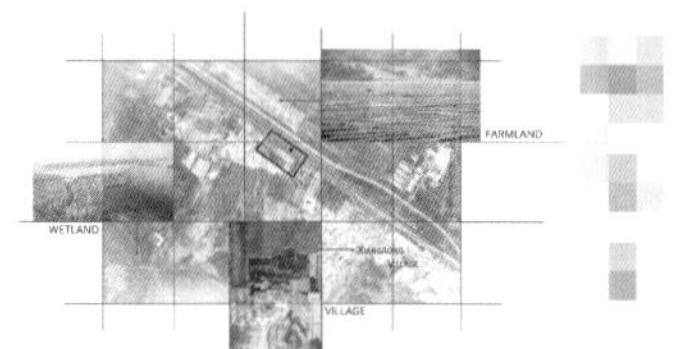

历史沿革 HISTORY OF CITY

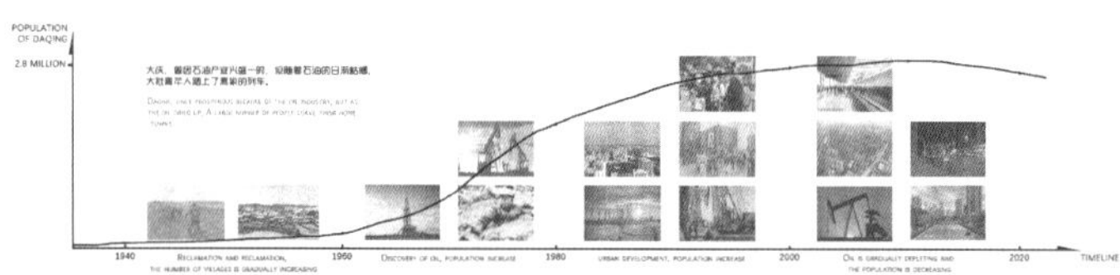

气候分析 CLIMATE ANALYSIS

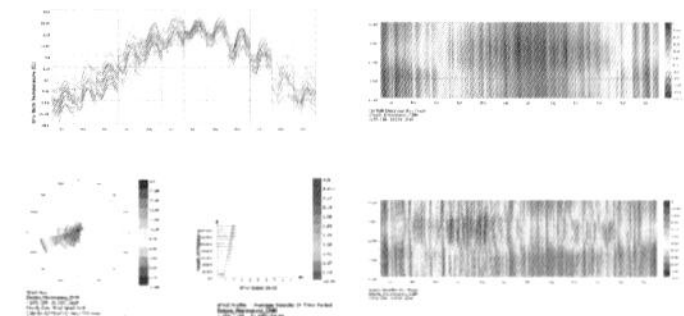

体块生成 BLOCK GENERATED

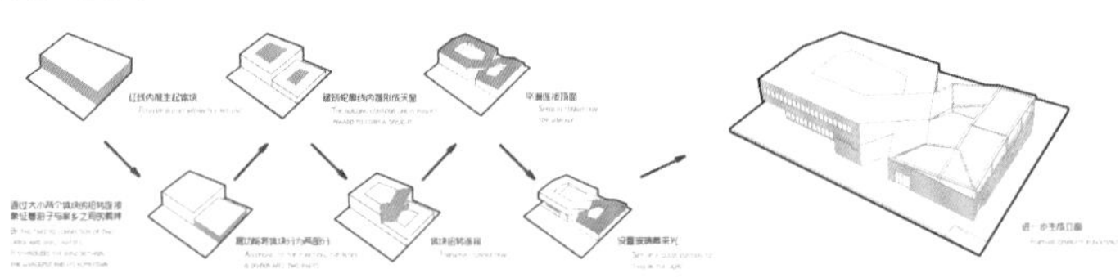

总平面图 GENERAL LAYOUT

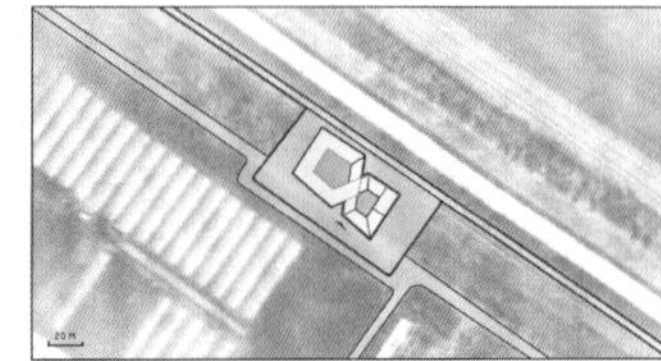

材质分析 MATERIAL ANALYSIS

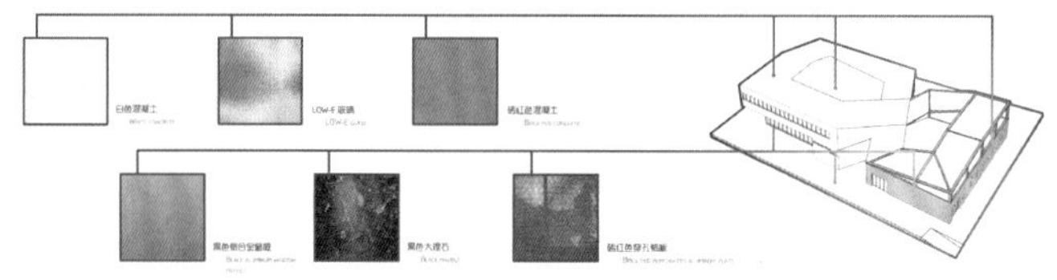

灵感来源 INSPIRATION SOURCE

“我相信有情感的建筑。‘建筑’的生命就是它的美。这对人类是很重要的。对一个问题如果有许多解决方法，其中的那种给使用者传达美和情感的就是建筑。”

“I BELIEVE IN EMOTIONAL ARCHITECTURE. THE LIFE OF ARCHITECTURE IS ITS BEAUTY. THIS IS VERY IMPORTANT TO HUMAN BEINGS. IS VERY IMPORTANT.IF THERE ARE MANY SOLUTIONS TO A PROBLEM, WHICH ONE IS COMMUNICATED TO THE USER,THE OF BEAUTY AND EMOTION IS ARCHITECTURE.”

——路易斯·巴拉干

LUIS BARRAGAN

各种色彩浓烈鲜艳的墙体的运用是巴拉干设计中鲜明的个人特色，后来也成为了墨西哥建筑的重要设计元素。在巴拉干游历欧洲的时候曾深深的被摩洛哥那种独特的地中海式气候下浓烈的色彩风格所打动。之后他便开始关注墨西哥民居中绚烂的色彩，并将其运用到了自己的众多作品中。

THE USE OF VARIOUS BRIGHTLY COLORED WALLS IS BALAKAN THE DISTINCTIVE PERSONAL FEATURES OF THE DESIGN LATER BECAME MEXIAN IMPORTANT DESIGN ELEMENT OF THE GOE BUILDING. TRAVELED TO EUROPE IN BALAKAN THE TIME WAS DEEPLY TOUCHED BY MOROCCO'S UNIQUE MEDITERRANEAN THE STRONG COLOR STYLE OF THE CLIMATE IS IMPRESSED. AFTER THAT HE WENT START PAYING ATTENTION TO THE BRILLIANT COLORS IN MEXICAN HOUSES AND BRING THEM TO LIFE IT HAS BEEN USED IN MANY OF HIS OWN WORKS.

立面生成 ELEVATION GENERATION

西南立面 WEST AND SOUTH ELEVATION

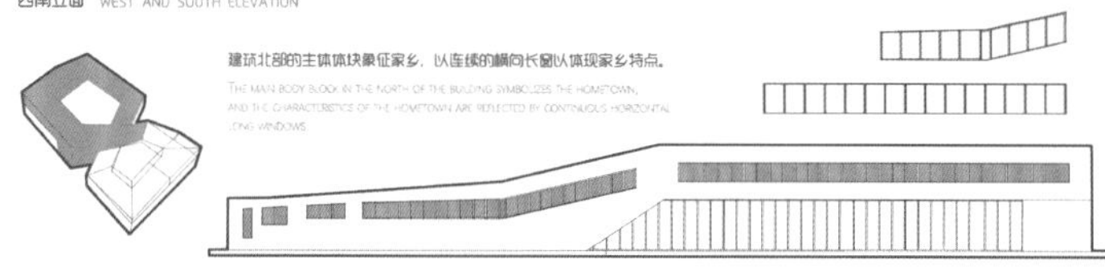

流线分析 TRAFFIC ANALYSIS

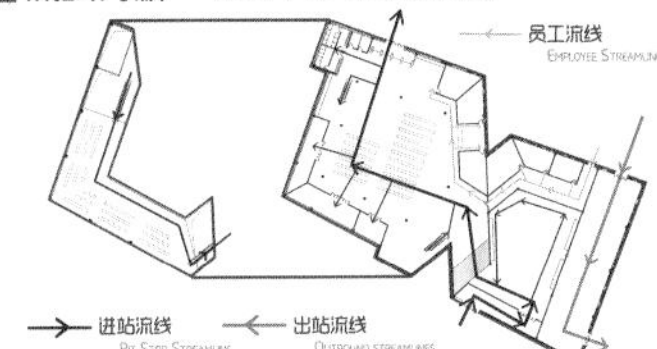

功能分区 FUNCTIONAL PARTITION

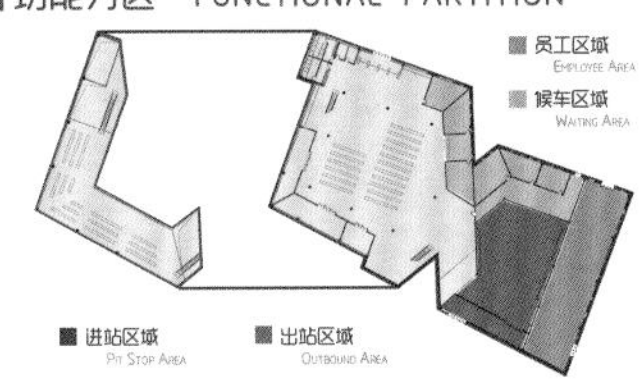

平面图 ICHNOGRAPHY

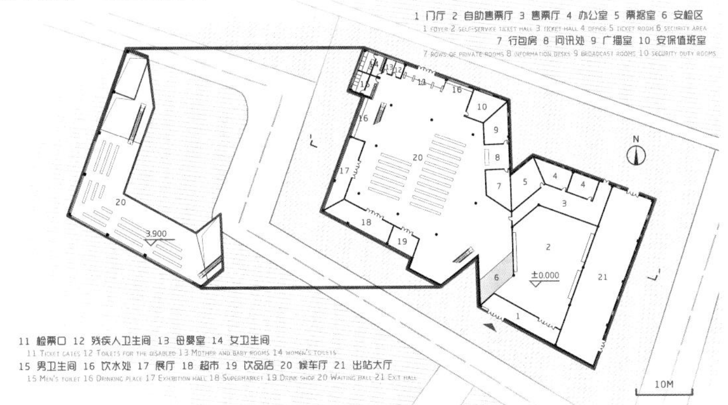

湿地资源 WETLAND RESOURCES

大庆不只有石油，更有全国城区内最大、保存最完整的湿地景观。湿地范围有120万公顷，占全国已知湿地总面积的3.12%，且发育比较成熟，湿地景观类型丰富。在转型升级的当今，我们应该重视湿地景观带来的旅游资源。通过在车站内设置小型湿地景观展厅，让更多的人了解到大庆秀美的自然风光，能为具有湿地景观的向荣村带来旅游资源，促进周边产业转型升级，促进经济发展。

Daqing is not only oil, but also the largest and best-preserved wetland lands-cape in the urban area of the country. wetland area is 1.2 million hectares, accounting for 3.12% of the total area of known wetlands in The country, and the development is relative mature, wetland landscape type is abundant. In today's transformation and upgrading, we should pay attention to wetland landscapes view of the tourism resources brought by the view. Let's get more by setting up a small wetland landscape exhibition hall inside the station Peo-ple learn about the beautiful natural scenery of Daqing, which can bring tourism to Xiangrong Village with wetland landscape Resources, promote the transformation and upgrading of surrounding industries, and promote economic development.

立面图 ELEVATION

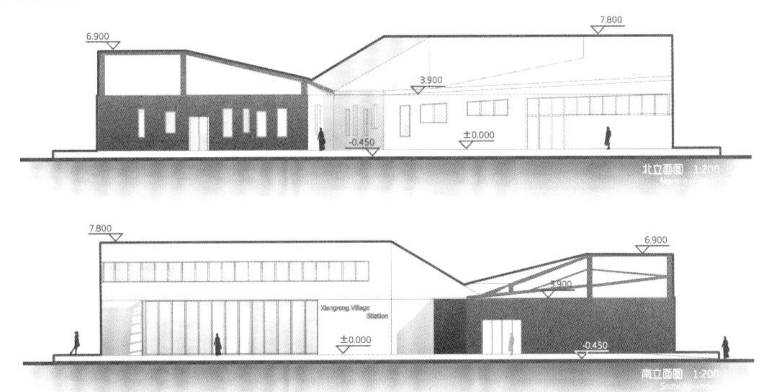

剖面图 PROFILE

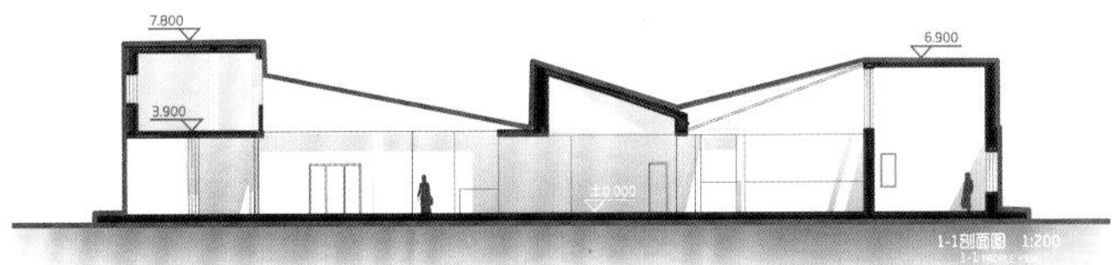

丝路·新纽带

Hakka Hub

参赛学校	University/College 嘉应学院 Jiaying University, China
指导教师	Supervisor(s) 张灵辉　ZHANG Linghui
参赛学生	Participant(s) 潘嘉颖　PAN Jiaying 丁文琳　DING Wenlin 管浩坤　GUAN Haokun

简介 Description

该设计我们以“丝路·新纽带”为主题设计了一座现代化的铁路客运站。该建筑的创作以丝绸为原型结合客家传统手工布艺上的纹理以及梅州地处山区的特点，形成了建筑的外立面。

在环境设计中采取生态优先、整体协调原则，通过在基地内置入大量的绿化、绿地空间，可收集雨水用于浇灌，从而达到节约水资源的目的。

This work is a modern railway station designed with the theme of “Silk Road, New Bond”. The shape of the building mimics silk, combined with the texture of traditional handmade fabrics of the Hakka people and the characteristics of Meizhou’s location in a mountainous region in south China. These are also reflected in the facade of the building.

The principle of prioritizing the biological environment and overall coordinating was adopted in the design of the greening of the station with abundant plants and green belts. Rain water is collected for irrigation to achieve water conservation.

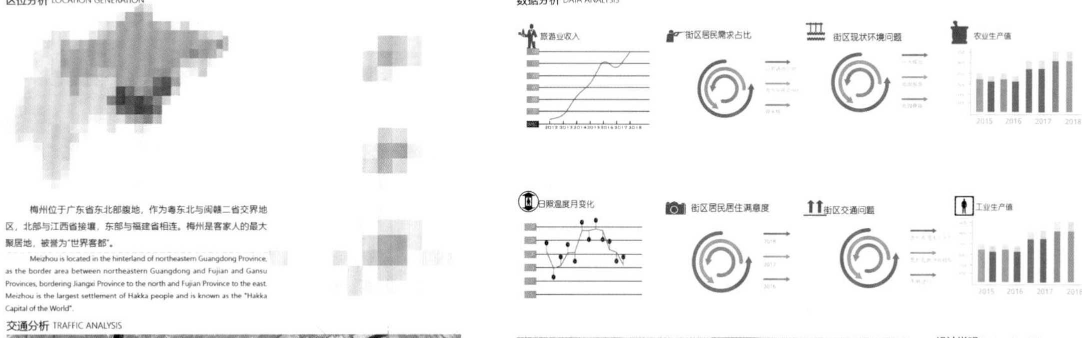
区位分析 LOCATION GENERATION
数据分析 DATA ANALYSIS
营器业收入
街区居民需求占比
街区现状环境问题
农业生产值
日照温度月变化
街区居民居住满意度
街区交通问题
工业生产值
梅州位于广东省东北部腹地，作为粤东北与闽赣二省交界地区，北部与江西省接壤，东部与福建省相连。梅州是客家人的最大聚居地，被誉为"世界客都"。
Meizhou is located in the hinterland of northeastern Guangdong Province, as the border area between northeastern Guangdong and Fujian and Gansu Provinces, bordering Jiangxi Province to the north and Fujian Province to the east. Meizhou is the largest settlement of Hakka people and is known as the "Hakka Capital of the World".

交通分析 TRAFFIC ANALYSIS

设计说明 Design description

在国家"一带一路"政策的大背景下，为加速推进着沿线各国经济、文化的合作交融，给古老的丝绸之路带来新的生机。本案在古老的丝绸之路上，以"丝路 · 新纽带"为主题设计了一座现代化的铁路客运站。以丝绸为原型创造出建筑的造型，结合客家传统手工布艺上的纹理，以及梅州地处山区的特点，设计了建筑的外立面。

在环境设计中采取生态优先、整体协调原则，通过在基地内置入大量的绿化、绿地空间。可收集雨水用于浇灌，从而达到节约水资源的目的。另外，可通过大量的地表植被净化地表水，达到以绿养水，以土养绿，以绿养土的良性循环。

Under the background of the national "Belt and Road" policy, in order to accelerate the economic and cultural cooperation and integration of countries along the route, it has brought new vitality to the ancient Silk Road. The case is on the ancient Silk Road, with the title of "Silk Road · "New Bond" is the theme of designing a modern railway passenger station. The shape of the building was created based on silk, combining the texture of traditional Hakka handmade fabrics and the characteristics of Meizhou's mountainous location to design the façade of the building.

In the environmental design, the principle of ecological priority and overall coordination is adopted, and a large number of green and green space is built into the base. Rainwater can be collected for irrigation, so as to achieve the purpose of saving water resources. In addition, surface water can be purified through a large amount of surface vegetation, so as to achieve a virtuous cycle of green water, green soil and green soil. This reflects the principles of sustainable development.

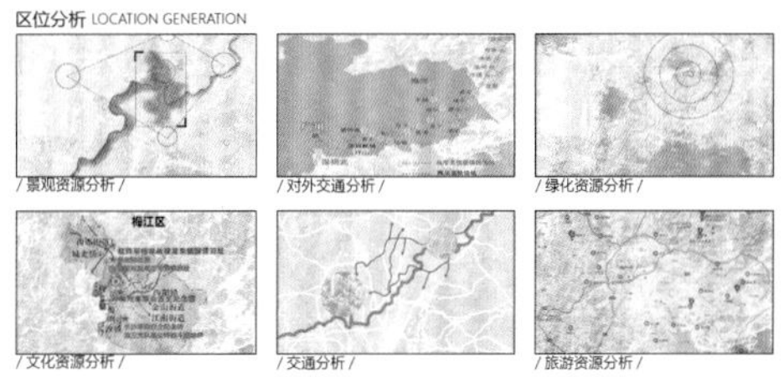
区位分析 LOCATION GENERATION
/ 景观资源分析 /
/ 对外交通分析 /
/ 绿化资源分析 /
梅江区
/ 文化资源分析 /
/ 交通分析 /
/ 旅游资源分析 /

丝路 · 新纽带 "客都枢纽"铁路客运站设计——2022"一带一路"国际大学生数字建筑设计 1

"Hakka Hub" Railway Passenger Station Design——2022 Belt and Road International Student Competition on Digital Architectural Design

在建筑的空间设计中采用了通高和错层的设计手法。顺应场地高差将建筑的空间分成两部分，一部分为售票办公空间，另一部分为候车区。为缓解南方地区夏季高温炎热的气候环境，采用了双层屋面的做法，在两层屋面之间形成空气间层，使空气通过时带走部分建筑热量的作用，由此达到被动式建筑节能。

In the spatial design of the building, the design method of pass height and split level is adopted. According to the height difference of the site, the space of the building is divided into two parts, one part is the ticket office space and the other part is the waiting area. In order to alleviate the hot and hot climate environment in the southern region in summer, the practice of double-layer roofing is adopted, and an air interlayer layer is formed between the two-storey roof, so that the air passes through the role of taking away part of the building heat, thereby achieving passive building energy saving.

西立面图 1：500 WEST ELEVATION 1:500

南立面图 1：500 SOUTH ELEVATION 1:500

1-1 剖面图 1：500 1-1 SECTIONAL DRAWING 1:500

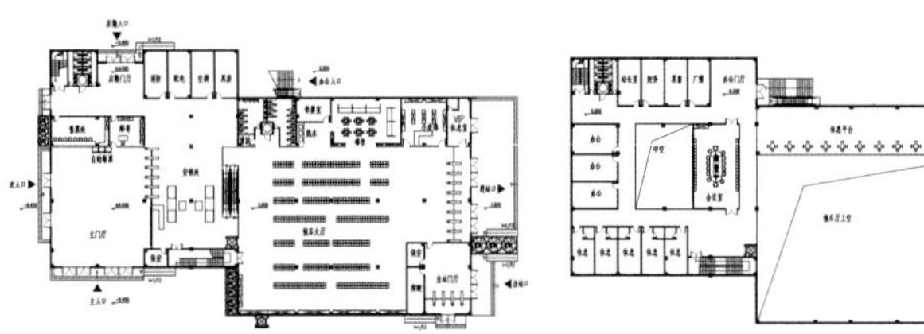

首层平面图 1：500 GROUND FLOOR PLAN 1:500

二层平面图 1：500 THE SECOND FLOOR PLAN 1:500

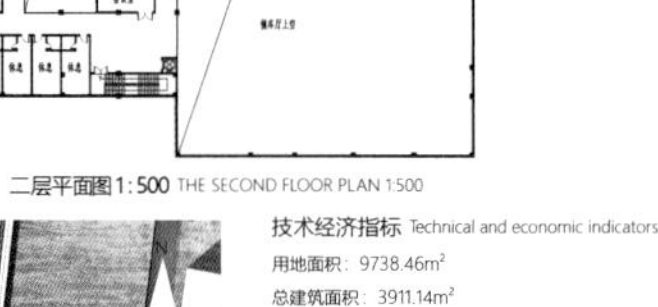

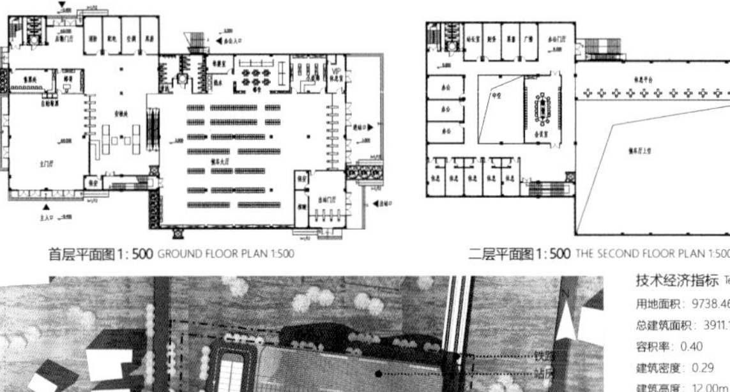

总平面图 1：1000

技术经济指标 Technical and economic indicators

用地面积：9738.46m²
总建筑面积：3911.14m²
容积率：0.40
建筑密度：0.29
建筑高度：12.00m
绿地率：37.24%
停车位：42
Site Area: 9738.46m²
Total GFA: 3911.14m²
FAR: 0.40
Building density: 0.29
Building Height: 12.00m
Green space rate: 37.24%
Parking space: 42

设计理念 DESIGN CONCEPT

以丝绸为原型创造出建筑的造型 The shape of the building was created based on silk

客家传统手工布艺上的纹理转译成建筑表皮

The texture on the traditional Hakka handmade fabric is translated into the architectural skin

提取山体轮廓，结合木格栅作为建筑的立面遮阳构件

Extract the contours of the mountain and combine the wooden grille as a shading component of the building's façade

景观小品设计 LANDSCAPE SKETCH DESIGN

丝路·新纽带 “客都枢纽”铁路客运站设计——2022“一带一路”国际大学生数字建筑设计 2
"Hakka Hub" Railway Passenger Station Design——2022 Belt and Road International Student Competition on Digital Architectural Design

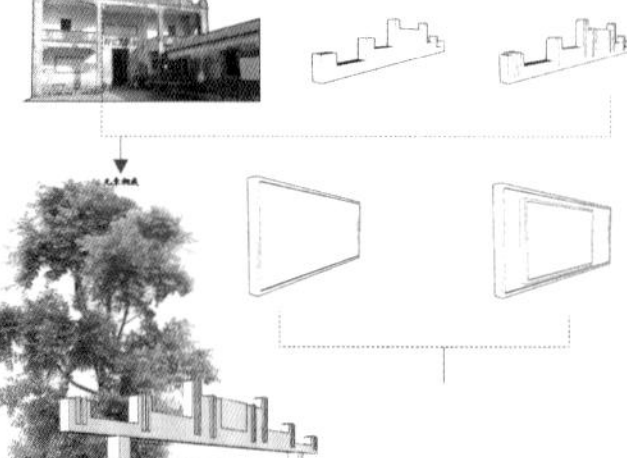

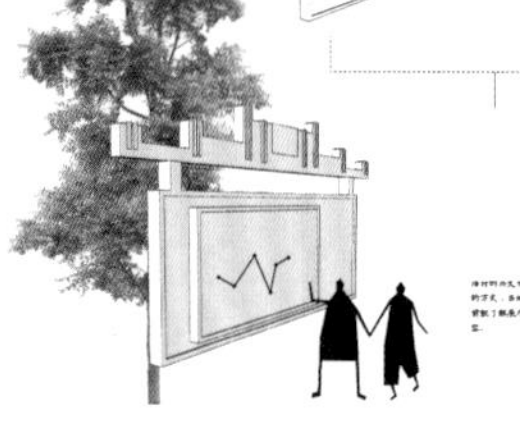

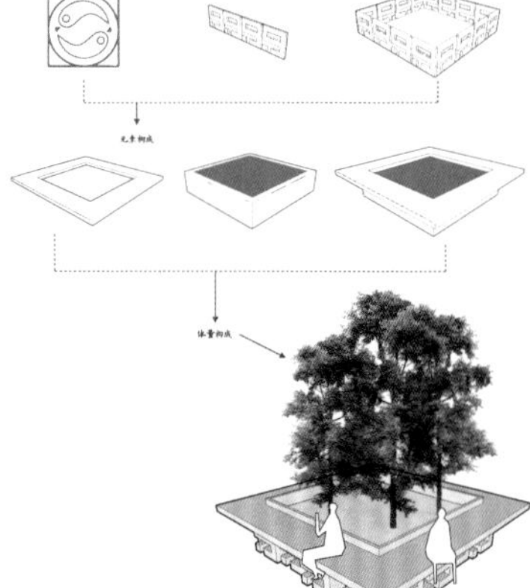

绿色节能设计 GREEN ENERGY-SAVING DESIGN

"一带一路"之城市记忆的延续

Continuation of the Urban Memory under the Belt and Road Initiative

参赛学校 University/College
江苏大学
Jiangsu University, China

指导教师 Supervisor(s)
李 晓 LI Xiao

参赛学生 Participant(s)
王明明 WANG Mingming
张思月 ZHANG Siyue
韩 旭 HAN Xu

简介 Description

铁路工业遗产是城市的记忆，铁路工业遗产的更新是城市快速发展的命脉。改革开放以来我国城市化进程不断加快，新老城区的冲突成为当前城市的主要矛盾，旧铁路的废弃不仅阻碍了城市的发展，也影响了城市的面貌。镇江宝盖山铁路工业遗址更新设计注重新老城区致密空间肌理冲突，融入城市梯度记忆，塑造城市未来空间发展的结构。以保护与更新为主题总结镇江宝盖山铁路工业遗产更新设计的思路，打造历史与时代交替的文化与旅游空间。

The heritage of the local railway industry is the memory of the city and its renewal is the lifeblood of a city with rapid development. Since Reform and Opening-up, China's urbanization process has been accelerating and the conflict between the new and old areas of a city has become prominent today. Desolated old railways not only hinder the further development of a city, but also make it unsightly. The design of renovating the Baogaishan railway site in Zhenjiang focuses on the texture conflict of the dense space in the new and old areas of the city, integrating the multiple layers of the city's memory and shaping the structure of future development of the city space. The ideas behind the renovation could be summarized as protection and revival with the aim of shaping a space for cultural and tourist activities featuring the overlapping of the past and present of the city.

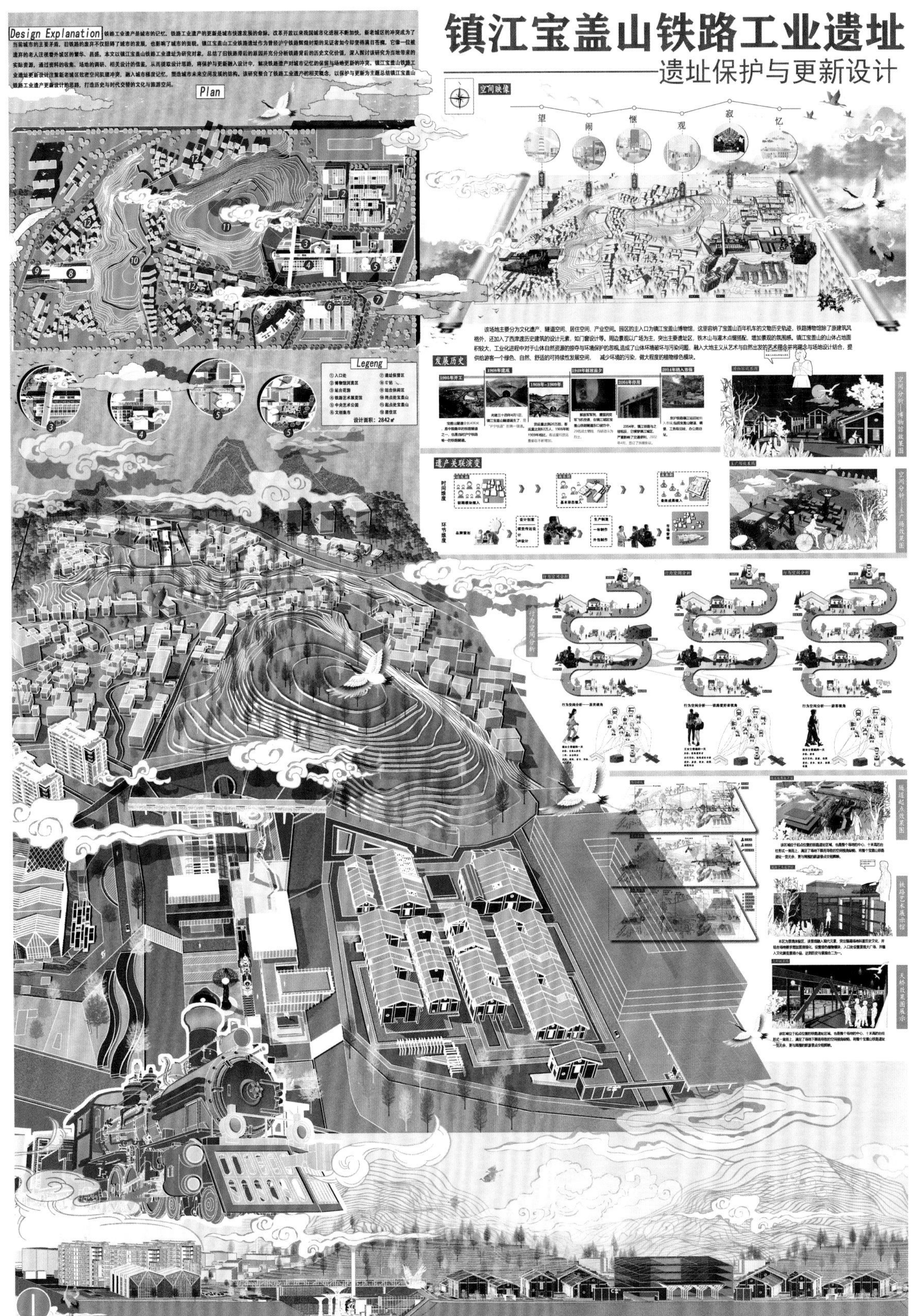
Design Explanation
铁路工业遗产是城市的记忆，铁路工业遗产的更新是城市快速发展的命脉。改革开放以来我国城市化进程不断加快，新老城区的冲突成为了当前城市的主要矛盾，旧铁路的废弃不仅阻碍了城市的发展，也影响了城市的面貌。镇江宝盖山工业铁路遗址作为曾经沪宁铁路辉煌时期的见证者如今却变得满目苍痍，它像一位被遗弃的老人注视着外城区的繁华、昌盛。本文以镇江宝盖山铁路工业遗址为研究对象，总结了旧铁路滞后的原因并充分分析铁路背后的历史文化价值，深入探讨该研究为当地带来的实际资源，通过资料的收集、场地的调研、相关设计的借鉴，从而提取设计思路，将保护与更新融入设计中，解决铁路遗产对城市记忆的保留与场地更新的冲突。镇江宝盖山铁路工业遗址更新设计注重新老城区致密空间肌理冲突，融入城市梯度记忆，塑造城市未来空间发展的结构。该研究整合了铁路工业遗产的相关概念，以保护与更新为主题总结镇江宝盖山铁路工业遗产更新设计的思路，打造历史与时代交替的文化与旅游空间。
Plan
镇江宝盖山铁路工业遗址
遗址保护与更新设计
空间映像
望 闹 惬 观 寂 忆
该场地主要分为文化遗产、隧道空间、居住空间、产业空间。园区的主入口为镇江宝盖山博物馆，这里容纳了宝盖山百年机车的文物历史轨迹，铁路博物馆除了原建筑风格外，还加入了西津渡历史建筑的设计元素，如门窗设计等。周边景观以广场为主，突出主要遗址区，铁木山与灌木点缀搭配，增加景观的氛围感。镇江宝盖山的山体占地面积较大，工业化进程中对于山体自然资源的掠夺与环境保护的忽视造成了山体环境破坏与污染问题，融入大地主义从艺术与自然出发的艺术理念并将理念与场地设计结合，提供给游客一个绿色、自然、舒适的可持续性发展空间，减少环境的污染，做大程度的植物绿色模块。
Legeng
①入口处
②博物馆浏览区
③站台花街
④铁路艺术展览馆
⑤中央艺术公园
⑥文创集市
⑦遗址保留区
⑧车站
⑨站台休闲区
⑩终点处宝盖山
⑪起点处宝盖山
⑫居住区
设计面积：2842㎡
发展历史
1905年开工
1908年建成
1908年-1909年
1949年解放前夕
2004年停用
2014年纳入市保
光绪三十四年4月1日，镇江宝盖山隧道诞生了，是“沪宁铁路”的第一隧道。
货运量达到20万担，客运量达到63万人，1908年和1909年相比，客运量和货运量都在不断增加。
2004年，镇江铁路与之错轨后，它横穿镇江城区，严重影响了交通便利。2002年4月，签订了拆除协议。
遗产关联演变
时间维度
初期模块植入
基本形态植入
最终成果植入

镇江宝盖山铁路工业遗址

遗址保护与更新设计

各功能比例分析图

约310，000平方米 居住区 70%
约110，000平方米 山区 25%
约1，962平方米 其他区 2%
约2，842平方米 隧道遗址区 3%

研究范围总占地面积：423，130平方米
改造遗址总长度：406米

铁路变化

沪宁旧线
沪宁新线

空间维度

基地分析

解决策略

设计概念

设计思路梳理

激活阶段、传承历史、展望未来
窗口展示、城市魅力
古尚新、旧尚潮
场地交汇
更新与保留
空间感、历史文化感

历史地貌　遗址营造　遗产传承　城市更新　国内现状　国外现状　区位特点　历史积淀　人文特色　更新选择　遗址规划　空间设计

相关概念界定　相关理论基础　国内外研究进展　区位概况　人文历史资源　业态规划管理　空间环境管理

历史遗产保护更新相关概念及研究进展　历史遗产保护更新案例分析研究

剖面图

UnderSpace

剖面图

场地介绍

空间分析

博物馆　车站　站台花园　历史展示

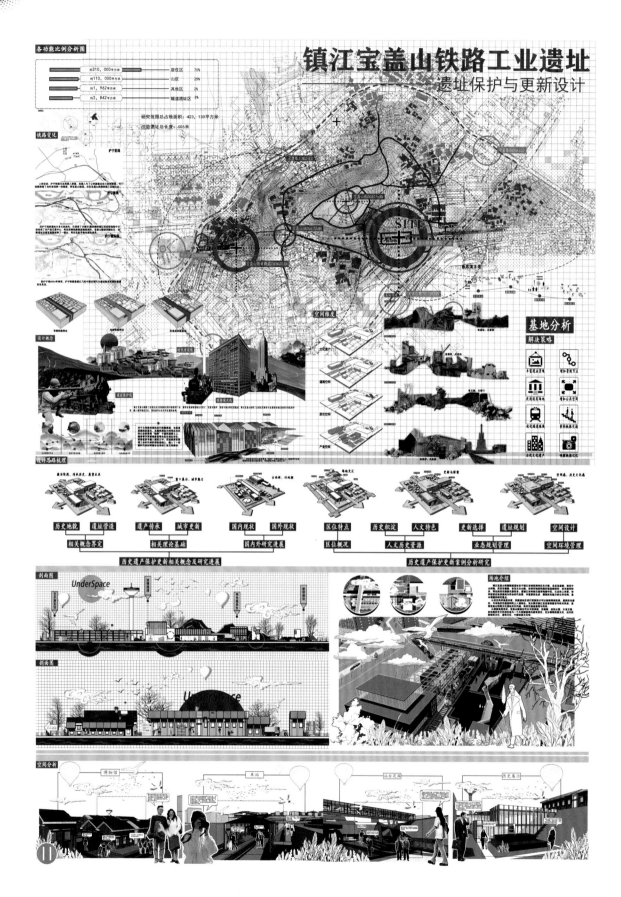

洄泝

Tracing and Recalling

参赛学校　University/College

兰州交通大学

Lanzhou Jiaotong University, China

指导教师　Supervisor(s)

王昱鸥　WANG Yu'ou

李振泉　LI Zhenquan

参赛学生　Participant(s)

刘沛雨　LIU Peiyu

崔金云　CUI Jinyun

朱童延　ZHU Tongyan

简介　Description

本设计地处甘肃天水，建筑具有浓郁的西北特色。天水作为丝绸之路的第三站，该现代化的铁路运输网络的设计，会是联结丝路沿线地区的新纽带之一。

设计从地域性的角度出发。建筑高低错落的屋顶轮廓是连绵起伏的山水的隐喻，也是对中国传统坡屋顶文化的继承与延续；形体两方向的延伸、跌落，形成丰富的体量穿插，整体造型似秀美的山峦，与场地周边环境相呼应；有自山下而仰山巅，谓之高远；自山前而窥山后，谓之深远；自近山而忘远山，谓之平远的深意，营造“洄泝经千里，烟波接两乡”的意境。所以以洄泝作为设计的主旨，洄泝指回旋的水，可引申为回顾或向上追求，也是渴望沿线各国能够在文化与经济上观往知来，合作共融。

The designed station is located in the city of Tianshui, Gansu Province in China. The building has apparent northwest characteristics. As the third station on the silk road, the design of this modern railway transportation network near Tianshui will be one of the new hubs connecting the areas along the silk road.

The design is based on regionalism. The contour of the roofs with different heights is a metaphor for the undulating landscape and also inheritance and continuation of the traditional Chinese culture of slope roof; the extension and falling of the roof bodies in two directions form a rich image full of intersections so that the overall shape is like a beautiful mountain ridge, echoing the surrounding environment of the site; looking from the bottom to the top, you will be amazed by its height and breadth; looking from the front to the back, you will be overwhelmed by its depth and farness; looking from the nearby to the far-away, you will experience the profoundness of flat and far that creates the artistic conception of “traveling back and forth for a thousand miles, the fog on waters connecting the two townships”. This is why “Tracing and Recalling” is the theme of this design. The basic meaning is whirling water, which can be extended to retrospect or a pursuit upwards, containing our hope that the countries along the road can cooperate in culture and economy and join hands through the past and the future.

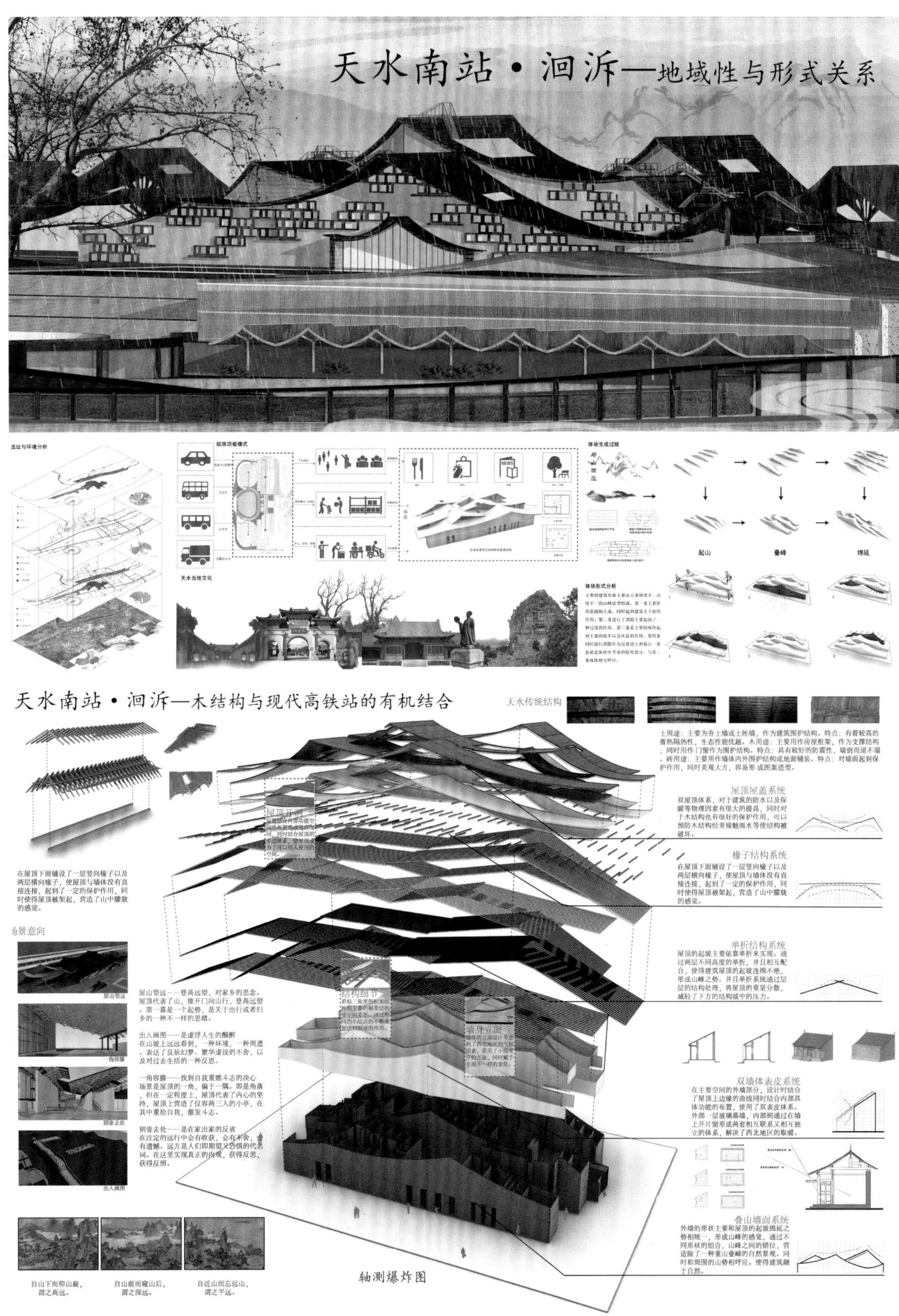
天水南站·洄泝——地域性与形式关系
选址与环境分析
站房功能模式
天水当地文化
体块生成过程
起山
叠峰
绵延
体块形式分析
天水南站·洄泝——木结构与现代高铁站的有机结合
天水传统结构
土用途：主要为夯土墙或土坯墙，作为建筑围护结构。特点：有着较高的蓄热隔热性，生态性能优越。木用途：主要用作房屋框架，作为支撑结构；同时用作门窗作为围护结构。特点：具有较好的防震性，墙倒而屋不塌。砖用途：主要用作墙体内外围护结构或地面铺装。特点：对墙面起到保护作用，同时美观大方，容易形成图案造型。
屋顶屋盖系统
双屋顶体系，对于建筑的防水以及保暖等物理因素有很大的提高，同时对于木结构也有很好的保护作用，可以预防木结构经常接触雨水等使结构被破坏。
椽子结构系统
在屋顶下面铺设了一层竖向椽子以及两层横向椽子，使屋顶与墙体没有直接连接，起到了一定的保护作用，同时使得屋顶被架起，营造了山中朦胧的感觉。
举折结构系统
屋顶的起坡主要依靠举折来实现。通过两层不同高度的举折，并且相互配合，使得建筑屋顶的起坡连绵不绝，形成山峰之势。并且举折系统通过层层的结构处理，将屋顶的重量分散，减轻了下方的结构城中的压力。
双墙体表皮系统
在主要空间的外墙部分，设计时结合了屋顶上边缘的曲线同时结合内部具体功能的布置，使用了双表皮体系。外部一层玻璃幕墙，内部则通过在墙上开片窗形成两套相互联系又相互独立的体系，解决了西北地区的取暖。
叠山墙面系统
外墙的形状主要和屋顶的起坡绵延之势相统一，形成山峰的感觉，通过不同形状的组合，山峰之间的错位，营造除了一种重山叠嶂的自然景观。同时和周围的山势相呼应。使得建筑融于自然。
屋顶开洞
结构细节
墙身立面
在屋顶下面铺设了一层竖向椽子以及两层横向椽子，使屋顶与墙体没有直接连接，起到了一定的保护作用，同时使得屋顶被架起，营造了山中朦胧的感觉。
景意向
屋山望远
一角容膝
别壶去处
出入画图
屋山望远——登高远望，对家乡的思念。屋顶代表了山，推开门向山行，登高远望。第一幕是一个起势，是关于出行或者归乡的一种不一样的思绪。
出入画图——是虚浮人生的醒醉
在山坡上远远看到，一种环境，一种周遭。表达了良辰幻梦、繁华虚设的不舍，以及对过去生活的一种反思。
一角容膝——找到自我重燃斗志的决心
场景是屋顶的一角，偏于一隅。即是角落，但在一定程度上，屋顶代表了内心的坚持，屋顶上营造了仅容两三人的小亭，在其中重拾自我，激发斗志。
别壶去处——是在家出家的反省
在注定的远行中会有收获，会有不舍，会有遗憾。远方是人们即期望又恐惧的代名词。在这里实现真正的内观，获得反思，获得反照。
自山下而仰山巅，谓之高远。
自山前而窥山后，谓之深远。
自近山而忘远山，谓之平远。
轴测爆炸图

天水南站·洄泝—平面功能和流线组织

一层平面图

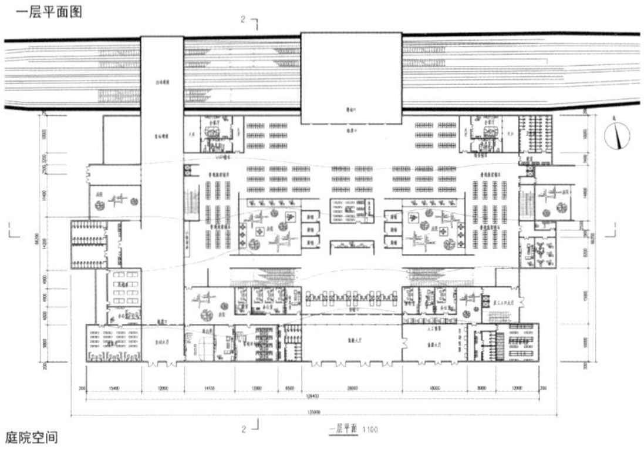

二层平面图

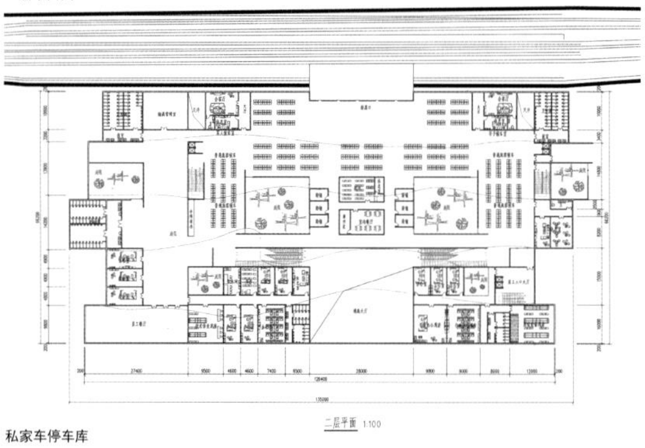

庭院空间

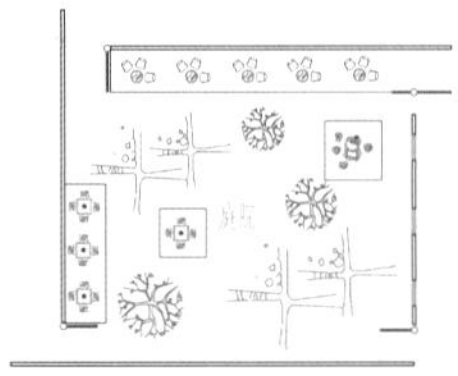

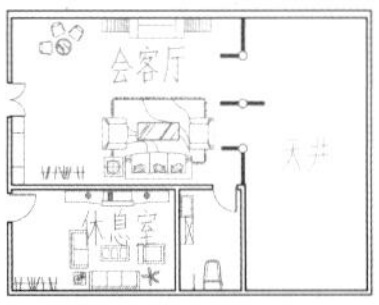

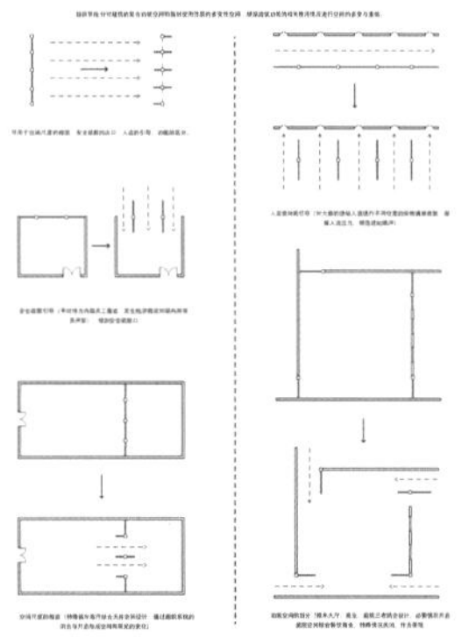

私家车停车库

特殊候车空间

出租车停车库

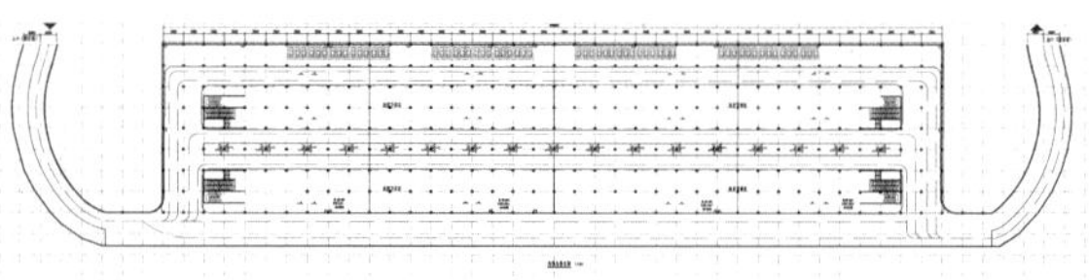

丝途·丝绸之路火车站房设计

Situ-Silk Road Railway Station Design

参赛学校 University/College
华北水利水电大学
North China University of Water Resources and Electric Power, China

指导教师 Supervisor(s)
张忆萌 ZHANG Yimeng
连鲁楠 LIAN Lunan

参赛学生 Participant(s)
王思俨 WANG Siyan
徐　凯 XU Kai
张旖珂 ZHANG Yike
秦凯博 QIN Kaibo
曹佳欣 CAO Jiaxin

简介 Description

该设计为四级火车站，选址位于浙江省金华市中浦江县郑家坞镇江滨东路。我们此次设计沿用了江南传统民居的屋顶形式，采用坡屋顶可以更好地达到排水的作用，整体结构采用的是网架结构，建筑中心以水为界，中间为一条长廊，既分割了两边的功能空间，又联合了两边的空间，使建筑具有你中有我，我中有你的风格特点，结合基地地形，设置摆放建筑，两个变形的“L”相互对应，针对金华市的气候，做了一些灰空间和对外开放空间的设计。

We designed a fourth-class railway station, sitting on the Jiangbin East Road in Zhengjiawu Town, Zhongpujiang County in Jinhua City, Zhejiang Province in China. We adopted the style of traditional residential houses in the south of the Yangtze River for the roof, which is a sloping one for better drainage. Overall, the building is in a grid structure with a watershed at its center. The long corridor in the middle divides the space into two areas with different functions, which are also connected to represent a highly intertwined relationship. The layout of the building is based on the ground conditions of the site. The two parts of the station look like two reshaped "L" standing facing each other from a bird's view. Given the climate conditions of Jinhua City, gray and open spaces were also included in the design.

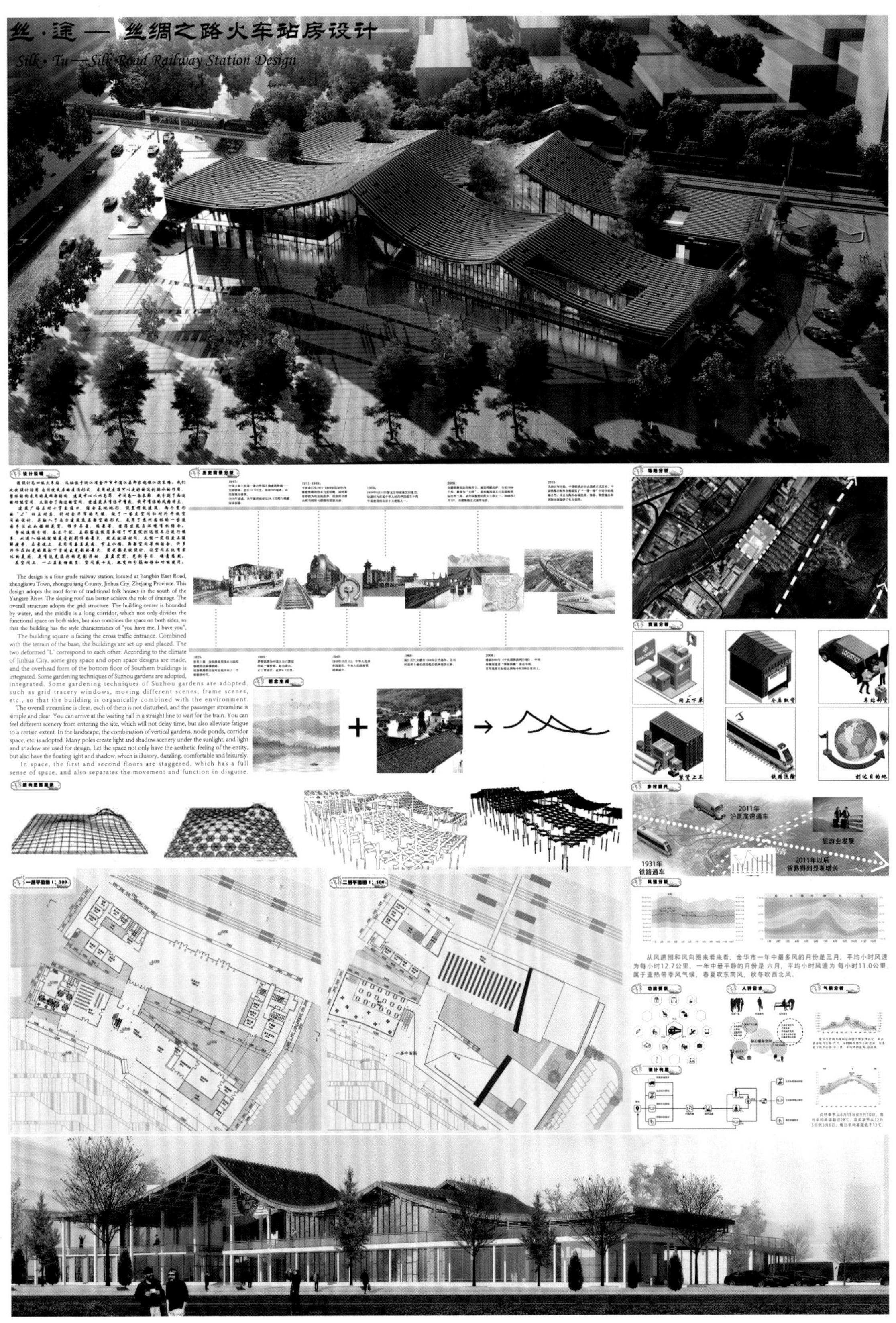
丝·途— 丝绸之路火车站房设计
Silk · Tu—Silk Road Railway Station Design
The design is a four grade railway station, located at Jiangbin East Road, zhengjiawu Town, zhongpujiang County, Jinhua City, Zhejiang Province. This design adopts the roof form of traditional folk houses in the south of the Yangtze River. The sloping roof can better achieve the role of drainage. The overall structure adopts the grid structure. The building center is bounded by water, and the middle is a long corridor, which not only divides the functional space on both sides, but also combines the space on both sides, so that the building has the style characteristics of "you have me, I have you",
The building square is facing the cross traffic entrance. Combined with the terrain of the base, the buildings are set up and placed. The two deformed "L" correspond to each other. According to the climate of Jinhua City, some grey space and open space designs are made, and the overhead form of the bottom floor of Southern buildings is integrated. Some gardening techniques of Suzhou gardens are adopted, integrated. Some gardening techniques of Suzhou gardens are adopted, such as grid tracery windows, moving different scenes, frame scenes, etc., so that the building is organically combined with the environment.
The overall streamline is clear, each of them is not disturbed, and the passenger streamline is simple and clear. You can arrive at the waiting hall in a straight line to wait for the train. You can feel different scenery from entering the site, which will not delay time, but also alleviate fatigue to a certain extent. In the landscape, the combination of vertical gardens, node ponds, corridor space, etc. is adopted. Many poles create light and shadow scenery under the sunlight, and light and shadow are used for design, Let the space not only have the aesthetic feeling of the entity, but also have the floating light and shadow, which is illusory, dazzling, comfortable and leisurely.
In space, the first and second floors are staggered, which has a full sense of space, and also separates the movement and function in disguise.
2011年
沪昆高速通车
旅游业发展
1931年
铁路通车
2011年以后
贸易得到显著增长
从风速图和风向图来看来看，金华市一年中最多风的月份是三月，平均小时风速为每小时12.7公里，一年中最平静的月份是 六月，平均小时风速为 每小时11.0公里。属于亚热带季风气候，春夏吹东南风，秋冬吹西北风。

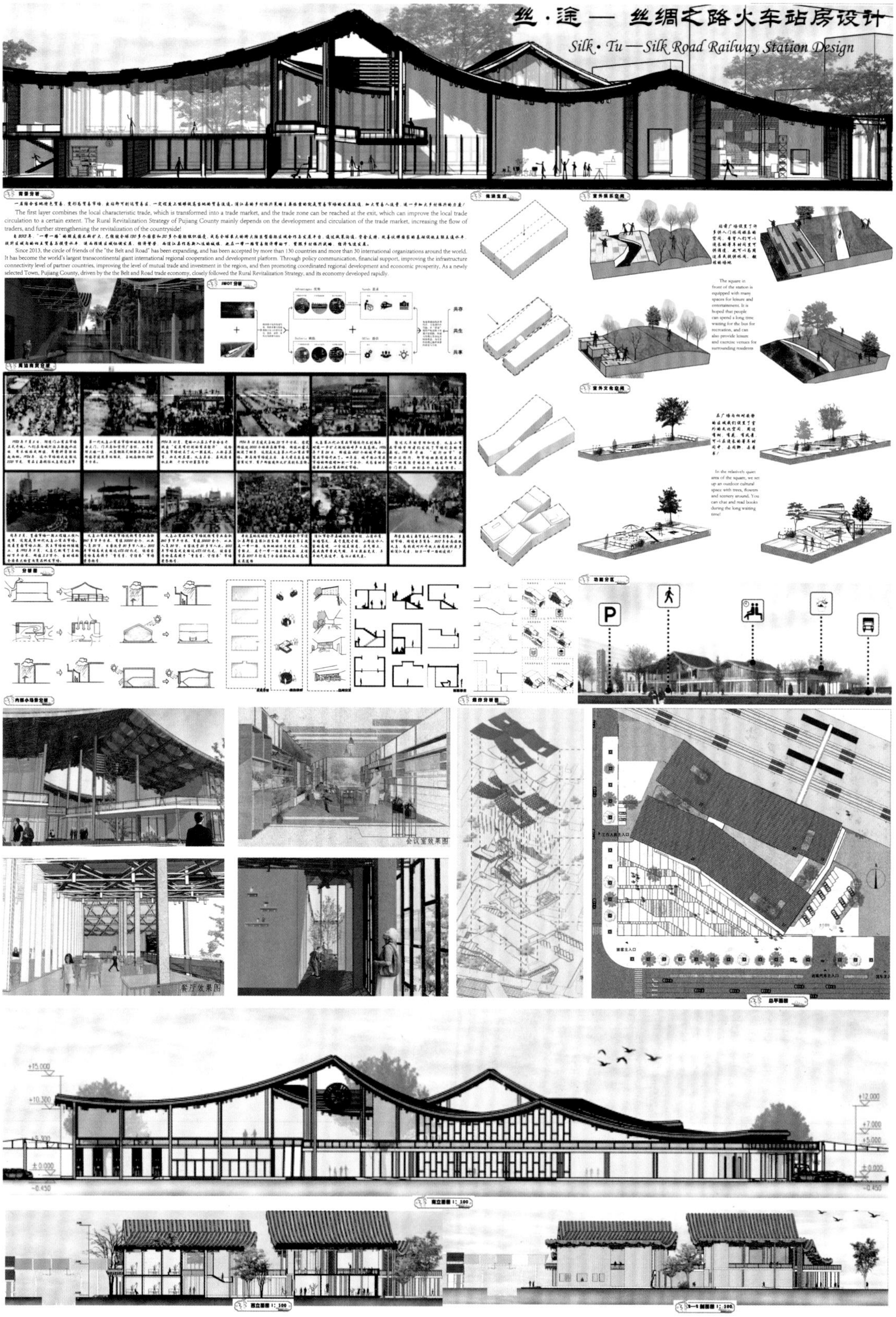

丝·途—— 丝绸之路火车站房设计

Silk • Tu—Silk Road Railway Station Design

背景分析

一层结合当地特色贸易，变形为贸易市场，在站外可到达贸易区，一定程度上能够提高当地的贸易流通。浦江县的乡村振兴策略主要依靠的就是贸易市场的发展流通，加大贸易人流量，进一步加大乡村振兴的力度!

The first layer combines the local characteristic trade, which is transformed into a trade market, and the trade zone can be reached at the exit, which can improve the local trade circulation to a certain extent. The Rural Revitalization Strategy of Pujiang County mainly depends on the development and circulation of the trade market, increasing the flow of traders, and further strengthening the revitalization of the countryside!

自2013年，“一带一路”的朋友圈不断扩大，已经被全球130多个国家和30多个国际组织接受，成为全球最大的跨大陆巨型国际区域合作与发展平台。通过政策沟通，资金支持，改善伙伴国家的基础设施互联互通水平，提升区域内的相互贸易与投资水平，进而推进区域协调发展，经济繁荣。而浦江县作为新入选的城镇，在一带一路贸易经济带动下，紧跟乡村振兴战略，经济飞速发展。

Since 2013, the circle of friends of the "the Belt and Road" has been expanding, and has been accepted by more than 130 countries and more than 30 international organizations around the world. It has become the world's largest transcontinental giant international regional cooperation and development platform. Through policy communication, financial support, improving the infrastructure connectivity level of partner countries, improving the level of mutual trade and investment in the region, and then promoting coordinated regional development and economic prosperity. As a newly selected Town, Pujiang County, driven by the the Belt and Road trade economy, closely followed the Rural Revitalization Strategy, and its economy developed rapidly.

SWOT 分析

共存 共生 共享

The square in front of the station is equipped with many spaces for leisure and entertainment. It is hoped that people can spend a long time waiting for the bus for recreation, and can also provide leisure and exercise venues for surrounding residents.

In the relatively quiet area of the square, we set up an outdoor cultural space with trees, flowers and scenery around. You can chat and read books during the long waiting time!

会议室效果图

餐厅效果图

扶摇乘风

Fufeng Railway Station

参赛学校 University/College
青岛理工大学
Qingdao University of Technology, China

指导教师 Supervisor(s)
岳乃华 YUE Naihua

参赛学生 Participant(s)
李　豪 LI Hao
苑清宇 YUAN Qingyu
张逸婷 ZHANG Yiting
杜伟宁 DU Weining
姜　腾 JIANG Teng

简介 Description

扶风县铁路客运站房位于陕西省宝鸡市扶风县，与古代丝绸之路起点长安紧密相邻。本次设计围绕丝绸之路这一主题，取飞鸟、丝绸、波纹等形象，为扶风县历史文化名城的新形象展示名片，车站充分考虑了建筑的工艺需求、当地的气候条件、场地的独特位置和使用者的环境需求，通过先进的设计理念，多元化的功能布局实现了历史文化与经济效益的平衡，诠释了新丝路开创新局的时代内涵。

The Fufeng railway station is located in Fufeng County, Baoji City, Shaanxi Province in China, closing to Chang'an which is the starting point of the ancient Silk Road. Focusing on the theme of the Silk Road, this design takes the images of birds, silk, ripples and so on to display a business card for the new image of Fufeng County as a historical and cultural city. Considering the techniques required by the building, the local climatic conditions, the unique location of the site and the environmental needs of users, the station realizes the balance between history, culture and economic benefits through advanced design concepts and diversified functional layout and explains the connotation of the new silk road leading to a new prospect of the time.

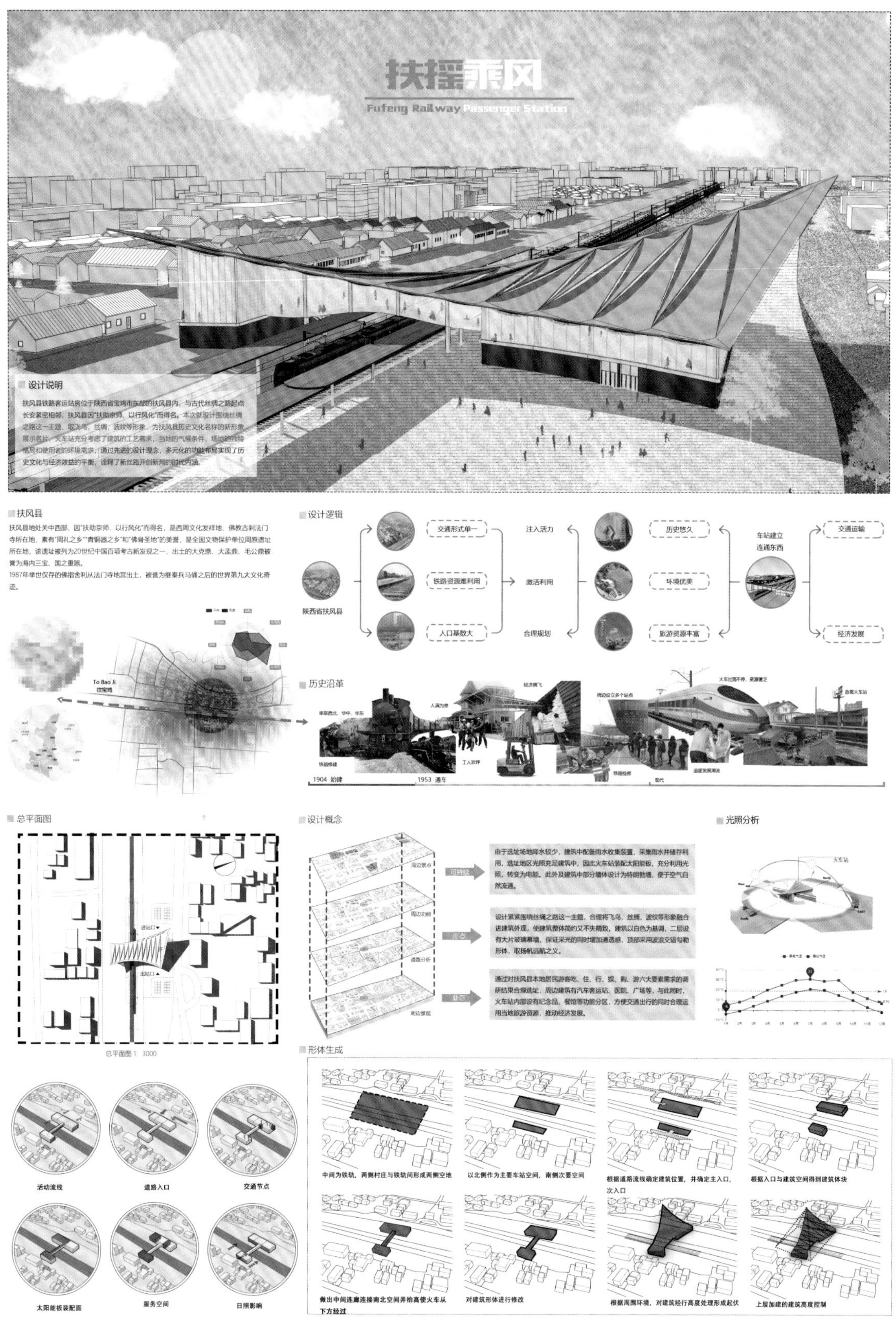
扶摇乘风
Fufeng Railway Passenger Station
设计说明
扶风县铁路客运站房位于陕西省宝鸡市东部的扶风县内，与古代丝绸之路起点长安紧密相邻，扶风县因"扶助京师、以行风化"而得名。本次就设计围绕丝绸之路这一主题，取飞鸟，丝绸，波纹等形象，为扶风县历史文化名称的新形象展示名片。火车站充分考虑了建筑的工艺需求，当地的气候条件，场地的独特格局和使用者的环境需求，通过先进的设计理念，多元化的功能布局实现了历史文化与经济效益的平衡，诠释了新丝路开创新局的时代内涵。
扶风县
扶风县地处关中西部，因"扶助京师、以行风化"而得名，是西周文化发祥地、佛教古刹法门寺所在地，素有"周礼之乡""青铜器之乡"和"佛骨圣地"的美誉，是全国文物保护单位周原遗址所在地，该遗址被列为20世纪中国百项考古新发现之一，出土的大克鼎、大盂鼎、毛公鼎被誉为海内三宝、国之重器。
1987年举世仅存的佛指舍利从法门寺地宫出土，被誉为继秦兵马俑之后的世界第九大文化奇迹。
To Bao Ji
往宝鸡
设计逻辑
陕西省扶风县
交通形式单一
铁路资源难利用
人口基数大
注入活力
激活利用
合理规划
历史悠久
环境优美
旅游资源丰富
车站建立
连通东西
交通运输
经济发展
历史沿革
铁路修建
人满为患
工人欢呼
铁路检修
1904 始建
1953 通车
现代
总平面图
进站口
出站口
总平面图 1: 3000
设计概念
周边景点
周边功能
道路分析
周边景观
可持续
形态
业态
由于选址场地降水较少，建筑中配备雨水收集装置，采集雨水并储存利用，选址地区光照充足建筑中，因此火车站装配太阳能板，充分利用光照，转变为电能。此外及建筑中部分墙体设计为特朗勃墙，便于空气自然流通。
设计紧紧围绕丝绸之路这一主题，合理将飞鸟，丝绸，波纹等形象融合进建筑外观，使建筑整体简约又不失精致。建筑以白色为基调，二层设有大片玻璃幕墙，保证采光的同时增加通透感，顶部采用波浪交错勾勒形体，取扬帆远航之义。
通过对扶风县本地居民游客吃、住、行、娱、购、游六大要素需求的调研结果合理选址，周边建筑有汽车客运站、医院、广场等，与此同时，火车站内部设有纪念品、餐馆等功能分区，方便交通出行的同时合理运用当地旅游资源，推动经济发展。
光照分析
火车站
活动流线
道路入口
交通节点
太阳能板装配面
服务空间
日照影响
形体生成
中间为铁轨，两侧村庄与铁轨间形成两侧空地
以北侧作为主要车站空间，南侧次要空间
根据道路流线确定建筑位置，并确定主入口，次入口
根据入口与建筑空间得到建筑体块
做出中间连廊连接南北空间并抬高使火车从下方经过
对建筑形体进行修改
根据周围环境，对建筑经行高度处理形成起伏
上层加建的建筑高度控制

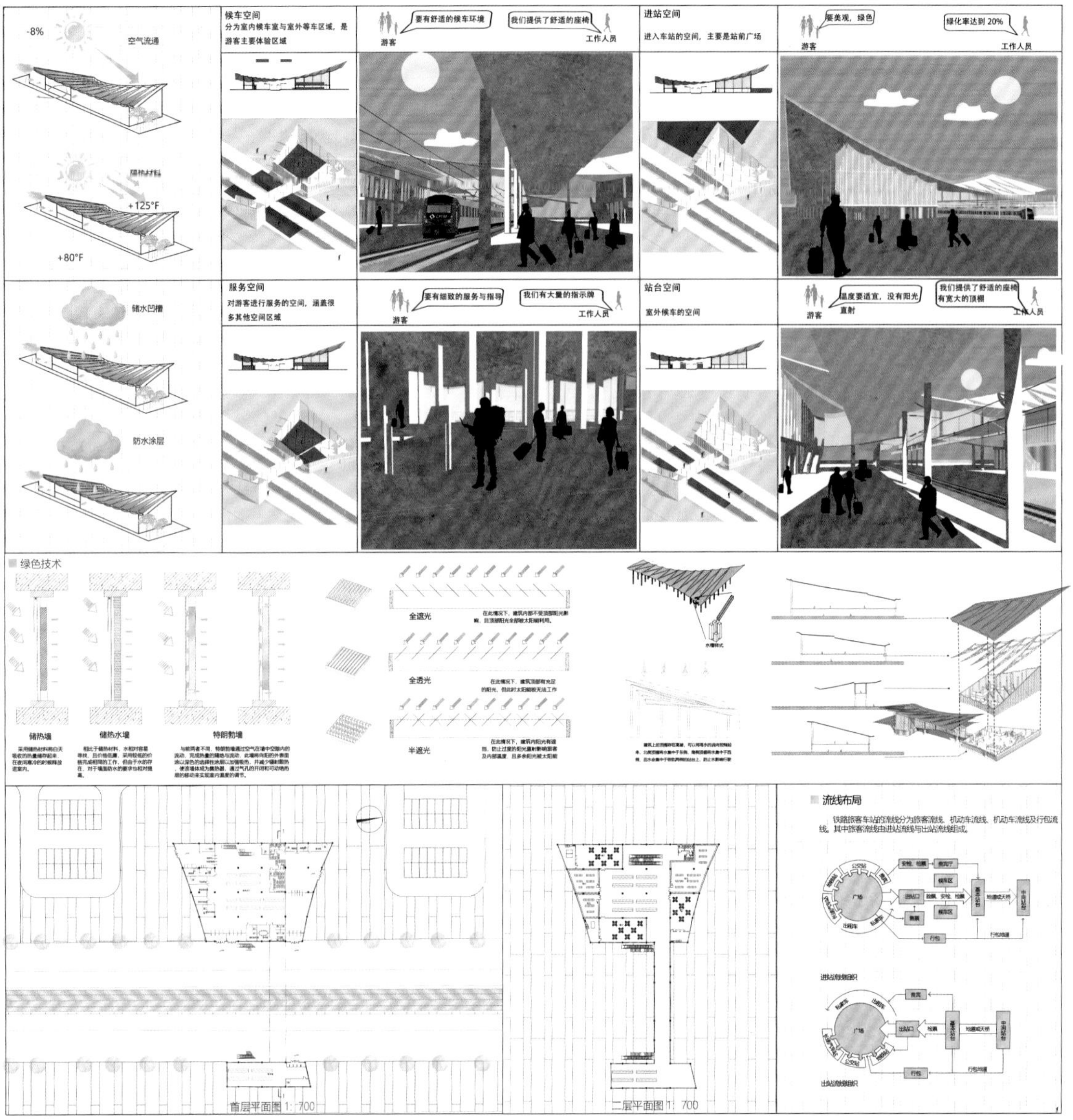

-8%
空气流通
+125°F
+80°F

候车空间
分为室内候车室与室外等车区域，是游客主要体验区域

游客：要有舒适的候车环境
工作人员：我们提供了舒适的座椅

进站空间
进入车站的空间，主要是站前广场

游客：要美观，绿色
工作人员：绿化率达到 20%

储水凹槽
防水涂层

服务空间
对游客进行服务的空间，涵盖很多其他空间区域

游客：要有细致的服务与指导
工作人员：我们有大量的指示牌

站台空间
室外候车的空间

游客：温度要适宜，没有阳光直射
工作人员：我们提供了舒适的座椅有宽大的顶棚

绿色技术

储热墙

采用储热材料将白天吸收的热量储存起来，在夜间寒冷的时候释放进室内。

储热水墙

相比于储热材料，水相对容易寻找，且价格低廉，采用较低的价格完成相同的工作，但由于水的存在，对于墙面防水的要求也相对提高。

特朗勃墙

与前两者不同，特朗勃墙通过空气在墙中空腔内的流动，完成热量的隔绝与流动，此墙将向阳的外表面涂以深色的选择性涂层以加强吸热，并减少辐射散热，使该墙体成为集热器，通过气孔的开闭和可动绝热层的移动来实现室内温度的调节。

全遮光

在此情况下，建筑内部不受顶部阳光影响，且顶部阳光全部被太阳能利用。

全透光

在此情况下，建筑顶部有充足的阳光，但此时太阳能板无法工作

半遮光

在此情况下，建筑内阳光有遮挡，防止过度的阳光直射影响旅客及内部温度，且多余阳光被太阳能

首层平面图 1：700

二层平面图 1：700

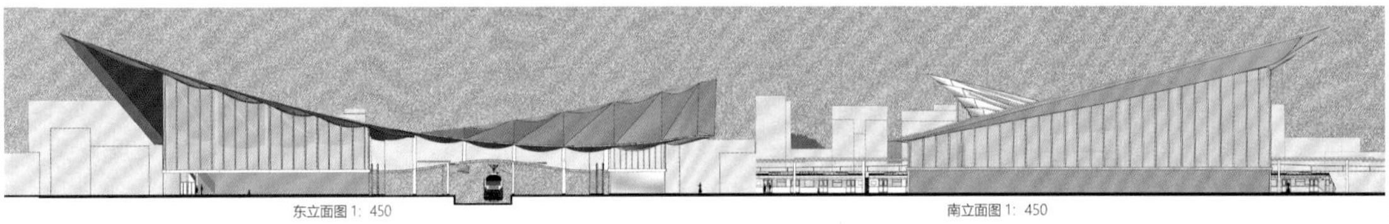

流线布局

铁路旅客车站的流线分为旅客流线、机动车流线、机动车流线及行包流线。其中旅客流线由进站流线与出站流线组成。

广场
公交站
出租车
私家车
安检、检票
贵宾厅
候车区
进站口
检票、安检、检票
售票
基本站台
地道或天桥
中间站台
行包
行包地道

进站流线组织

广场
私家车
出租车
贵宾
出站口
检票
基本站台
地道或天桥
中间站台
公交站
行包
行包地道

出站流线组织

东立面图 1：450

南立面图 1：450

归来·失落站台

Return - Lost Platform

参赛学校 University/College
华北理工大学轻工学院
Qing Gong College, North China University of Science and Technology, China

指导教师 Supervisor(s)
唐晨辉 TANG Chenhui

参赛学生 Participant(s)
蔡云龙 CAI Yunlong

简介 Description

归来，从现代到过去，在空间上的跳跃，以一种物质化的东西表达时空之间的介质，就好比宇宙中的黑洞一般存在，失落站台是以山洞的形式，但又有现代的玻璃，以感受时间上的离开和到来，当你通过站台去往别的地方，回来是若干年后，而此处的站台不再是普通的站台。而是你时间和空间上的见证，当周边建筑在百年之后成为废墟，失落站台却悄然融入到周边环境中，成为悄然的山洞，一种拥有时空穿越魅力的自然体。

Return means traveling from modern times to the past and a leap in space. It expresses the connection between time and space with a materialized medium. Mimicking a black hole in the universe, the Lost Platform is designed as a cave with the modern material of glass

to symbolize the transient time. You leave the platform and may not come back until years later, finding the platform here is no longer just an ordinary platform for you but a witness of time and space. When the surrounding buildings turned into ruins after a hundred years, the platform is gradually integrated into the environment, very similar to a cave that has always been silent——a natural existence with the magic of traveling through time and space.

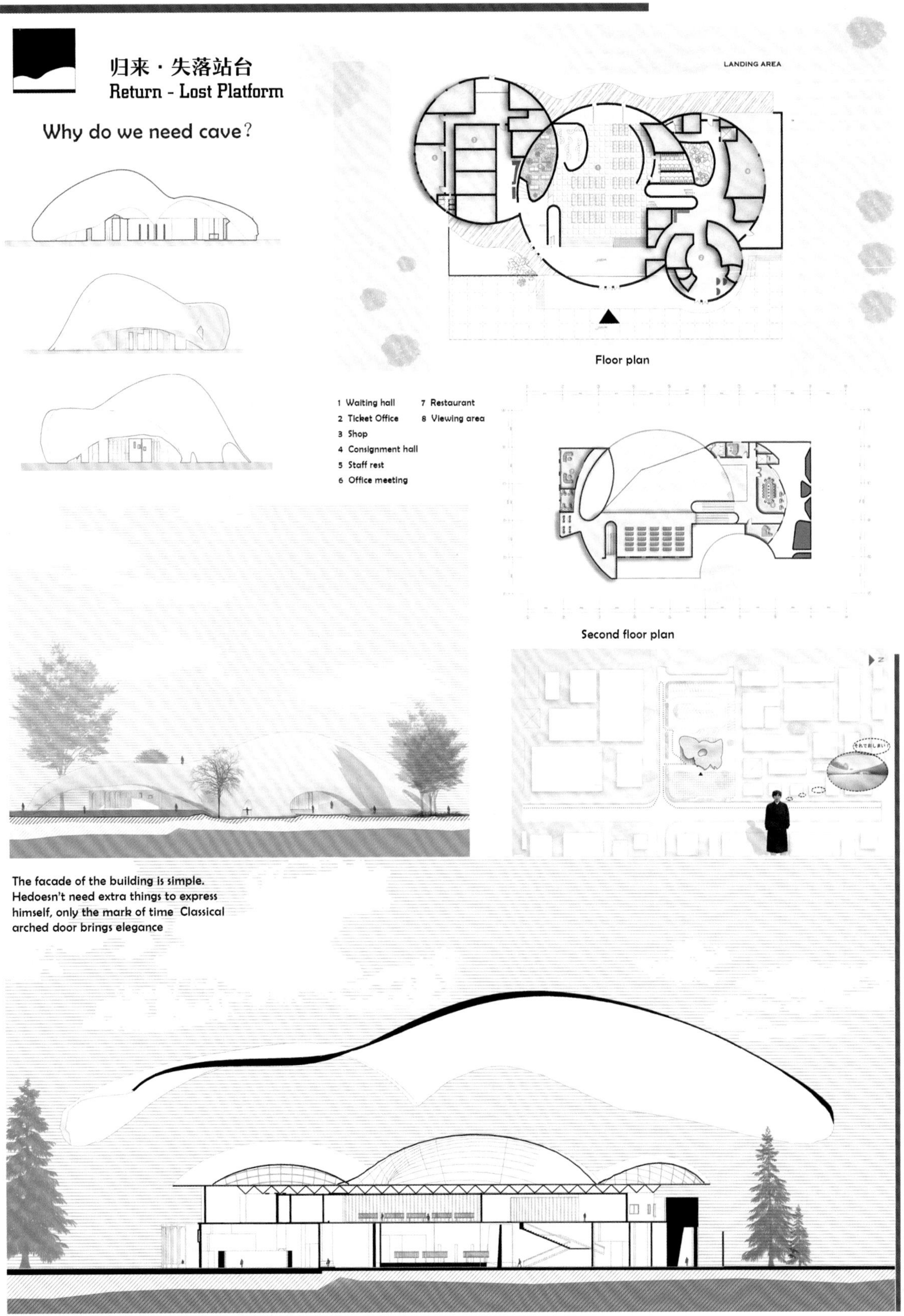
归来 · 失落站台
Return - Lost Platform
Why do we need cave?
LANDING AREA
Floor plan
1 Waiting hall
2 Ticket Office
3 Shop
4 Consignment hall
5 Staff rest
6 Office meeting
7 Restaurant
8 Viewing area
Second floor plan
The facade of the building is simple.
Hedoesn't need extra things to express
himself, only the mark of time Classical
arched door brings elegance

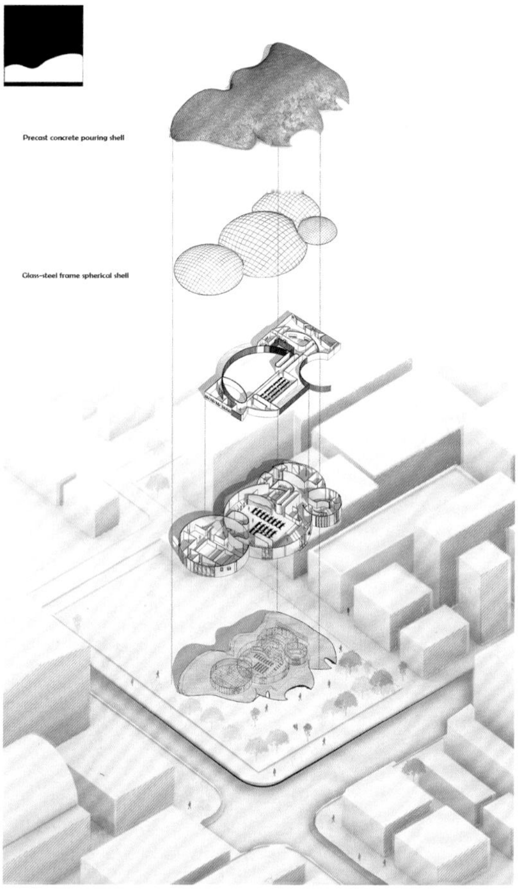

Precast concrete pouring shell

Glass-steel frame spherical shell

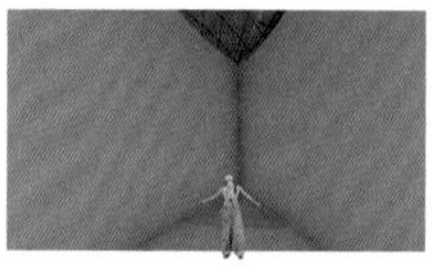

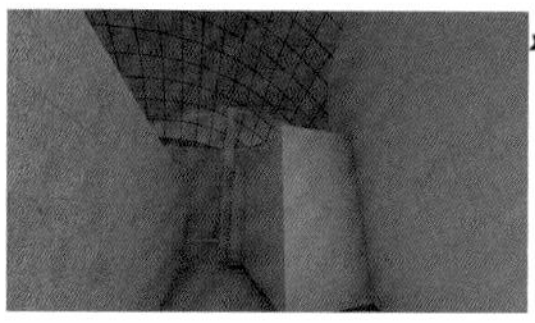

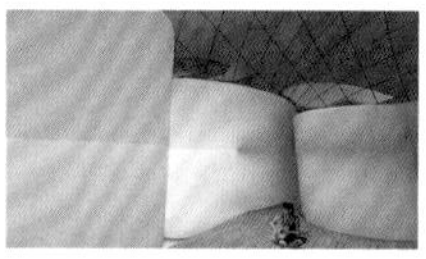

Because !

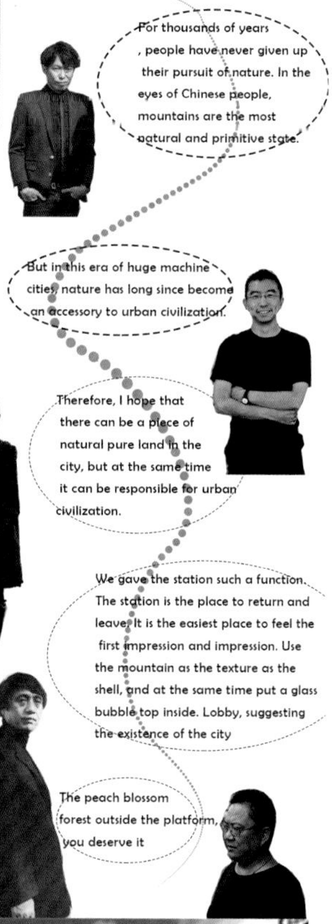

For thousands of years, people have never given up their pursuit of nature. In the eyes of Chinese people, mountains are the most natural and primitive state.

But in this era of huge machine cities, nature has long since become an accessory to urban civilization.

Therefore, I hope that there can be a place of natural pure land in the city, but at the same time it can be responsible for urban civilization.

We gave the station such a function. The station is the place to return and leave. It is the easiest place to feel the first impression and impression. Use the mountain as the texture as the shell, and at the same time put a glass bubble top inside. Lobby, suggesting the existence of the city

The peach blossom forest outside the platform, you deserve it

梁赞交通运输中转枢纽

Transport Interchange Hub in Ryazan

参赛学校 University/College
俄罗斯莫斯科理工大学梁赞分校
Ryazan Institute, Moscow Polytechnic University, Russia

指导教师 Supervisor(s)
Osina Natalya Alexandrovna

参赛学生 Participant(s)
Zhuravleva Alexandra Georgievna

简介 Description

根据全球铁路运输的战略版图，多条运输路线途径俄罗斯，包括从中国到欧洲的新丝绸之路纬向走廊及从印度到圣彼得堡的南北经向走廊。为贯彻上述战略布局，需要在梁赞的现代领土上建立一个交通运输中转枢纽（HUB），建立一个强大的铁路客运综合体。该枢纽最重要的任务在于提供一个综合解决方案，连接梁赞与所有城市区域，包括新城区及历史定居区。因此，该枢纽是梁赞市及整个区域客流的主要集散地。

建筑项目的目标包括：

- 扩大城市道路网络；
- 重建市内铁路交通系统；

- 重建历史上的梁赞 1 号车站及其铁路运输线路;
- 联通城市的新旧领土。

建筑项目的效果包括:

- 为梁赞现有的交通运输网减负;
- 简化物流;
- 为梁赞地区旅游业发展提供新机遇;
- 促进外来劳动力的涌入。

According to the strategic development of railway transport in the world, there are a number of transport corridors passing through Russia. These are the latitudinal corridors of the New Silk Road, going from China to Europe, and the North-South meridional corridor, going from India to St. Petersburg. To comply with the declared status, it is necessary to build a transport interchange hub (HUB) on the territory of the modern part of Ryazan with a powerful railway passenger complex. The most important indicator of the work of the transport hub is a comprehensive solution that connects the object with all urban areas: on the territory of a new city and a historical settlement. Thus, the HUB acts as the main entrance and distributor of passenger flows not only in the city, but also in the region.

Project objectives:

- expansion of the city's road network;
- resumption of railway communication within the city;
- reconstruction of the historical station Ryazan-1 and its railway lines;
- organization of a link between the modern and historical territories of the city.

Project effects:

- unloading the existing transport network of Ryazan;
- simplification of logistics;
- opportunities for the development of tourism in the Ryazan region;
- influx of new workers.

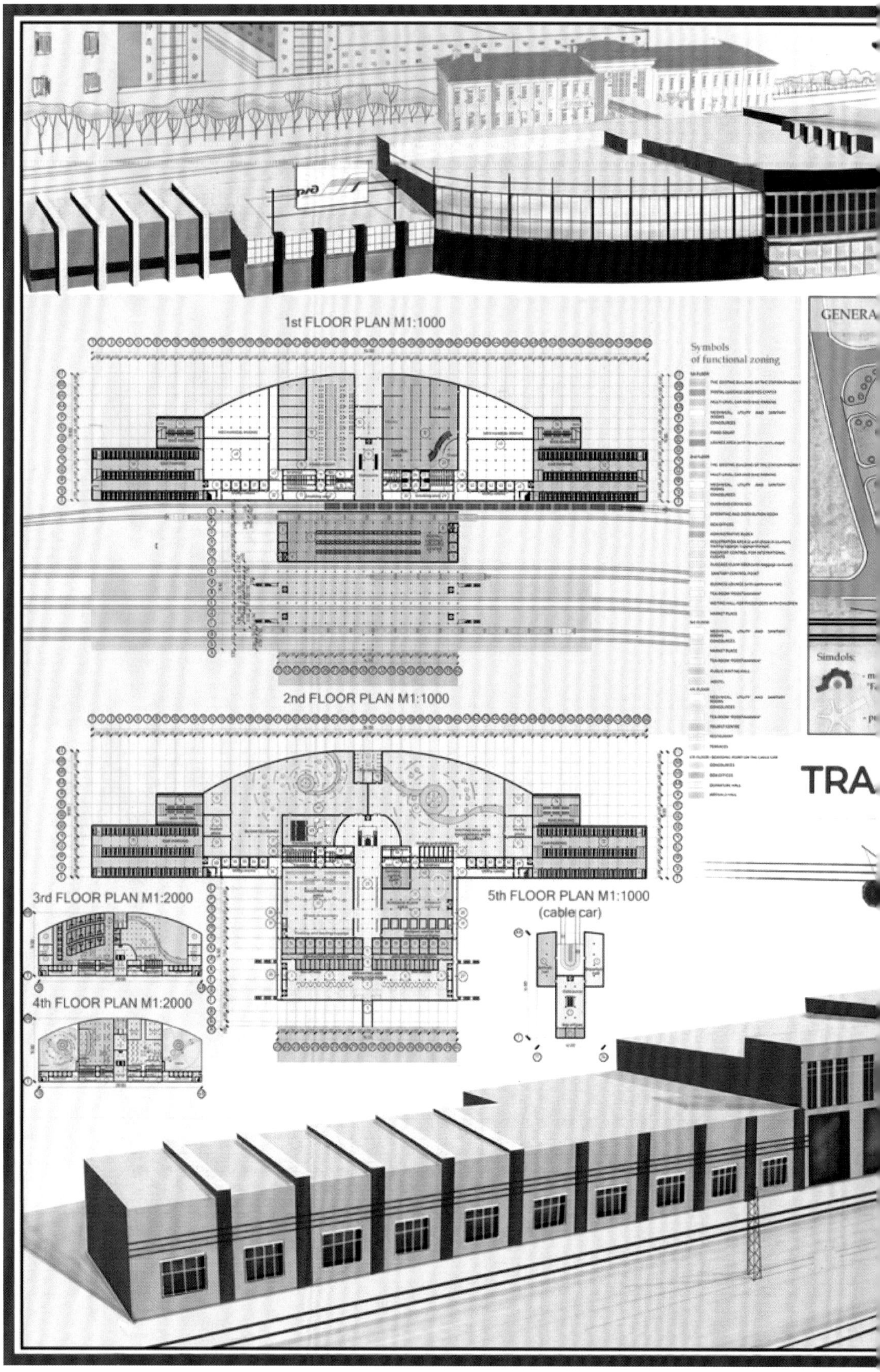

1st FLOOR PLAN M1:1000

Symbols
of functional zoning

GENERA

Simdols

2nd FLOOR PLAN M1:1000

TRA

3rd FLOOR PLAN M1:2000

4th FLOOR PLAN M1:2000

5th FLOOR PLAN M1:1000
(cable car)

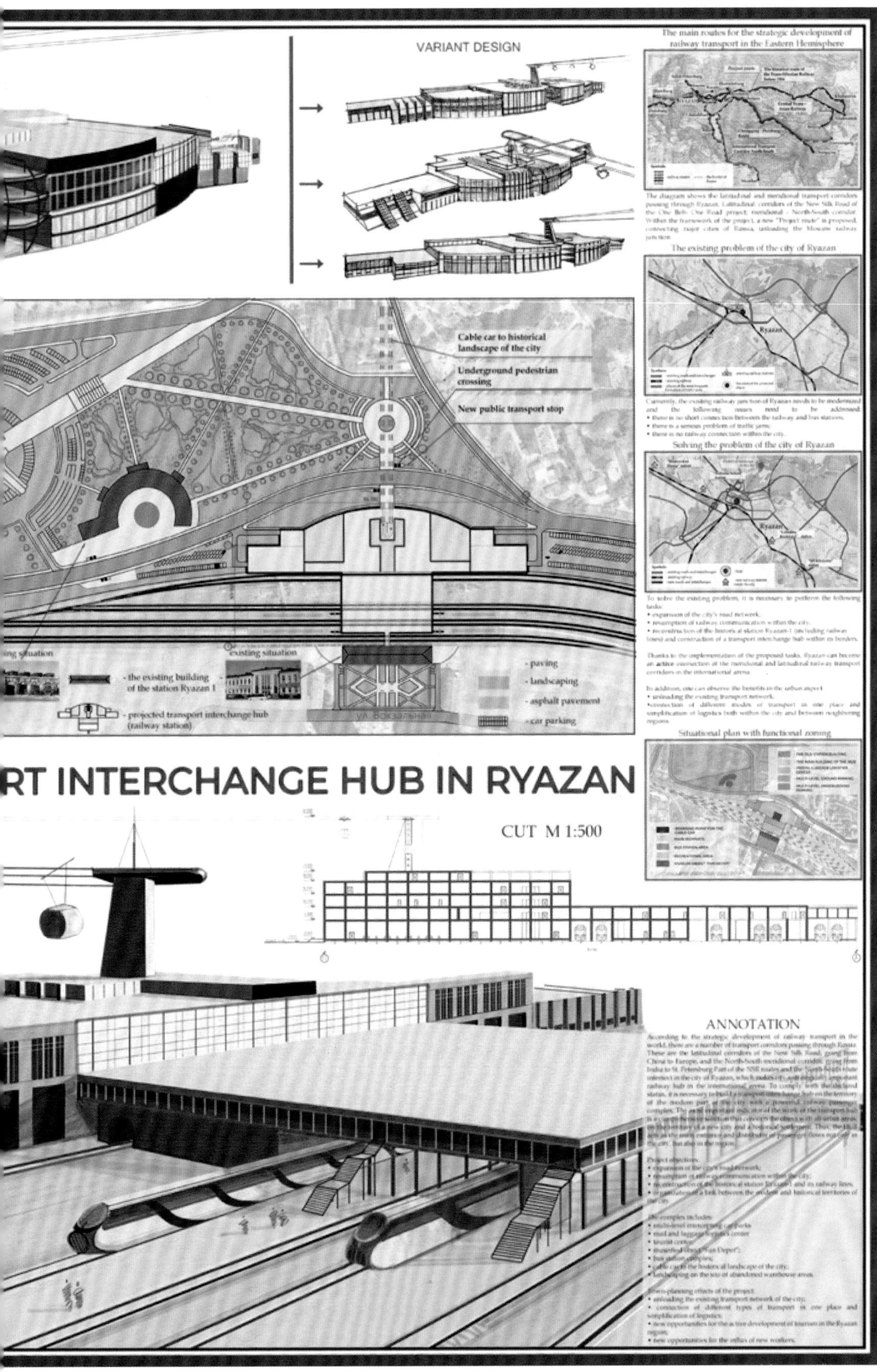
VARIANT DESIGN
The main routes for the strategic development of railway transport in the Eastern Hemisphere
The existing problem of the city of Ryazan
Solving the problem of the city of Ryazan
Cable car to historical landscape of the city
Underground pedestrian crossing
New public transport stop
existing situation
- the existing building of the station Ryazan 1
- projected transport interchange hub (railway station)
- paving
- landscaping
- asphalt pavement
- car parking
Situational plan with functional zoning
RT INTERCHANGE HUB IN RYAZAN
CUT M 1:500
ANNOTATION

梁赞火车站改建工程

Reconstruction of the Building of the Railway Station in Ryazan

参赛学校 University/College
俄罗斯莫斯科理工大学梁赞分校
Ryazan Institute, Moscow Polytechnic University, Russia

指导教师 Supervisor(s)
Karetnikova Svetlana Veniaminovna

参赛学生 Participant(s)
Titenok Alena Sergeevna

简介 Description

梁赞火车站始建于 1864 年，历史悠久，20 世纪 60 年代末经历重建。车站设有 3 个客运站台，由人行天桥连接。目前，车站建筑已经过时，不符合现代要求。

该项目的主要目标是对现有建筑进行重新修缮，更换四坡顶，在站台上安装圆顶顶棚，更换景观元素、涂料类型和小型建筑物的外观。建筑物外观的配色方案保留原始版本。火车站建筑的原地形参数保持不变。

项目计划使用钢方管和钢化玻璃窗格设计大跨度的圆顶顶棚。建筑中心部分的屋顶将替换为半透明涂层。

大楼的入口为行动不便的人配备斜坡。

建筑的尺寸为 115 米 ×20 米，建筑面积为 4200 平方米。

项目实施过程中，使用 AutoDesk Revit 软件包完成了对现有建筑的数字化建模。

The historic building of the railway station was built in 1864, reconstructed in the late 1960s. The station has 3 passenger platforms connected by a footbridge. At the moment, the station building is obsolete and does not meet modern requirements.

The main goal of the project is the reconstruction of the building with the replacement of the hip roof, the installation of a domed canopy over the platforms, the replacement of landscaping elements, types of coatings and small architectural forms. The color scheme of the external appearance of the building will be preserved in its original version. The topographic parameters of the site subject to reconstruction will not change.

When designing a large-span dome cover, the use of steel square pipes and tempered glass panes is envisaged. The roof of the central part of the building was replaced with a translucent coating.

Entrances to the building are additionally equipped with ramps for people with limited mobility.

The dimensions of the building are 115m×20m, the area of the building is $4200m^2$.

During the implementation of the project, the existing building was digitized in the AutoDesk Revit software package.

Reconstruction of the railway station building in Ryazan

1 Ground floor plan

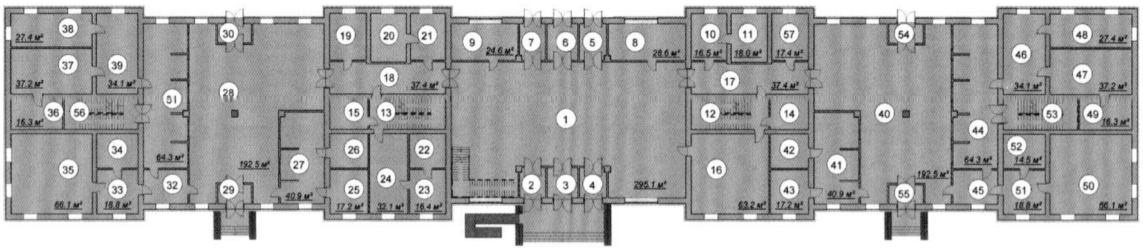

2 Second floor plan

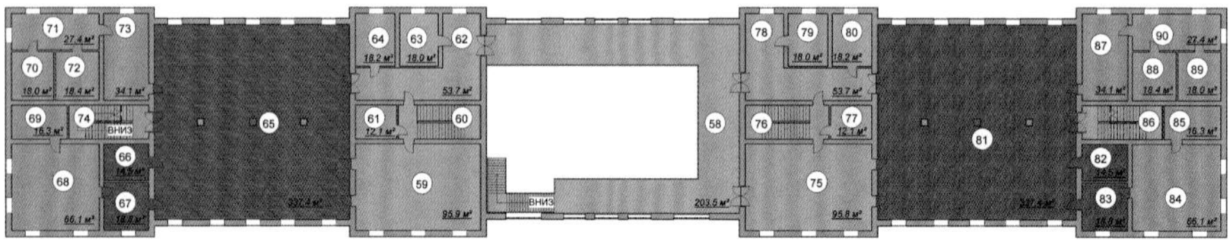

3 Third floor plan

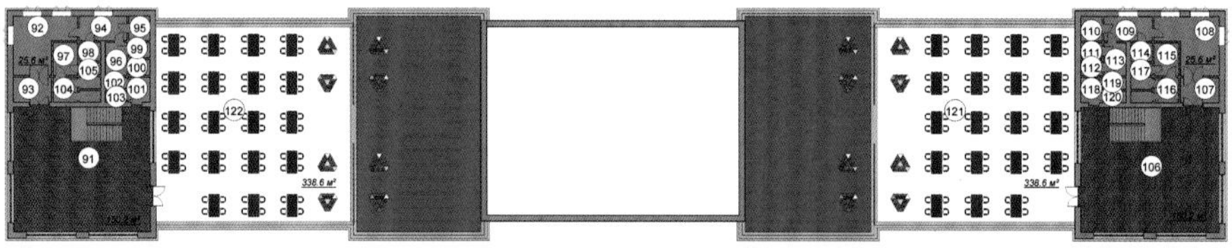

Explication of the 1st floor premises	
№	Name of premises
1	Lobby
2-7	Tambour
8	Commercial premises
9	Commercial premises
10	Security room
11	Server
12	Staircase
13	Staircase
14	Stock
15	Stock
16	Commercial premises
17	Hall
18	Hall
19	Commercial premises
20	Commercial premises
21	Female rest room
22	PBX room
23	Communication room
24	ITR room
25	Staff quarters
26	Staff quarters
27	Short distance ticket offices
28	Waiting hall
29	Tambour
30	Tambour
31	Short distance ticket offices
32	Cass Tambour
33	Kitchen
34	Wardrobe
35	Staff lounge
36-39	Archive room
40	Waiting hall
41	Long-distance ticket offices
42	Stock
43	Cabinet
44	Long-distance ticket offices
45	Cass Tambour
46	Staff quarters
47	Shower room
48	WC
49	Staircase
50	Staff quarters
51	Administrative premises
52	WC
53	Staircase
54	Tambour
55	Tambour
56	Staircase
57	Male rest room

Explication of the 2,3st floor premises	
№	Name of premises
58	The second light
59	Economic department
60	Staircase
61	Stock
62	Hall
63	Male rest room
64	Female rest room
65	Hall for visitors
66	Office
67	Secretary's room
68	Station master's office
69	Restroom
70	Stock
71	Chief Engineer's Office
72	Archive room
73	ITR room
74	ITR room
75	Accountant's office
76	Staircase
77	Stock
78	The corridor
79	Female rest room
80	Male rest room
81	Hall for visitors
82	KUI room
83	Secretary's room
84	Distance manager's office
85	Restroom
86	Staircase
87	ITR room
88	Archive room
89	Wardrobe
90	ITR room
91	Cafe dining room
92	Kitchen
93	Distribution zone
94	Kitchen
95	Washing room
96	Product warehouse
97	Cloakroom
98	Warehouse utensils
99,100	Cooling chamber
101	WC
102	Male rest room
103	Female rest room
104	Shower room
105	WC
106	Cafe dining room
107	Distribution
108	Kitchen
109	Kitchen
110	Washing room
111,112	Cooling chamber
113	Product warehouse
114	Warehouse utensils
115	Wardrobe
116	Shower room
117	WC
118	WC
119	Male rest room
120	Female rest room
121,122	Open cafe area

Within the framework of the international program "Railway: a new connection on the Silk Road", a concept was developed for the object of reconstruction: "The building of the railway station in Ryazan", aimed at updating the external appearance of the city of Ryazan.
The historic building of the Ryazan railway station was opened on August 27, 1864. In the late 1960s it was rebuilt. It is located in the historical cluster of the city and is one of the main stations of the transport network of passenger and freight container transportation of the central administrative district of the Russian Federation.
When developing the project, it was decided not to change the topographic and geological parameters of the land plot subject to reconstruction. The color scheme of the external appearance of the building is preserved in its original version. When facing the facade, porcelain stoneware slabs of white limestone color with a thickness of 20 mm are used.
The project provides for the reconstruction of the cover of the station building, as well as the construction of a spatial shell of complex shape, covering the station building, three main railway tracks and platforms for boarding and disembarking passengers. The shape of the shell is a sinusoid line with variable height marks on two tiers. The shell is made of rectangular steel pipes forming honeycombs, which are filled with double-glazed windows made of tempered glass. The hip roof of the existing building is being dismantled. In the central part of the building, part of the coating is dismantled and replaced with a translucent dome. It is planned to arrange an operated roof with landscaping elements and various recreation areas.
When implementing the reconstruction project, the needs of people with limited mobility were taken into account, and a ramp was provided.
The space-planning solution of the building is made taking into account the principles of forming the most useful use of the space of the internal environment, the passenger service area, the short- and long-distance ticket office areas, the waiting area and other auxiliary premises are designed. As part of the general improvement of the territory of the station site, all the asphalt concrete pavement of the sidewalks was replaced with paving stones; the installation of lighting poles and small architectural forms in the waiting area is envisaged.
The implementation of the above design solutions will update the appearance and will breathe new life into the historic building of the railway station and, which is the face of the city of Ryazan.

					2022	62 - 07/22		
Measur.	Plots	Sheet	№ Doc.	Signature	Date	Reconstruction of the railway station building in Ryazan		
Developed		[illegible]				The building of the railway station in Ryazan	Stage: PD	Sheet: A1
Checked		[illegible]					Sheets:	
Control		[illegible]				Ground floor plan, Second floor plan, Third floor plan, Visual range	RI (f) MPU	

A1 format

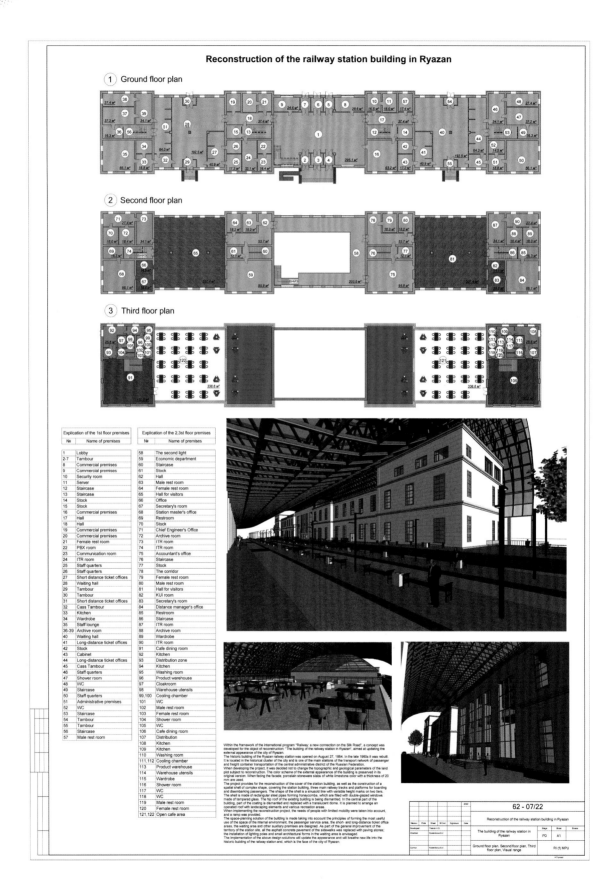

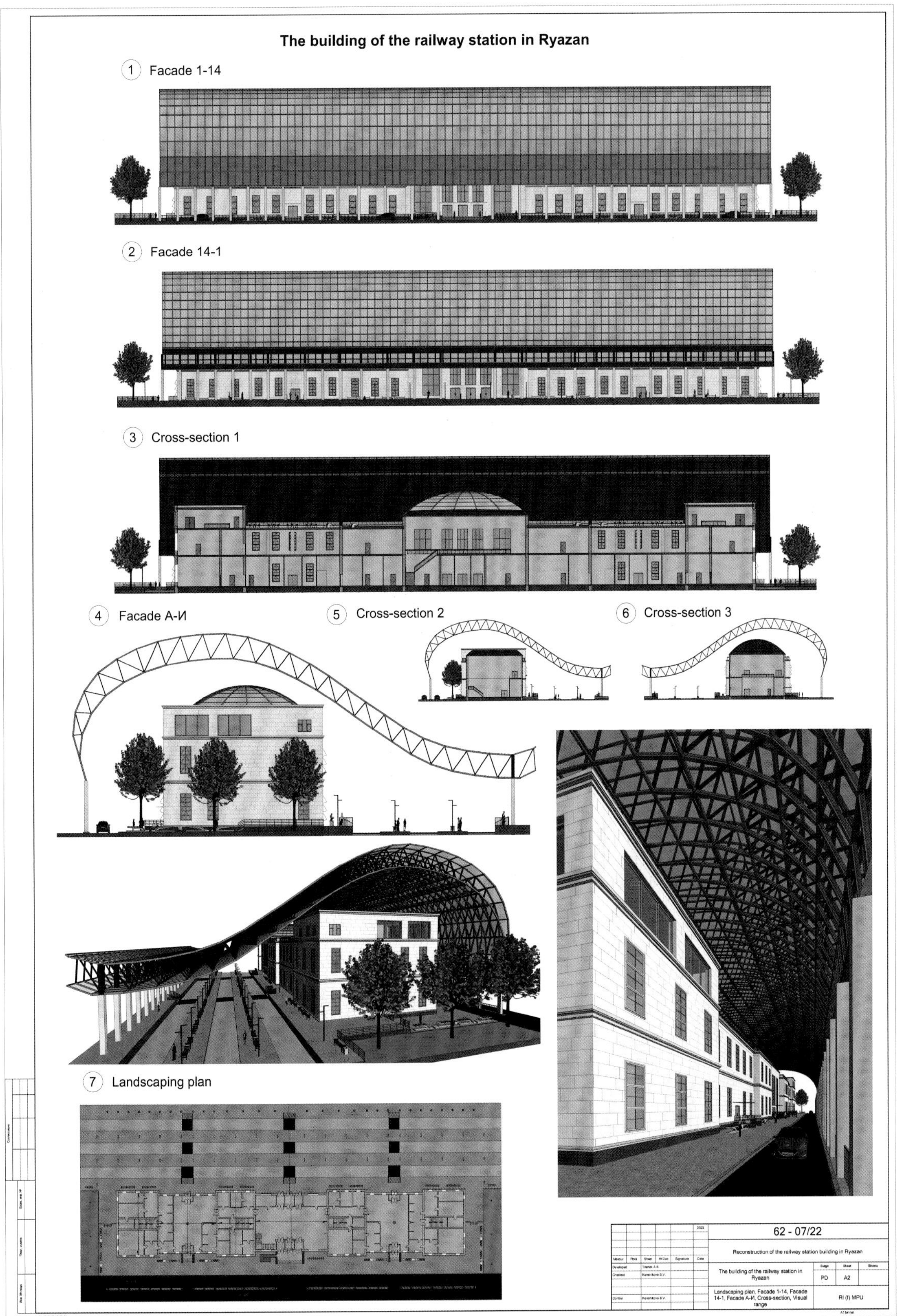
The building of the railway station in Ryazan
1 Facade 1-14
2 Facade 14-1
3 Cross-section 1
4 Facade A-И
5 Cross-section 2
6 Cross-section 3
7 Landscaping plan
62 - 07/22
Reconstruction of the railway station building in Ryazan
The building of the railway station in Ryazan
PD
A2
Landscaping plan, Facade 1-14, Facade 14-1, Facade A-И, Cross-section, Visual range

疫情下“Green-Architecture”客运站建筑设计

“Green-Architecture” Railway Station Adaptive to COVID-19

参赛学校 University/College
山东建筑大学
Shandong Jianzhu University, China

指导教师 Supervisor(s)
周　倩　ZHOU Qian
赵胜中　ZHAO Shengzhong

参赛学生 Participant(s)
杨一琦　YANG Yiqi
丁　绮　DING Qi
徐　繁　XU Fan
赵　帅　ZHAO Shuai
赵超宇　ZHAO Chaoyu

简介 Description

以疫情下“Green-Architecture”客运站建筑为主题所设计的“毕家站”位于吉林省长春市东北部毕家店附近的京哈铁路沿线上，旨在通过设立本站带动周边村落及企业发展。该地区地势平坦，交通便利。有着毗邻102国道、东北亚机械城、吉林工程技术师范学院等优势条件。

“毕家站”采用铁路贯穿式及对称设计。鉴于长春地区矿产资源丰富，本站点将碳元素结构模型应用于墙体及窗等部位。屋顶设立太阳能光伏板，优化首层及二层的分区。对站前广场的功能划分、用地布局、配套设施、绿化设计、环境保护等方面进行调控，实现对客流量的控制与区域防疫的和谐统一。

The Bijia station, designed under the theme of “Green-Architecture” Railway Station adaptive to COVID-19, is located along the Beijing-Harbin Railway, near Bijiadian, northeast of Changchun City in Jilin Province. The region enjoys flat terrain and convenient transportation. Combined with the location advantage of adjacency to the National Highway 102, Northeast Asia International Machinery City and Jilin Engineering Normal University, the station is designed to support the development of the surrounding villages and enterprises.

The design of the Bijia Station is symmetrical with the railway going through it. Given the abundant mineral resources in this area, a model of carbon structure is applied to the walls and windows of the station. PV panels are installed on the roof to optimize the division of the ground and the second floor. The function partitioning of different parts of the station square, the use of land, supportive facilities, greening design, environmental protection and other aspects are properly arranged to coordinate the control of passenger flows and regional COVID-19 prevention.

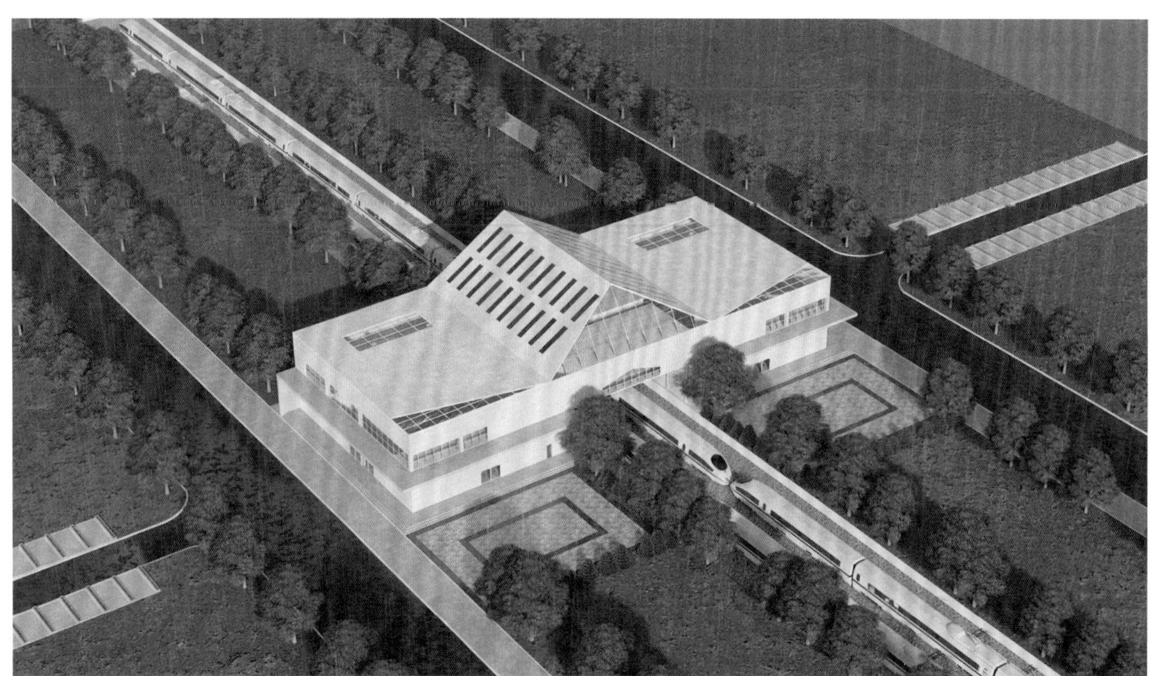

吉林毕家站建筑设计

The Green-Architecture Design of The Bijia Station

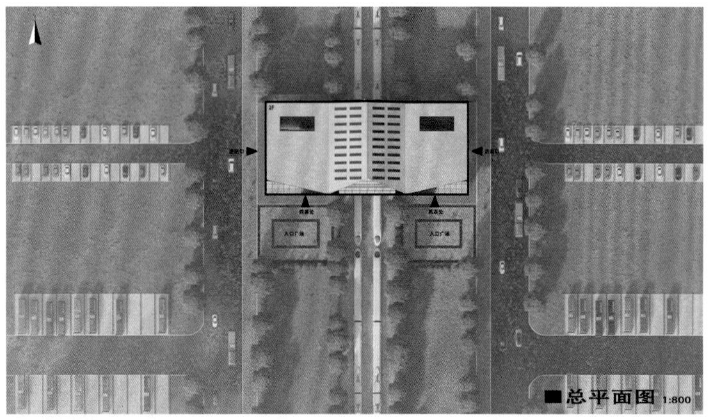

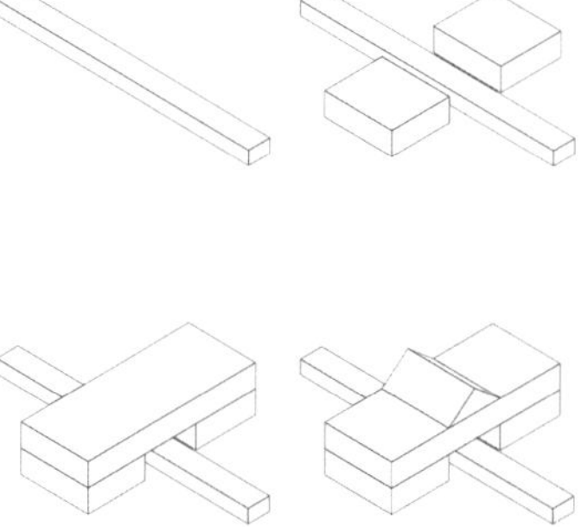

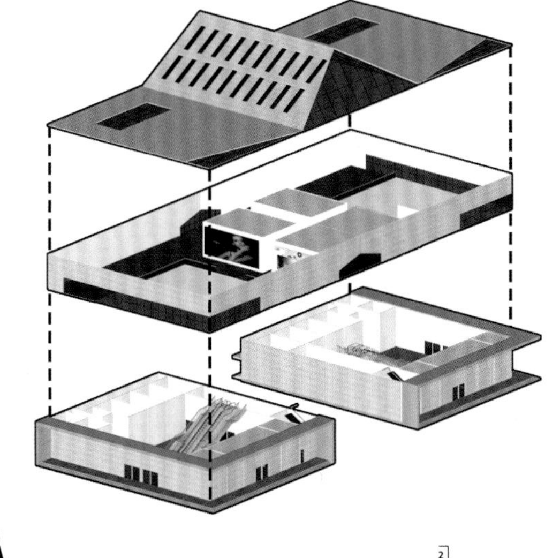

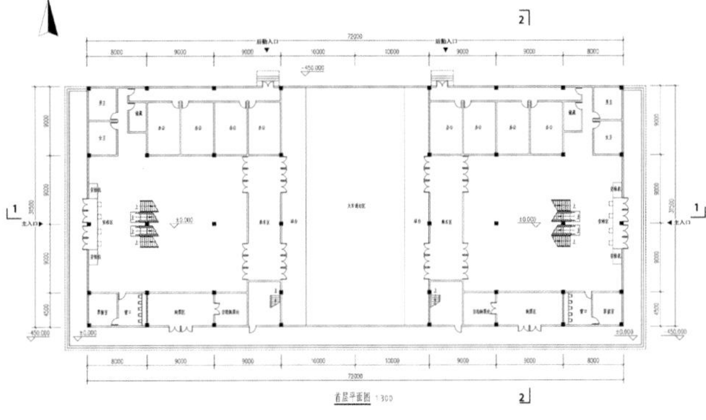
首层平面图 1:300

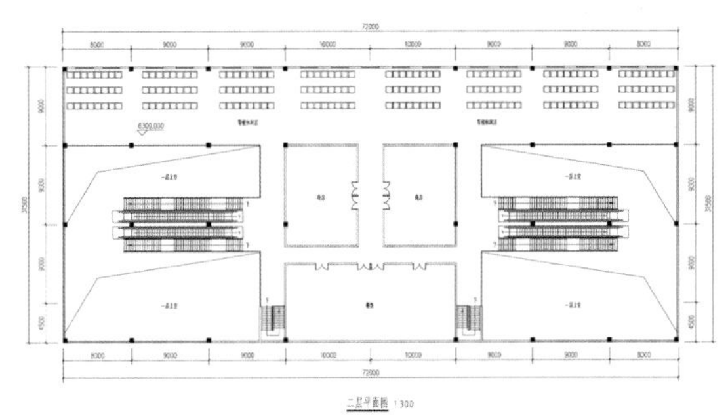
二层平面图 1:300

剖面图1-1 1：300
剖面图2-2 1：300

万物生——济南铁路文化体验中心

The Birth of All the Things—Jinan Railway Culture Experience Center

参赛学校 University/College
山东建筑大学
Shandong Jianzhu University, China

指导教师 Supervisor(s)
张莉莉 ZHANG Lili
苏允桥 SU Yunqiao

参赛学生 Participant(s)
苏贝勒 SU Beile
刘士硕 LIU Shishuo
姚子琪 YAO Ziqi
宋 莹 SONG Ying
张 秦 ZHANG QIN

简介 Description

万物衍生在空间的轮回，却保持了不变的延生之相，万物为灵，万物造物之所能，本在世界，也在心界。

本方案的选址位于济南市四门塔景区附近，毗邻泰山站与济南站。设计结合中国传统文化，以“铁路——架起‘丝路’新纽带”为表现重点，打造自然景观与中国式美学的视觉盛宴。

All things derive from the samsara of space but maintain the same phase of extended life. All things are spirits and the power of all things' creation is both in the world and in the heart.

The site of this plan is located near the Four-Gates Pagoda scenic spot in Jinan, adjacent to the Taishan station and the Jinan Station. The design combines traditional Chinese culture and focuses on "Railway-supporting the new links of the Silk Road" to create a visual feast of natural landscape and Chinese aesthetics.

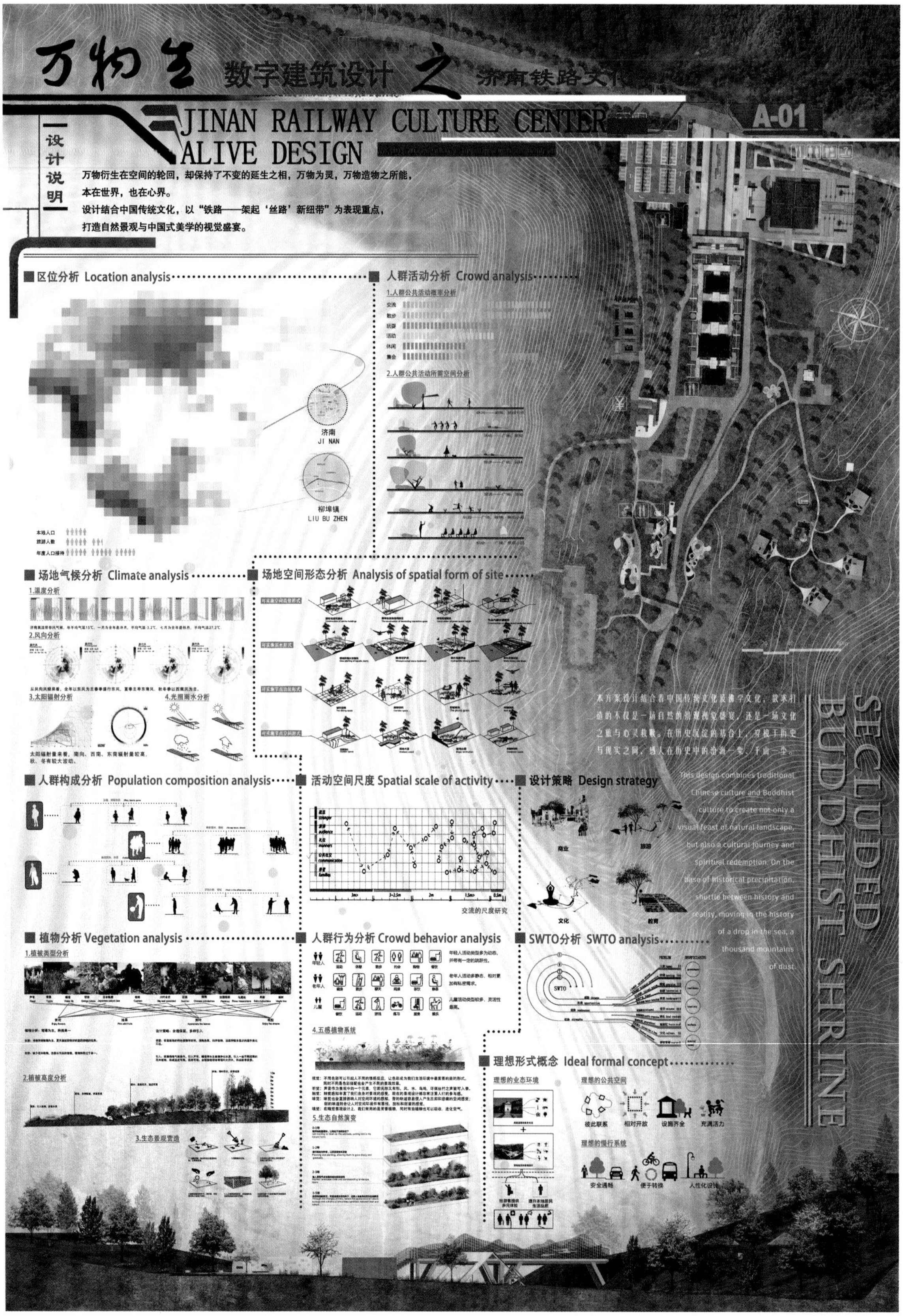

万物生 数字建筑设计之济南铁路文化

JINAN RAILWAY CULTURE CENTER ALIVE DESIGN

A-01

设计说明

万物衍生在空间的轮回，却保持了不变的延生之相，万物为灵，万物造物之所能，本在世界，也在心界。

设计结合中国传统文化，以“铁路——架起‘丝路’新纽带”为表现重点，打造自然景观与中国式美学的视觉盛宴。

区位分析 Location analysis

济南 JI NAN

柳埠镇 LIU BU ZHEN

本地人口

旅游人数

年度人口接待

人群活动分析 Crowd analysis

1.人群公共活动概率分析

交流 散步 玩耍 活动 休闲 集会

2.人群公共活动所需空间分析

场地气候分析 Climate analysis

1.温度分析

2.风向分析

3.太阳辐射分析

4.光照雨水分析

场地空间形态分析 Analysis of spatial form of site

人群构成分析 Population composition analysis

活动空间尺度 Spatial scale of activity

交流的尺度研究

设计策略 Design strategy

商业 旅游 文化 教育

植物分析 Vegetation analysis

1.植被类型分析

2.植被高度分析

3.生态景观营造

人群行为分析 Crowd behavior analysis

4.五感植物系统

5.生态自然演变

SWTO分析 SWTO analysis

理想形式概念 Ideal formal concept

理想的业态环境

理想的公共空间

彼此联系 相对开放 设施齐全 充满活力

理想的慢行系统

安全通畅 便于转换 人性化设计

本方案设计结合着中国传统文化及佛学文化，欲求打造的不仅是一场自然的景观视觉盛宴，还是一场文化之旅与心灵救赎。在历史沉淀的基台上，穿梭于历史与现实之间，感人在历史中的沧海一粟、千山一尘。

This design combines traditional Chinese culture and Buddhist culture to create not only a visual feast of natural landscape, but also a cultural journey and spiritual redemption. On the base of historical precipitation, shuttle between history and reality, moving in the history of a drop in the sea, a thousand mountains of dust.

SECLUDED BUDDHIST SHRINE

A-02
游客中心 Visitor Center
丝路风情广场 Cultural Square
接待禅窑 Meditation Homestay
景观桥 Landscape bridge
涌泉茶舍 Yongquan Tea House
文化展厅 Museum of Culture
休闲广场 Leisure Plaza

白日梦蓝

The Daydream Memories of Blue

参赛学校 University/College

华南农业大学

South China Agricultural University, China

指导教师 Supervisor(s)

周彝馨 ZHOU Yixin

许媛媛 XU Yuanyuan

参赛学生 Participant(s)

陈俊廷 CHEN Junting

潘楚乔 PAN Chuqiao

李旭才 LI Xucai

宋琳珠 SONG Linzhu

简介 Description

该项目位于广东省佛山市顺德区北郊镇桃村。基地内呈梯形布置，建筑面积3246平方米，用地面积45000平方米，包含交通、休闲、娱乐等功能。基地西侧为村庄聚落，衔接好站体与村落之间的联系，是设计的一大重点。建筑概念从“一带一路”的象征物丝绸飘带出发，并结合当地北滘的意向“百滘”进行设计。站房形体由两条飘带缠绕汇聚而成，分为入站房和出站房两个部分，两者通过二层的飘带通来进行连接。以飘带一体式设计串连起轨道、站房和广场，并以此设计了景观通廊，使市民得以在其中穿梭和休闲。

The project is located in Taocun Village, North suburb of Shunde District, Foshan City, Guangdong Province in China. The trapezoidal-shaped building covers a construction area of 3246 square meters and a land area of 45000 square meters. The station has multiple functions, including traffic, leisure, entertainment, etc. On the west of the site there are some villages. A key to this design is to connect the station and the villages. The design is based on silk ribbons, the symbol of the "Belt and Road". The place is also called "Baijiao", which means a hundred streams. This is also introduced to the design. The station itself looks like two flowing ribbons wound together with the two parts being the entrance area and the exit area that are connected by a corridor on the second floor. The two ribbons belong to a single station building joint with the railways and the square. In line with the building, there is also a sightseeing corridor for people to walk through and relax.

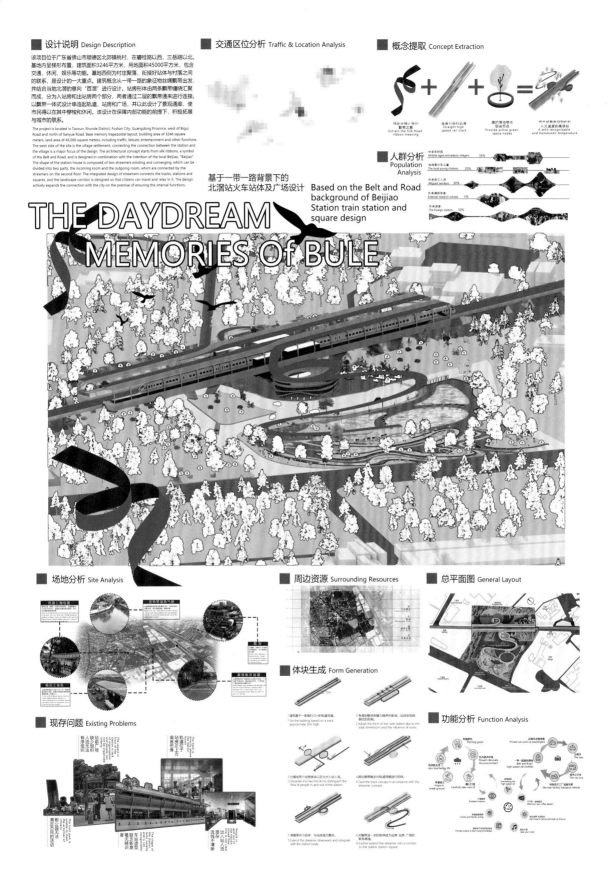

设计说明 Design Description

该项目位于广东省佛山市顺德区北郊镇桃村，在碧桂路以西，三岳路以北。基地内呈梯形布置，建筑面积3246平方米，用地面积45000平方米，包含交通，休闲，娱乐等功能。基地西侧为村庄聚落，衔接好站体与村落之间的联系，是设计的一大重点。建筑概念从一带一路的象征物丝绸飘带出发，并结合当地北滘的意向“百滘”进行设计。站房形体由两条飘带缠绕汇聚而成，分为入站房和出站房两个部分，两者通过二层的飘带通来进行连接。以飘带一体式设计串连起轨道，站房和广场，并以此设计了景观通廊，使市民得以在其中穿梭和休闲。该设计在保障内部功能的前提下，积极拓展与城市的联系。

The project is located in Taocun, Shunde District, Foshan City, Guangdong Province, west of Bigui Road and north of Sanyue Road. Base memory trapezoidal layout, building area of 3246 square meters, land area of 45,000 square meters, including traffic, leisure, entertainment and other functions. The west side of the site is the village settlement, connecting the connection between the station and the village is a major focus of the design. The architectural concept starts from silk ribbons, a symbol of the Belt and Road, and is designed in combination with the intention of the local Beijiao, "Baijiao". The shape of the station house is composed of two streamers winding and converging, which can be divided into two parts, the incoming room and the outgoing room, which are connected by the streamers on the second floor. The integrated design of streamers connects the tracks, stations and squares, and the landscape corridor is designed so that citizens can travel and relax in it. The design actively expands the connection with the city on the premise of ensuring the internal functions.

交通区位分析 Traffic & Location Analysis

概念提取 Concept Extraction

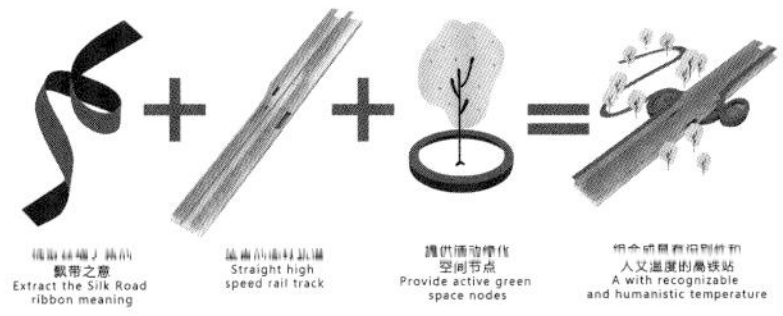

飘带之意
Extract the Silk Road ribbon meaning

Straight high speed rail track

空间节点
Provide active green space nodes

人文温度的高铁站
A with recognizable and humanistic temperature

人群分析 Population Analysis

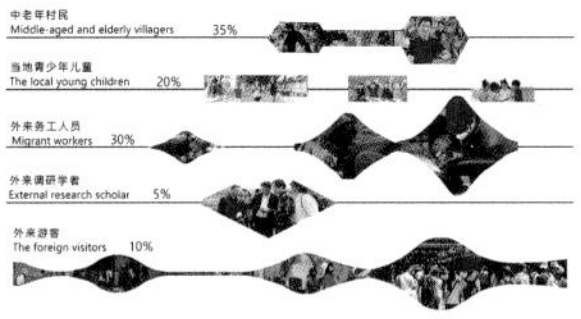

中老年村民	Middle-aged and elderly villagers	35%
当地青少年儿童	The local young children	20%
外来务工人员	Migrant workers	30%
外来调研学者	External research scholar	5%
外来游客	The foreign visitors	10%

基于一带一路背景下的北滘站火车站体及广场设计

Based on the Belt and Road background of Beijiao Station train station and square design

THE DAYDREAM MEMORIES Of BULE

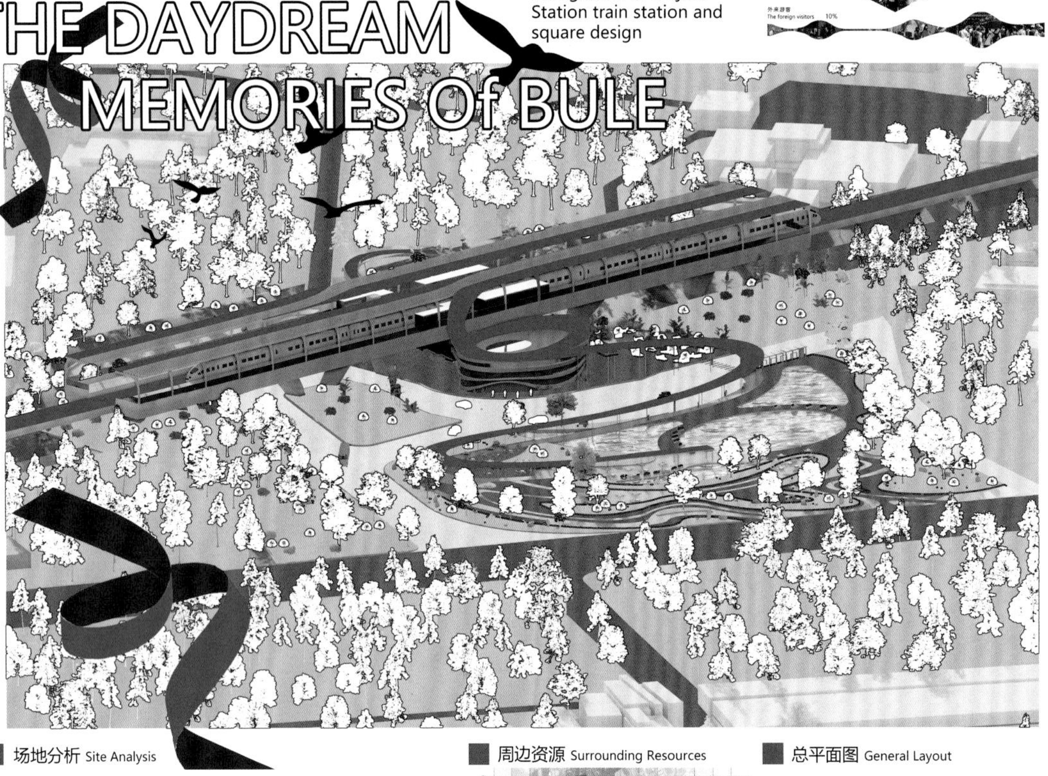

场地分析 Site Analysis

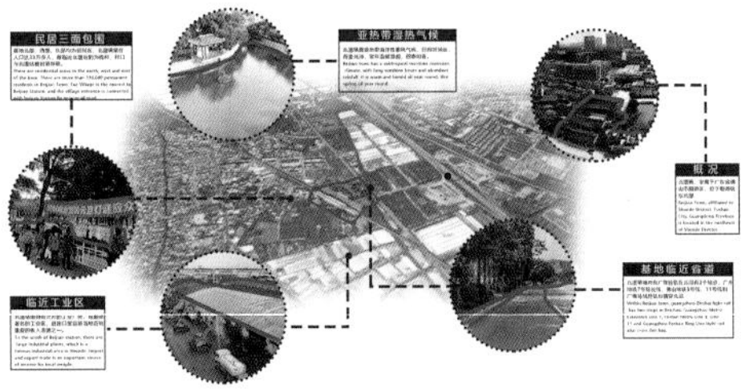

临近工业区

基地临近省道

周边资源 Surrounding Resources

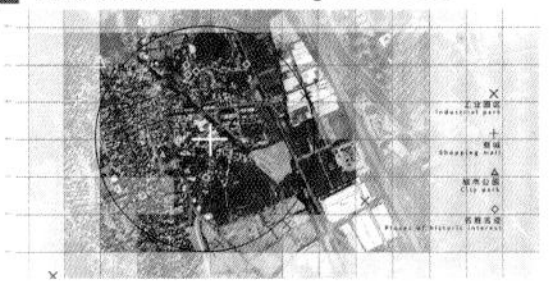

总平面图 General Layout

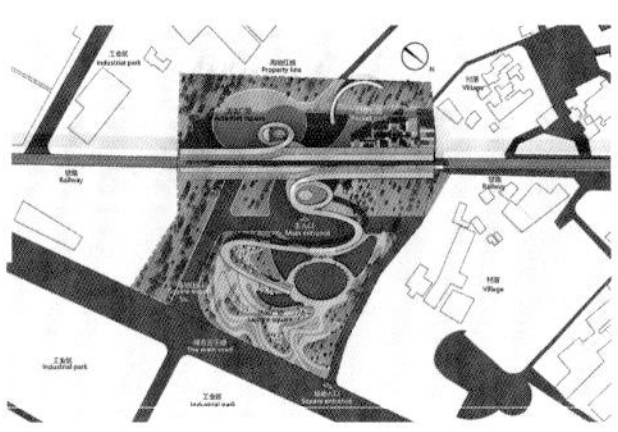

体块生成 Form Generation

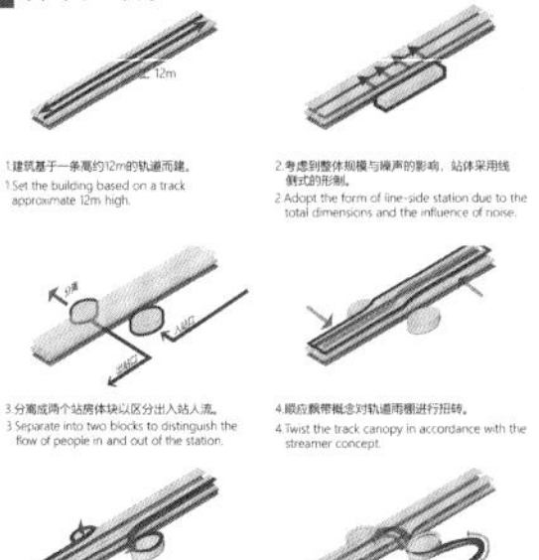

1.建筑基于一条高约12m的轨道而建。
1.Set the building based on a track approximate 12m high.

2.考虑到整体规模与噪声的影响，站体采用线侧式的形制。
2.Adopt the form of line-side station due to the total dimensions and the influence of noise.

3.分离成两个站房体块以区分出入站人流。
3.Separate into two blocks to distinguish the flow of people in and out of the station.

4.顺应飘带概念对轨道雨棚进行扭转。
4.Twist the track canopy in accordance with the streamer concept.

5.将飘带向下延伸，与站体结为整体。
5.Extend the streamer downward and integrate with the station body.

6.对飘带进一步的延伸成为站房-站房-广场的联系廊道。
6.Further extend the streamer into a corridor to link station-station-square.

现存问题 Existing Problems

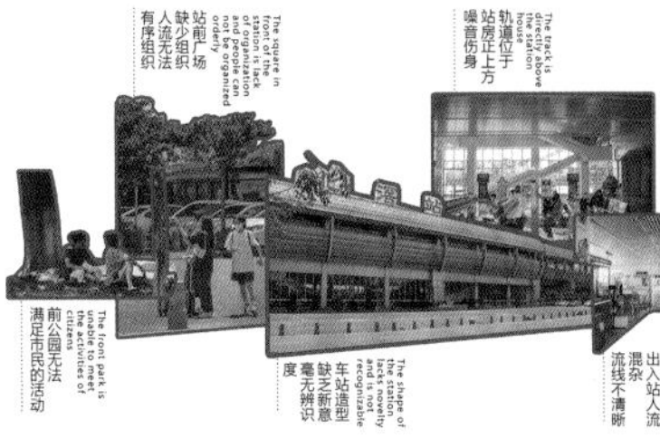

站前广场缺少组织人流无法有序组织
The square in front of the station is lack of organization and people can not be organized orderly

轨道位于站房正上方噪音伤身
The track is directly above the station house

前公园无法满足市民的活动
The front park is unable to meet the activities of citizens

车站造型缺乏新意毫无辨识度
The shape of the station lacks novelty and is not recognizable

出入站人流混杂流线不清晰
People in and out of the station mixed flow line is not clear

功能分析 Function Analysis

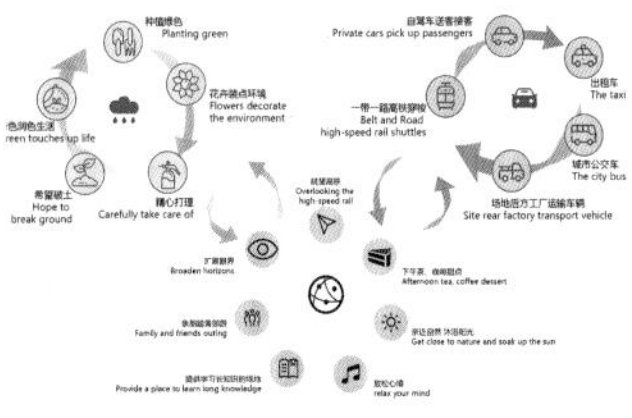

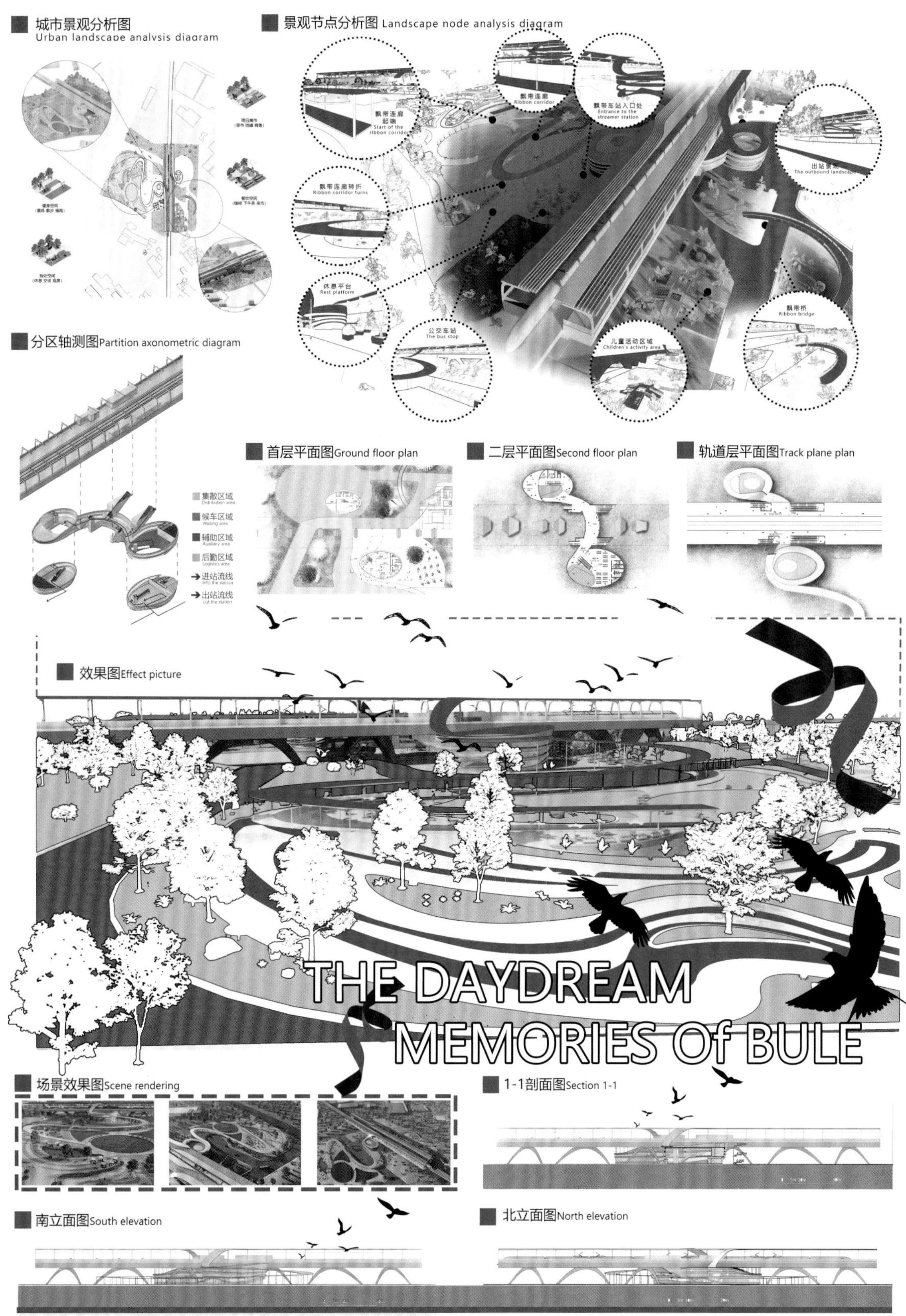
城市景观分析图
Urban landscape analysis diagram
景观节点分析图 Landscape node analysis diagram
飘带连廊起端
Start of the ribbon corridor
飘带连廊
Ribbon corridor
飘带车站入口处
Entrance to the streamer station
出站景观
The outbound landscape
飘带连廊转折
Ribbon corridor turns
休息平台
Rest platform
公交车站
The bus stop
儿童活动区域
Children's activity area
飘带桥
Ribbon bridge
分区轴测图Partition axonometric diagram
集散区域
候车区域
辅助区域
后勤区域
进站流线
出站流线
首层平面图Ground floor plan
二层平面图Second floor plan
轨道层平面图Track plane plan
效果图Effect picture
THE DAYDREAM
MEMORIES Of BULE
场景效果图Scene rendering
1-1剖面图Section 1-1
南立面图South elevation
北立面图North elevation

穿越碧江，古今对话——高铁客运站设计

Crossing the Bijiang River to Enable an Ancient and Modern Dialogue—A Design of High-speed Railway Station

参赛学校 University/College
华南农业大学
South China Agricultural University, China

指导教师 Supervisor(s)
周彝馨 ZHOU Yixin
许媛媛 XU Yuanyuan

参赛学生 Participant(s)
李远哲 LI Yuanzhe
彭卓婷 PENG Zhuoting
朱冰冰 ZHU Bingbing
周咏欣 ZHOU Yongxin

简介 Description

随着城市化进程的加快，交通压力变大，道路堵塞严重，环境污染加剧，居民休闲可达性降低，因此人们越来越重视实现城乡空间的融合发展，尤其是与环境的融合，与传统文化的融合和历史的融合。当年，碧江振响楼成为无数爱国学子汲取力量的重要战地，在这里，碧江的红色“星火”照亮了一颗颗励志救亡图存的赤子之心。如今，

以此为灵感，设计者运用作环抱姿势的弧线去融合历史，联系文脉之路——碧江大道，以列车为镞穿越碧江，沟通古今。

With the acceleration of urbanization, traffic pressure is increasing, road congestion is serious, environmental pollution is intensifying and residents' access to leisure is reduced. Therefore, people pay more and more attention to the integrated development of urban and rural spaces, especially with the environment, the traditional culture and the history. In the past, the Zhenxiang building in Bijiang became an important battlefield for countless patriotic students where they could draw strength. Here, the red relic of Bijiang is like a "spark" lighting up the unstained aspiration to save the nation from extinction. Today, inspired by the red history of Bijiang, designers use the arch of the Bijiang Avenue as an embracing posture to refer back to the history and connect the dots of Bijiang's cultural spirit. The train looks like an arrow flying across Bijiang and connecting the past and the future.

穿越碧江 古今对话
——高铁客运站设计

基地选址：
中国广东省佛山市顺德区北滘镇碧江村长宁路

基地特色：
碧江村历史悠久，富有浓厚的文化氛围和保存较好的古建筑，文人辈出，也是红色革命的发展点之一。

设计说明：
随着城市化进程的加快，环境污染加剧，人们越来越重视实现城乡空间的融合发展，尤其是与环境的融合，与传统文化的融合。当年，碧江振响楼成为无数爱国学子汲取力量的重要战地，在这里，碧江的红色“星火”照亮了一颗颗励志救亡图存的赤子之心。如今，以此为灵感，设计者运用作环抱姿势的弧线去融合历史，联系文脉之路——碧江大道，以列车为镞穿越碧江，沟通古今。

古建资源丰富，保存较好

东南角小角度俯视

优势1——历史文化氛围浓厚，古建资源丰富

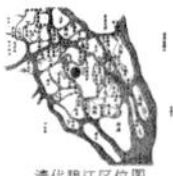
清代碧江区位图

清咸丰《顺德县志》中载碧江属龙头堡，民夹水而居，百货辐辏，为广州货物一大中转站。龙头山、伏龙、浮石、温鱼、狮诸岗等山岗皆在碧江内，陈氏昆岗在碧江西北，古人谓此乃“五兽当关，九龙入洞”的风水格局。地理位置的优越对碧江的经济发展起到了重要作用。

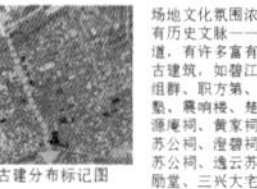
古建分布标记图

场地文化氛围浓厚，持有历史文脉——碧江大道，有许多富有特色的古建筑，如碧江金楼及组群、职方第、亦渔书塾、震响楼、楚珍祠、源庵祠、黄家祠、峭岩苏公祠、澄碧祠、丛兰苏公祠、逸云苏公祠、励堂、三兴大宅等，保护较为完善。

优势2——交通路网密集，周边有轨道站点和覆盖公交巴士

碧江及碧江大桥

京澳线、广珠西线高速等重要交通要道组织有序，沿线合理设置公共交通站点，公共出行方便。

劣势1——基地附近交通流量大，人车流线错杂

碧江大道与京澳线交叉路口

碧江大道车流情状况

场地交通车流量大，人车流线交叉，连接天桥少，交通情况复杂，碧江大道正对工业区入口，穿行车流量大。

劣势2——周边无公共场所，绿化建设缺少

场地上空，无大型公共空间

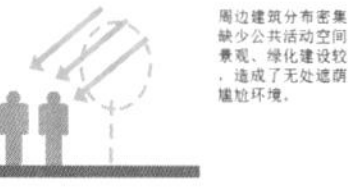
绿化建设较少，路面气温高

周边建筑分布密集，缺少公共活动空间；景观、绿化建设较少，造成了无处遮荫的尴尬环境。

挑战1——整合交通流线，优化进出站方式

原碧江站旧址建设在高架轨道下层，京澳线辅路之上，乘客需要走上天桥横跨京澳线才能连接到站房

原址的站房空间只能采取侧面采光，进深大，因此导致采光面积不足，内部空间整体较暗，灯光资源消耗大，不适合低碳发展

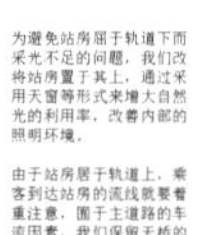

为避免站房居于轨道下而采光不足的问题，我们改将站房置于其上，通过采用天窗等形式来增大自然光的利用率，改善内部的照明环境。

由于站房居于轨道上，乘客到达站房的流线就要着重注意，囿于主道路的车流因素，我们保留天桥的传达形式，并对采用公共交通前来的人流做合理的组织调整，优化进出站的方式，方便出行。

挑战2——设计公共广场，营造宜人闲适共享空间

碧江金楼组群内部园林

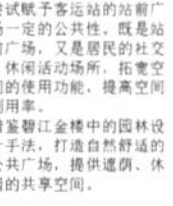

为改善当地无公共休闲空间、无绿化隐蔽空间，将尝试赋予客运站的站前广场一定的公共性，既是站前广场，又是居民的社交、休闲活动场所，拓宽空间的使用功能，提高空间利用率。
借鉴碧江金楼中的园林设计手法，打造自然舒适的公共广场，提供遮荫、休闲的共享空间。

现状道路| Road Analysis
城市路网 轨道
基地路网有序，场地中包含京澳线以及高架铁路，面对历史文脉碧江大道。东有碧江大桥，西有广珠西线高速

绿地水系| Greenbelt Water System
绿地 水系
基地东临碧江，周边绿化分布零散，基地范围无大面积的绿化空间，景观局限。

建筑肌理| Architechtural Texture
现有建筑
碧江保留了部分村落肌理，保存了祠堂、馆舍、书塾、民居和园林等较为丰富的历史文化遗产。建筑密集分布，但又是有序排布。以传统民居形式为主体。工业厂房大范围分布。

建筑类型分布| Distribution of Building Types
居住区 工业区 商业区
用地所处是京澳线上，京澳线以北是工业用区，以南是以居住区为主，有部分的商业建筑沿路分布，但是基本属于旧楼，目前使用较少。

基地范围| Base Area
基地选址文长方形地块，受当地道路限制，场地内包含了一条贯穿碧江的京澳线，其分隔了工业区和居住区，规划分明。沿京澳线东上就是碧江大桥。

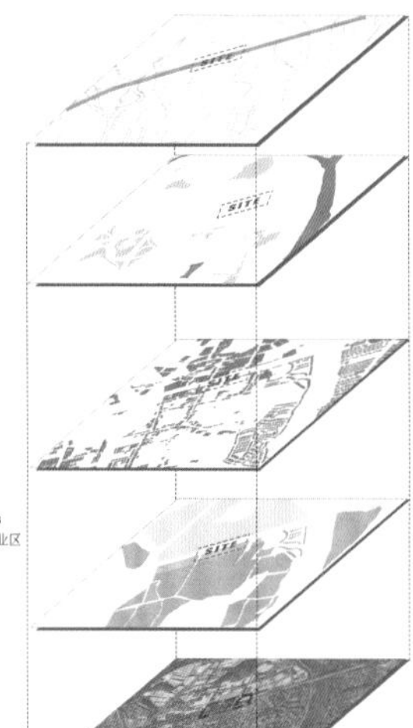

概念提取及演变

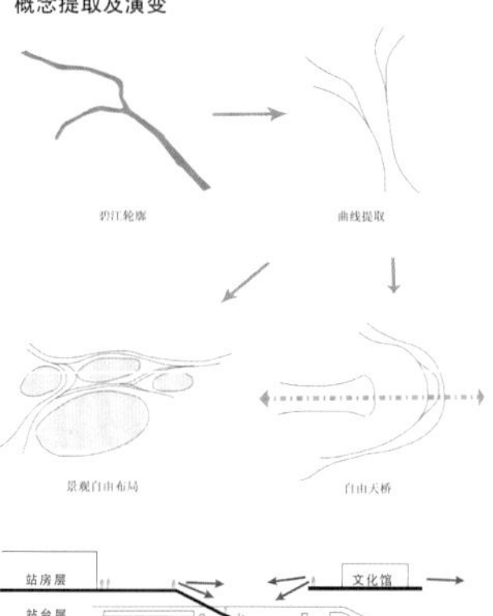

视线分析

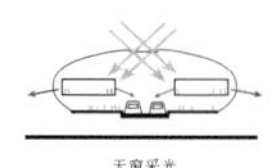
天窗采光

视线分析示意图

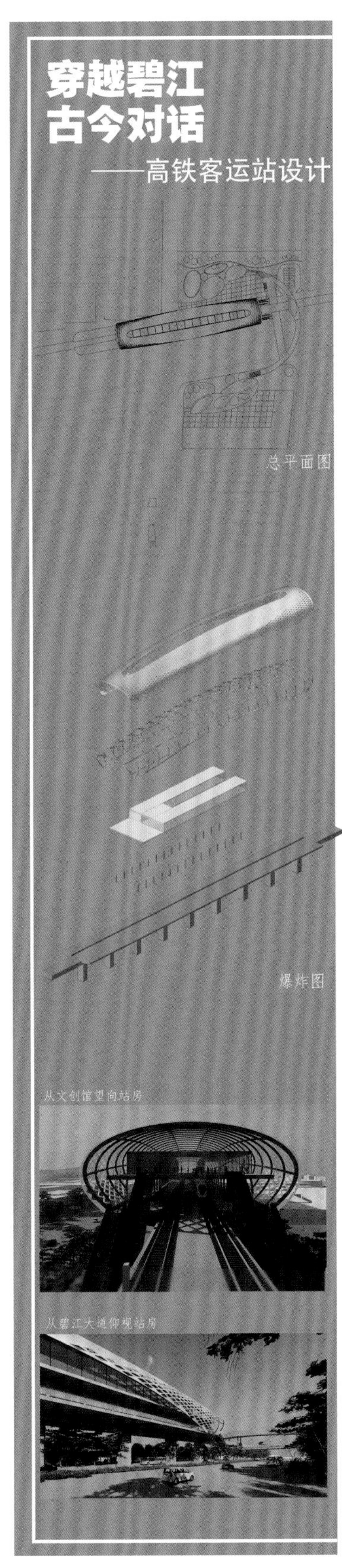

站前广场景观效果图

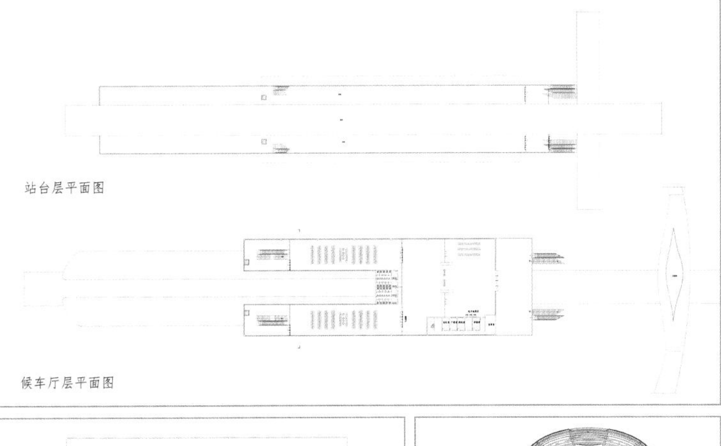

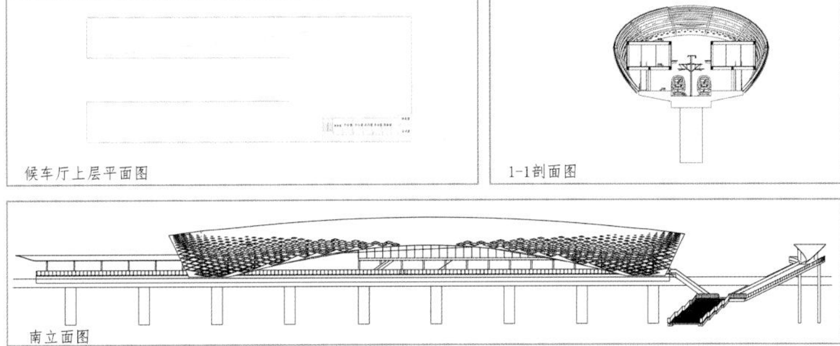

从站台望向候车厅

织错接市——成都红牌楼站枢纽设计

Woven Streets: Hongpailou Station in Chengdu, China

参赛学校 University/College

东南大学

Southeast University, China

指导教师 Supervisor(s)

胡碧琳 HU Bilin

周 霖 ZHOU Lin

参赛学生 Participant(s)

陈宝睿 CHEN Baorui

刘卓尔 LIU Zhuoer

袁丽婷 YUAN Liting

黄子杰 HUANG Zijie

刘存浩 LIU Cunhao

简介 Description

本方案位于曾作为汉藏交流枢纽的红牌楼，跨越铁路，组织起织补城市的慢行平台，为成都市民提供多样而全天候的活动空间，亦结合周边站点进行城市设计来推动整体开发。步道组织起第二层城市肌理，也在进入车站和步行通过之间取得微妙平衡。乘客借助步道系统进入进站广场后，便可在自然光和风包围下抵达站台。

在结构设计上，巨构 Y 型柱支撑起简洁的大跨空间，容纳下三层立体车站。横向上富有秩序的门架形象也致敬了场地的历史传统。

The planned site is located in Hongpailou, Chengdu, China, which used to be the hub of communication between the Han and Tibetan peoples. Across the railway, the design creates a platform for slow traveling to supplement the city. The station provides a space for the citizens in Chengdu where they can enjoy rich activities that is available 24/7. The design plan covers the whole city by combining the features of surrounding stops to fuel an overall development. The walkway bolsters the second layer of the urban texture and strikes a delicate balance between walking into the station and walking by. After entering the station plaza through the walkway, passengers can reach the platform in the open natural light and wind. In terms of structural design, the giant Y-shaped column supports the simple large-span space to accommodate the three-story three-dimensional station. The order of the horizontal gantry pays tribute to the historical tradition of the site.

设计说明

以轨道交通站点为中心，10分钟步行路程以内，同时满足居住、工作、购物、娱乐、出行、休憩等需求的多功能公园城市社区。以轨道交通站点为中心的TOD开发，聚集人流商流，带动新区发展，推动城市更新，是重构城市形态、优化城市空间的重要支撑，也是革新资源配置，重塑经济地理的重要抓手，更是提升空间品质，推动场景营城的重要载体。

基于城市设计的原则打造公共空间，以行人尺度为依据优化城市体验，为居民社区建立新的通勤习惯，核心区将补充城市功能，构建完整的城市生活服务圈，在“回家路上”解决80%的日常生活需求，呈现高度融合的通勤体系。

方案选址位于汉藏交流枢纽红牌楼，以环城生态带、武侯政务圈、武侯电商产业圈、武侯品质居住圈围绕的城市新中心，依托政务、产业、居住、自然环境核心地缘资源，形成“人城境业”和谐统一的TOD活力生活圈。

车站主体建筑借助传统建筑的表现形式，以红牌楼为灵感，在红牌楼的过去和未来之间架起桥梁，表达了其在项目中兼顾传承与现代化的愿景。通过使用适应性参数化设计，得以创造出拥有多个变体的设计，同时又能够维持大区范围内的统一感。项目位于四川省成都市武侯区，抗震设防烈度为7度（场地基本地震动峰值加速度0.1g），铁路客运车站应提高一度进行抗震设计并有相应性能设计要求，车站主体采用混凝土框架+异形钢支撑组合，兼顾经济性与结构安全性，Y形立柱设计在满足结构跨度的同时，减少竖向构件对建筑使用空间的入侵，带来更为宽敞的站厅平面开放空间，其造型也更加利于BRB阻尼器等减隔震构件的布置。Y形立柱保留钢骨架原貌，突出其本身结构之美，体现车站“向下生长”的力量感与生命力。

绿化公园作为红牌楼TOD15分钟生活圈的绿芯，是铁路红牌楼站及地铁太平园站区域及周边海绵城市建设的主要载体。公园设计倡导“四性”，即在地性、开放性、共享性及智慧性，从功能与文化的角度为周边居民提供了复合化的公园空间。

区位图

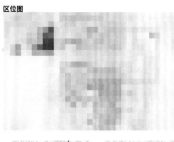

规划布局分析

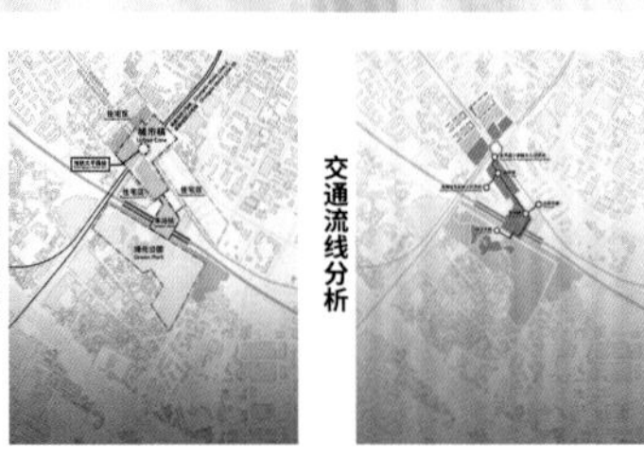

交通流线分析

爆炸功能图

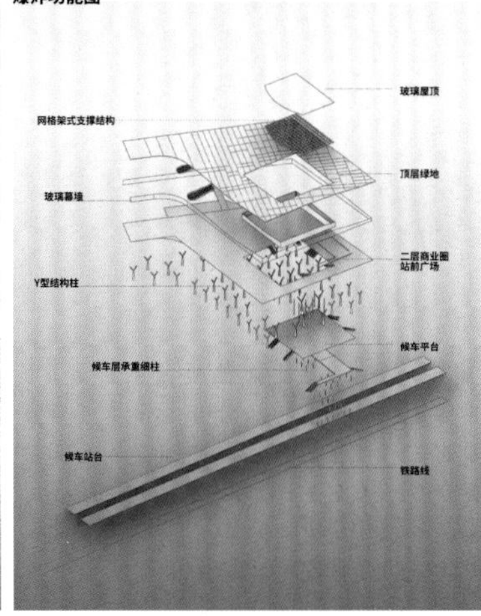

剖面图

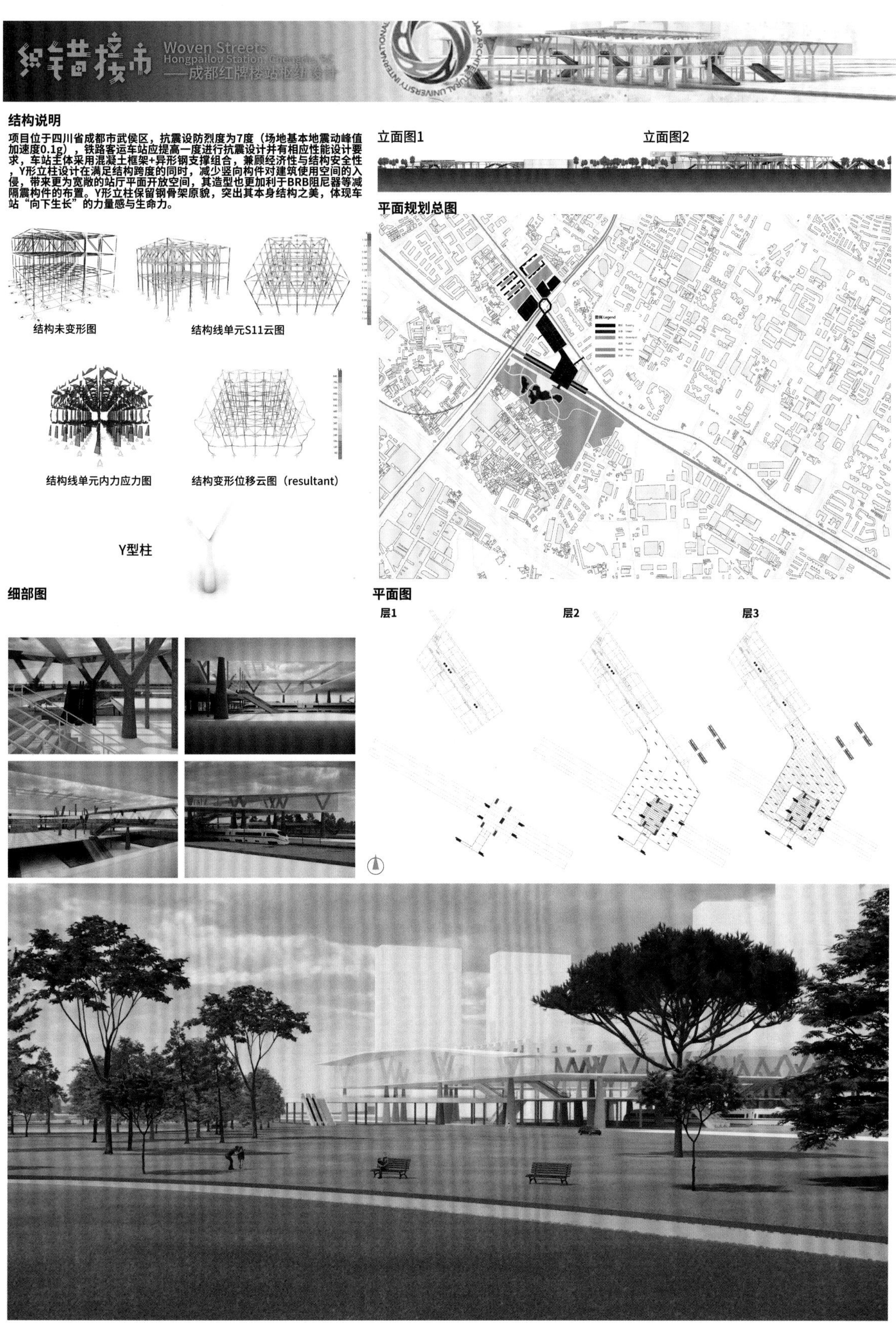
织错接市
Woven Streets
Hongpailou Station
——成都红牌楼站
结构说明
项目位于四川省成都市武侯区，抗震设防烈度为7度（场地基本地震动峰值加速度0.1g），铁路客运车站应提高一度进行抗震设计并有相应性能设计要求，车站主体采用混凝土框架+异形钢支撑组合，兼顾经济性与结构安全性，Y形立柱设计在满足结构跨度的同时，减少竖向构件对建筑使用空间的入侵，带来更为宽敞的站厅平面开放空间，其造型也更加利于BRB阻尼器等减隔震构件的布置。Y形立柱保留钢骨架原貌，突出其本身结构之美，体现车站“向下生长”的力量感与生命力。
结构未变形图
结构线单元S11云图
结构线单元内力应力图
结构变形位移云图（resultant）
Y型柱
细部图
立面图1
立面图2
平面规划总图
平面图
层1
层2
层3

普里奥焦尔斯克的多功能社区中心——客运枢纽站

Multifunctional Community Center with a Passenger Terminal in Priozersk

参赛学校 University/College
俄罗斯圣彼得堡国立建筑大学
St. Petersburg State University of Architecture and Civil Engineering, Russia

指导教师 Supervisor(s)
Olga Kokorina
Fedor Perov

参赛学生 Participant(s)
Lemudkina Alena

简介 Description

本项目展现了在火车站周围实现各种公共休闲功能的可能性，项目地点位于省级城市的中心地带，其重要性往往被人们低估。

普里奥焦尔斯克建于 1294 年，是一座有着丰富的历史和文化遗产的独特城市。它曾是被俄罗斯和芬兰争夺多年的要塞，因此保留了两个国家的文化痕迹。此外，这座城市还有湖泊和河流环绕。凭借其独特的文化遗产和自然风光，普里奥焦尔斯克成为一个旅游胜地。如今，作为一个坐落在圣彼得堡附近的小城市，普里奥焦尔斯克正在发展木材加工业。

为创造良好的周边环境、培育景观特征，项目选择将火车站建在一幢历史性建筑附近，周围有草木环绕。

该建筑可同时实现三类功能。

- 休闲：健身房、餐厅、展览馆、音乐厅。

- 工作：共享办公空间、图书馆。

- 火车站：候车厅、收款处、职工。

省城的公共多功能建筑不仅可以吸引游客，还可以留住他们，从而凝聚城市人口、促进城市繁荣。

This project demonstrates the possibility of combining various functions for public leisure around the railway station-an underestimated and very important center of a provincial town.

Priozersk is a unique city founded in 1294 with a rich history and heritage. For many years, the city played the role of a fortress and was a disputed territory between Russia and Finland, so it contains signs of both cultures. Besides, it is located between lakes and rivers. Because of its culture and nature, the city is a popular destination for tourists. Currently, Priozersk is a small town near St. Petersburg with a developing woodworking industry.

The location of the station was chosen near a historic building surrounded by greenery in order to create a favorable environment around the building and develop the view characteristics of the site.

The function of the complex is divided into 3 groups:

-Leisure: a fitness centre, a restaurant, an exhibition centre, a concert hall.

-Work: a coworking space, a library.

-Railway station: a waiting hall, cash desks, staffs.

A public multifunctional building in a provincial town can not only attract tourists, but also make them stay there, rally the population of the city and contribute to its further prosperity.

City of lakes: Priozersk

Priozersk (name comes from «by the lakes») is a unique city founded in 1294 with a rich history and heritage. For many years, the city played the role of a fortress and was a disputed territory between Russia and Finland, so it contains signs of both cultures. In addition, it is located between lakes and rivers, and because of its culture and nature, the city is a popular destination for tourists. Currently, Priozersk is a small town near St Petersburg with a developing woodworking industry.

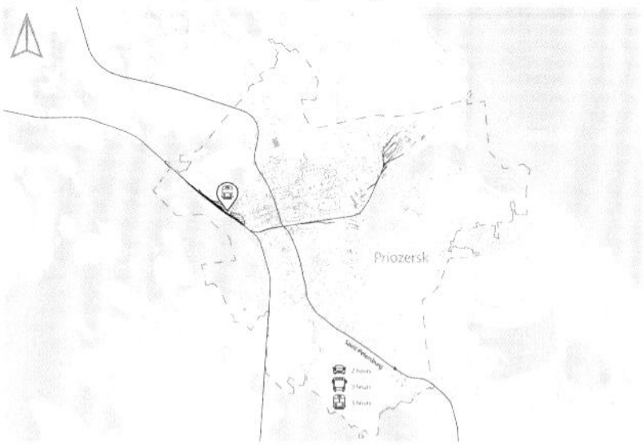

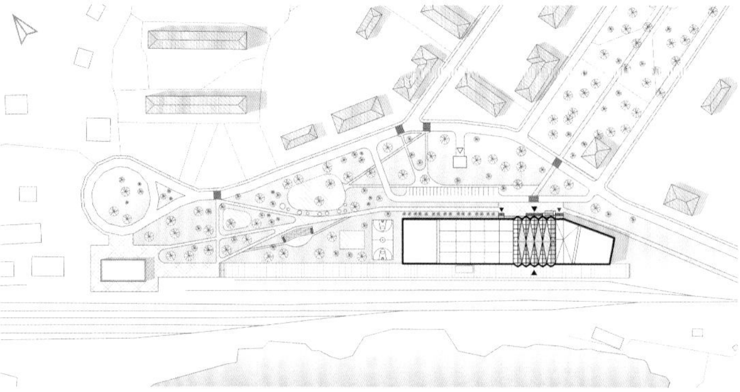

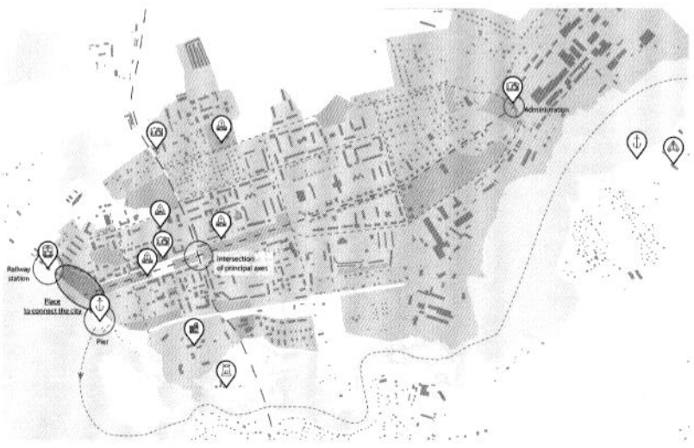

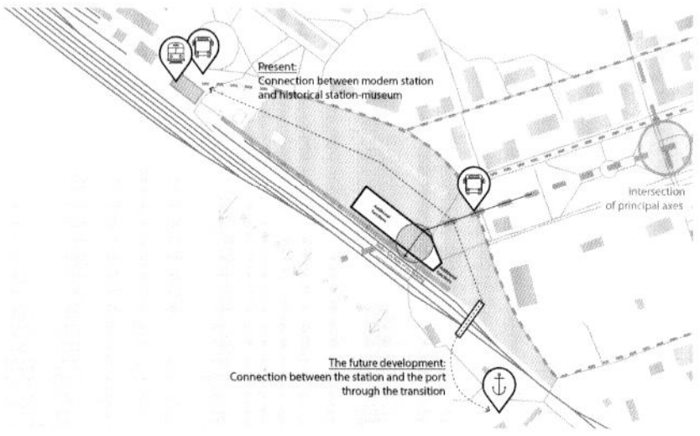

Present: Connection between modern station and historical station-museum

Intersection of principal axes

The future development: Connection between the station and the port through the transition

SWOT analisis

Strengths	Weaknesses
Medical resort Mainly seasonal tourism Heritage (Russian+Finnish) Closeness to unique nature The scale of the city is close to human	No city center No modern workplaces Less of city baeutification No organized landscaping sistem No modern places for city community

Opportunities	Treats
Create new workplaces Make tourism year-around Improve green spaces in the city Include waterway to tourism programm Attract new residents by modern improvments	Urban decline Loss of heritage Industrial growth Population outflow

Territory use:

- Multi-apartment development
- Individual residential buildings
- Territories of administration
- Industrial areas
- Green spots

- Railway station
- Bus station
- Pier
- Glamping
- Point of interest
- Church
- Fortress
- Sanatorium

The main idea of the formation of the environment is to preserve its original structure, to make the station building in harmony with it, to emphasize the city's structure, completing its main axis with the station building. Besides, the goal was to create a public space around the station, for which this site was perfect: a green area overlooking the lakes.

The building is located along the railway tracks, which emphasizes its purpose and uses the view characteristics for built-in functions. It also protects people in the park from the sun and wind from the river.

SWOT analysis helps to understand the current state of the city and what it needs at the moment. In my case, in order to use the benefits and prevent threats, I highlighted the necessary additional functions for the station:

- concert hall
- coworking
- fitness Centre.

Identity of Priozersk

Lakes · Fortress Korela · Heritage · Kirkha · Cathedral of the Nativity of the Virgin · European housing construction

Railway station · Water tower · Factory Waldhof · Courtyard of the Valaam Monastery · Panel linear building

Key Features

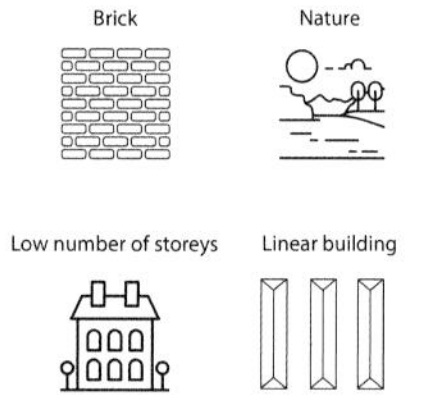

Brick · Nature · Low number of storeys · Linear building

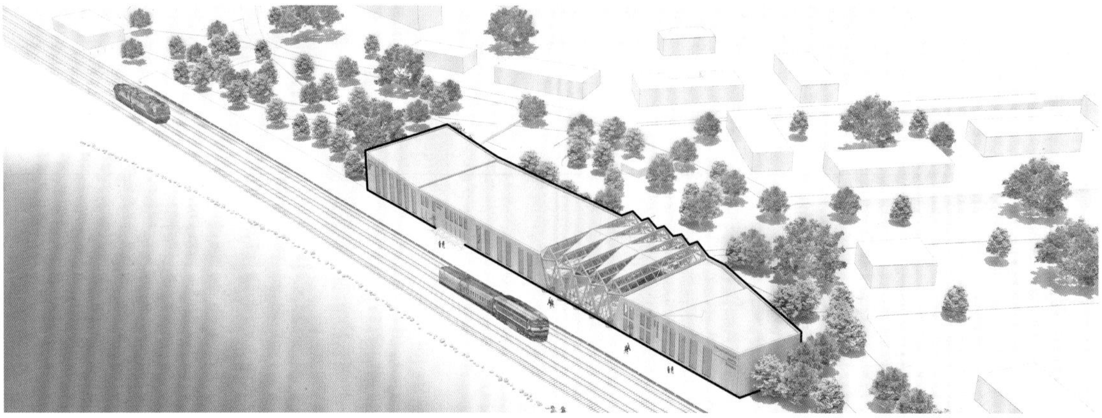

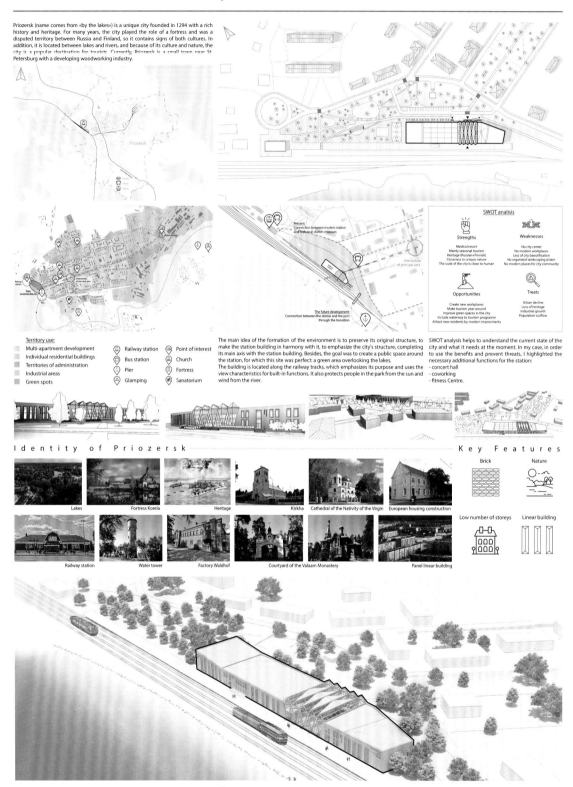

Railway station - multifunctional complex

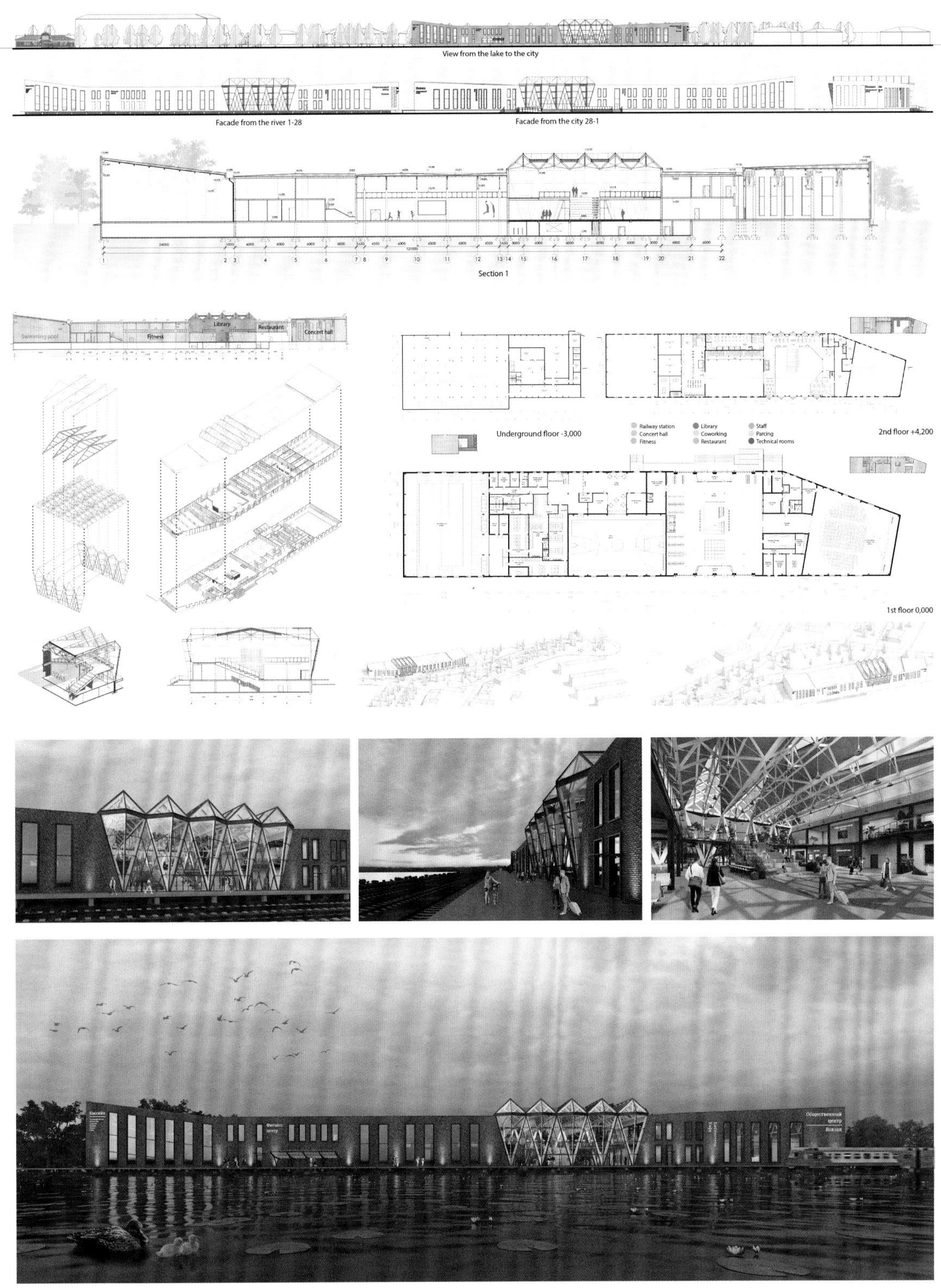

俄罗斯“瓦尔代”小城铁路客运站房设计

Design of the Railway Station Building in the Small Town of “Valdai” in Russia

参赛学校 University/College

俄罗斯圣彼得堡国立建筑大学

St. Petersburg State University of Architecture and Civil Engineering, Russia

指导教师 Supervisor(s)

朱晓菲 ZHU Xiaofei

Korzhempo Yan Alexandrovich

参赛学生 Participant(s)

张富瑞 ZHANG Furui

常 征 CHANG Zheng

阎晓凡 YAN Xiaofan

卢程遥 LU Chengyao

Anna Stytsenko

简介 Description

1. 广场空间设计

设计保留原有广场空间基本形式，车站大厅和候车厅部分与原有建筑空间体量相当，并作为场所背景，保留人们的原始城市印象，突出广场雕塑，保留城市历史。

2. 建筑空间布局

考虑到建筑未来对城市的影响，该建筑应成为城市中心公共领域的凝聚点，成为公共精神和物质载体。在空间功能上分为三个部分：

（1）城市公共交通属性部分（火车站交通属性功能）；

（2）城市公共聚集属性部分（餐饮、阅读、洽谈、休闲功能）；

（3）城市文化展示部分（制钟文化展示以及城市景观展示部分）。

这三部分在交通流线上彼此联系但功能上相对独立，互不影响。

1. Square Space Design

The design retains the basic form of the original square space. The station hall and waiting hall are of the same size as the original building space. They are also used as the background of the venue to preserve people's original impression of the city. The sculpture on the square is highlighted to record the history of the city.

2. Building Space Layout

Considering its impact on the city's future, this building should become the focal point of the public realm in the city center, and the physical carrier of public spirit. The space has three different functions:

(1) Urban public transportation (the traffic function of the railway station);

(2) Urban public gathering (catering, reading, negotiation, relaxation);

(3) Urban culture display (display of the clock-making culture and urban landscape display).

These three parts are inter-connected in traffic flows but independent in their respective functions.

俄罗斯“瓦尔代“小城铁路客运站房设计

DESIGN OF RAILWAY PASSTNGTR STAT ... THE SMALL TOWN OF "VALDAI" IN RUSSIA

场地周边环境要素
Elements of the surrounding environment of the site

设计说明 Design description

瓦尔代位于莫斯科与圣彼得堡之间的邮政路线上，是古老丝绸之路节点上的小城市。本市建于1770年，在历史上以铸造教堂的各种钟而闻名与整个俄罗斯。城市得名于市内风光秀丽的瓦尔代湖，建市前属于艾弗斯基修道院，其修道院位于瓦尔代湖中心的岛上。瓦尔代长居人口不足两万人，属于县级市，隶属于诺夫哥罗德州，是当地著名的旅游度假胜地。当地经济主要靠制钟业以及相关手工业和旅游业支撑，近年来围绕旅游与制钟的文化传统吸引不少外地游客到此观光，政府为其城市旅游业的发展，现需原址重建瓦尔代火车站，以提升城市魅力以及城市文化价值。瓦尔代城市空间布局缺少核心公共空间，城市布局分散，缺少标志性建筑，并且缺少城市文化符号。为解决以上城市问题，在瓦尔代火车站的原址重建项目中主要考虑以下几个方面

1、增加城市活力，提供更多的城市公共聚集空间，丰富城市公共文娱活动。

考虑到建筑为未来城市影响，该建筑应成为城市中心公共领域的凝聚点和展示城市文化魅力的平台。又因当地制钟的文化属性，在空间功能上分为三个部分：1、城市公共交通属性部分（火车站交通属性功能）；2、城市公共聚集属性部分（城市展厅、餐饮、阅读、洽谈、休闲功能）；3、城市文化展示部分（制钟文化展示以及城市景观展示部分）。

2、增强城市文化符号性，加大城市文化聚集性，体现城市文化标志性，尊重场地原始文化记忆性。

以高耸的钟楼为火车站文化符号的代表，建立新的城市天际线。加强当地的文化符号，增加城市空间的凝聚力，凸显当地公共领域的重要性。另外设计保留原有广场空间基本形式，并将建筑作为场所背景，保留人们对城市的原始k印象，突出广场雕塑，保留城市历史。

Located on the postal route between Moscow and St. Petersburg, Valdai is a small city on the ancient Silk Road. Founded in 1770, the city is historically famous throughout Russia for casting various church bells for churches. The city gets its name from the beautiful Lake Valdai in the city, which formerly belonged to the Ivorsky Monastery, which is located on an island in the center of Lake Valdai. Valdai has a population of less than 20,000. It belongs to a county-level city and belongs to Novgorod Oblast. it is a famous local tourist resort. The local economy is mainly supported by the clock-making industry, its handicrafts and tourism. In recent years, the cultural tradition of tourism and clock-making has attracted many foreign tourists to visit here. The government needs to rebuild the Valdai Railway Station on its original site for the development of urban tourism in order to enhance the charm of the city and its cultural value. The urban spatial layout of Valdai lacks core public space, the urban layout is scattered, the landmark buildings and high-rise buildings are lacking, and the urban cultural symbols are lacking. In order to solve the above problems of the city, the following aspects are mainly considered in the original site reconstruction project of Valdai Railway Station. **1. Increase urban vitality, provide more urban public gathering space, and enrich urban public cultural and recreational activities.** Considering the impact of thebuilding on the future city, the building should become a meeting point for the public realm in the city center and a platform for displaying the charm of the city's culture. Taking into account the cultural attributes of local clock making, the space function is divided into three parts. 1. The urban public transportation attribute part (the railway station traffic attribute function); 2. The urban public gathering attribute part (catering, reading, negotiation, leisure function); 3. the urban culture display part (the clock-making culture display and the urban landscape display part). **2. Zeng strengthens the symbol of urban culture, increases the concentration of urban culture, and reflects the iconicity of urban culture.** With the towering clock tower as the representative of the cultural symbol of the railway station, a new city skyline will be established. Strengthen local cultural symbols, increase the cohesion of urban space, and highlight the importance of the local public realm.

经济技术指标			
建筑用地面积	12666.79m²	候车厅总面积	1236m²
建筑总面积	3896.56m²	售票厅面积	326m²
建筑基底面积	2583.64m²	办公部分面积	243m²
容积率	0.31	钟楼文化馆面积	365m²
绿化率	45.6%	城市景观厅面积	184m²
建筑密度	0.23	技术设备房总面积	1542m²

区位分析
District Analysis

总平面图
General layout

场地现状分析
Site status analysis

火车站建筑元素类型学分析
The typological analysis of architectural elements of railway station

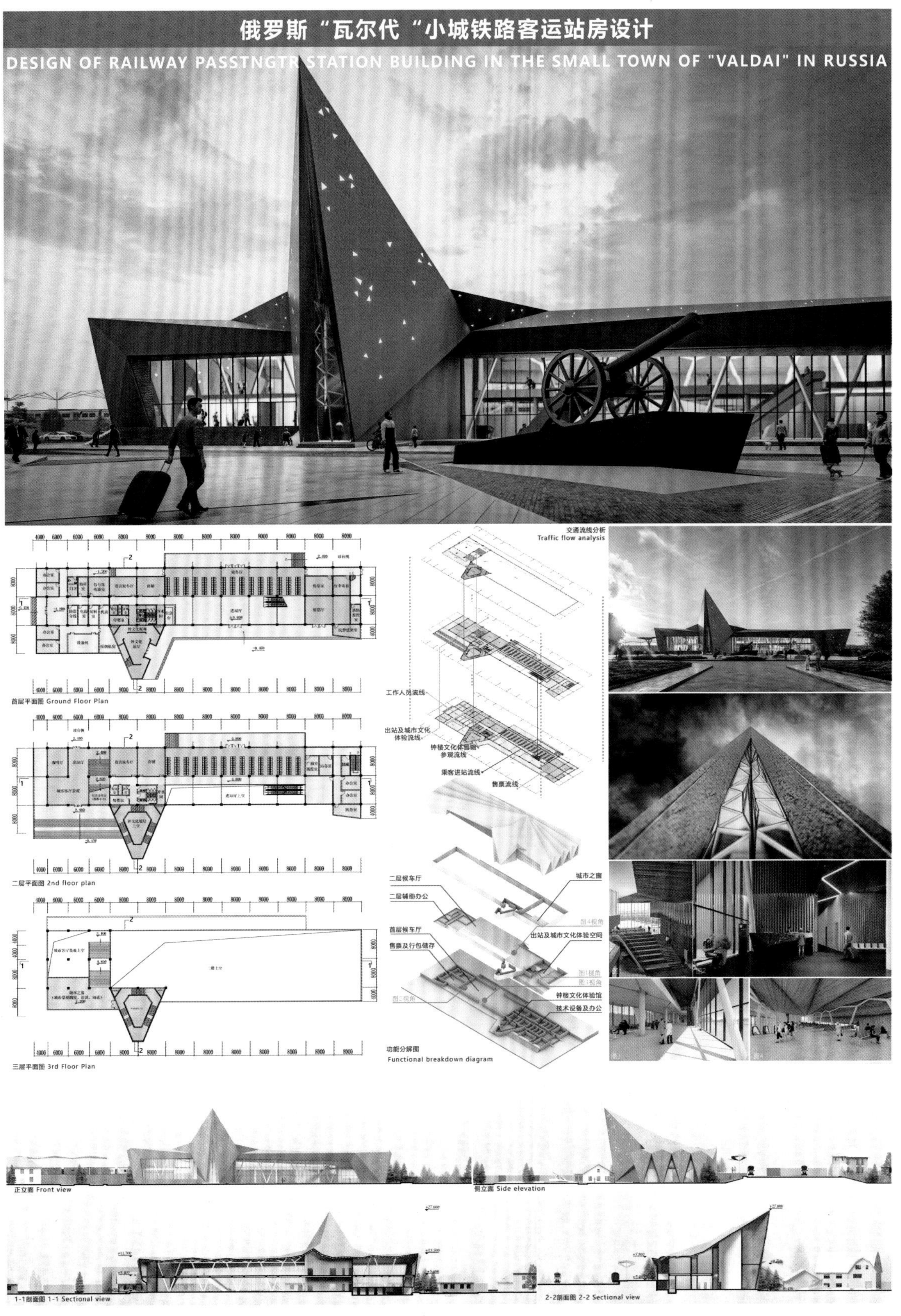
俄罗斯"瓦尔代"小城铁路客运站房设计
DESIGN OF RAILWAY PASSTNGTR STATION BUILDING IN THE SMALL TOWN OF "VALDAI" IN RUSSIA
首层平面图 Ground Floor Plan
二层平面图 2nd floor plan
三层平面图 3rd Floor Plan
交通流线分析
Traffic flow analysis
工作人员流线
出站及城市文化体验流线
钟楼文化体验馆参观流线
乘客进站流线
售票流线
二层候车厅
二层辅助办公
城市之窗
首层候车厅
售票及行包储存
出站及城市文化体验空间
钟楼文化体验馆
技术设备及办公
功能分解图
Functional breakdown diagram
正立面 Front view
侧立面 Side elevation
1-1剖面图 1-1 Sectional view
2-2剖面图 2-2 Sectional view

丝路驿站

Stop and Admire of the Silk Road

参赛学校 University/College

苏州科技大学

Suzhou University of Science and Technology, China

指导教师 Supervisor(s)

蔡新江 CAI Xinjiang

张 芳 ZHANG Fang

参赛学生 Participant(s)

闫亚琳 YAN Yalin

张雨晨 ZHANG Yuchen

韦婷婷 WEI Tingting

都晗琦 DU Hanqi

祖润亚 ZU Runya

简介 Description

该铁路客运站房位于丝绸之路沿边城镇，建筑面积约为 3000 平方米，主要承重结构为大尺寸柱子，以营造较大空间，不仅如此，柱与屋顶有机结合，带来连绵贯通的空间感受。屋顶主要采用膜结构，具有一定的透光率，白天可减少照明强度和时间，能很好地节约能源，也不易附着灰尘，常年保持外观的洁净。造型为流线型，体现丝绸的顺滑之感，又具有韵律和层次感的动态，兼具时尚的美感，使过路者感受到中国古今文化融合的魅力。

The railway station is located in the town along the Silk Road, with a construction area of about 3000 square meters. The main load-bearing structure consists of large columns to create a broader space. In addition to that, the columns are organically combined with the roof to arouse a feeling of stretching space. The roof mainly adopts a membrane structure with a certain light transmittance to reduce the lighting intensity and time in the daytime. This design saves energy well and effectively keeps away dust, so the appearance of the building looks clean all year round. The contour of the building is streamlined to resemble the smoothness of silk. The station is also of dynamic and diversified rhythms, which adds to it the beauty of modernity and allows passers-by to appreciate the charming integration of ancient and modern Chinese culture.

丝路驿站

Stop and Admire of the Silk Road

兰州

总平面图 General layout

设计说明 Design description

该铁路客运站房位于丝绸之路沿边城镇，建筑面积约为3000㎡，主要承重结构为大尺寸柱子，以营造较大空间，不仅如此，柱与屋顶有机结合，带来连绵贯通的空间感受。屋顶主要采用膜结构，具有一定的透光率,白天可减少照明强度和时间,能很好地节约能源，也不易附着灰尘，常年保持外观的洁净。造型为流线型，体现丝绸的顺滑之感，又具有韵律和层次感的动态，兼具时尚的美感，使过路者感受到中国古今文化融合的魅力。

The railway passenger station building is located in the town along the Silk Road, with a construction area of about 3000 square meters. The main bearing structure is large-size columns to create a larger space. Not only that, the columns are organically combined with the roof to bring a continuous spatial feeling. The roof mainly adopts membrane structure, which has a certain light transmittance. In the daytime, the lighting intensity and time can be reduced, which can save energy well. And it is not easy to attach dust. It keeps the appearance clean all the year round. The shape is streamlined, which reflects the smooth feeling of silk. It also has the dynamic sense of rhythm and hierarchy, which reflects fashionable beauty. This makes passers-by feel the charm of the integration of ancient and modern Chinese culture.

一层平面图 One floor plan 1:200

二层平面图 Two-story floor plan 1:200

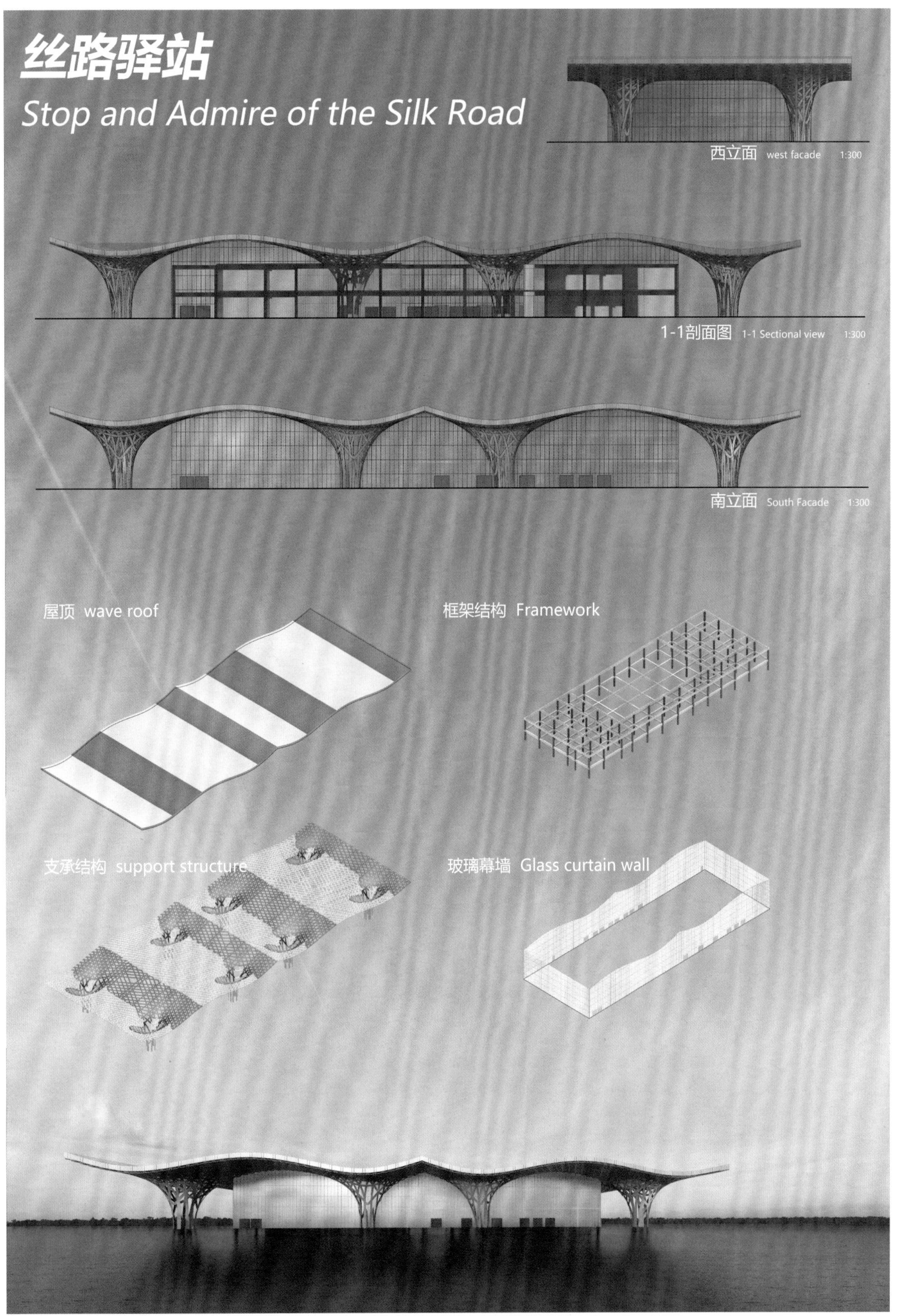
丝路驿站
Stop and Admire of the Silk Road
西立面 west facade 1:300
1-1剖面图 1-1 Sectional view 1:300
南立面 South Facade 1:300
屋顶 wave roof
框架结构 Framework
支承结构 support structure
玻璃幕墙 Glass curtain wall

回到切尔诺贝利

Back to Chernobyl

参赛学校 University/College
苏州科技大学
Suzhou University of Science and Technology, China

指导教师 Supervisor(s)
冯　进　FENG Jin
董志国　DONG Zhiguo

参赛学生 Participant(s)
林嘉俊　LIN Jiajun
孙亦心　SUN Yixin
朱　璇　ZHU Xuan
黄文琪　HUANG Wenqi
孙　心　SUN Xin

简介 Description

方案选取切尔诺贝利这个特殊的场地，以光锥为概念，设计可观景的车站，警示人们注意核安全，避免核事故再次发生。上部代表“未来”，随着双螺旋梯人们可以俯瞰全景。下半部分代表“过去”，设置候车厅。设计过程中使用了 grasshopper 全参数化建模以便调整模型形态，多处建筑形态采用函数控制（三叶玫瑰曲线函数，双曲函数等）使得建筑形态复杂且可控。同时，对异型形态结构的处理也使用了全参数化建模的方式，从而使结构能够完美契合异型建筑。

The project selects the special site of Chernobyl to design a station that allows for viewing by introducing the concept of light cone to warn people of nuclear safety to avoid the recurrence of nuclear accidents. The upper part represents "the future" where people could see the panorama on a double spiral staircase. The lower part represents "the past" with the waiting hall. In the design process, Grasshopper full parametric modeling was used to adjust the model form. Many architectural forms are controlled by functions (three-leaf rose curve function, hyperbolic function, etc.), which make them complicated and controllable. At the same time, the full parametric modeling method was also adopted to deal with the abnormal-shaped structure so that the structure can perfectly fit the building in abnormal shapes.

BACK TO CHERNOBYL I

—— New Yaniv Train Station design based on humanistic care

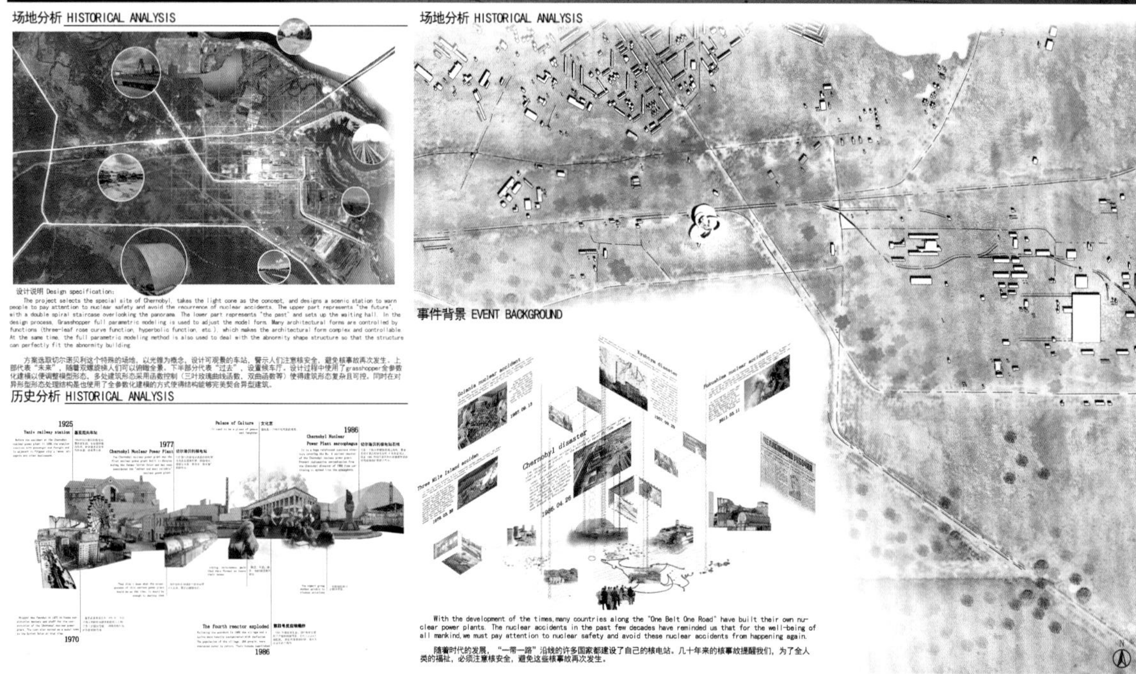

场地分析 HISTORICAL ANALYSIS
场地分析 HISTORICAL ANALYSIS
设计说明 Design specification:
The project selects the special site of Chernobyl, takes the light cone as the concept, and designs a scenic station to warn people to pay attention to nuclear safety and avoid the recurrence of nuclear accidents. The upper part represents "the future", with a double spiral staircase overlooking the panorama. The lower part represents "the past" and sets up the waiting hall. In the design process, Grasshopper full parametric modeling is used to adjust the model form. Many architectural forms are controlled by functions (three-leaf rose curve function, hyperbolic function, etc.), which makes the architectural form complex and controllable. At the same time, the full parametric modeling method is also used to deal with the abnormity shape structure so that the structure can perfectly fit the abnormity building.
方案选取切尔诺贝利这个特殊的场地，以光锥为概念，设计可观景的车站，警示人们注意核安全，避免核事故再次发生。上部代表“未来”，随着双螺旋梯人们可以俯瞰全景，下半部分代表“过去”，设置候车厅。设计过程中使用了grasshopper全参数化建模以便调整模型形态。多处建筑形态采用函数控制（三叶玫瑰曲线函数、双曲函数等）使得建筑形态复杂且可控。同时在对异形型形态处理结构是也使用了全参数化建模的方式使得结构能够完美契合异型建筑。
历史分析 HISTORICAL ANALYSIS
1925
Yaniv railway station
1977
Chernobyl Nuclear Power Plant
切尔诺贝利核电站
Palace of Culture
文化宫
1986
Chernobyl Nuclear Power Plant sarcophagus
切尔诺贝利核电站石棺
The fourth reactor exploded
1970
1986
事件背景 EVENT BACKGROUND
Chernobyl disaster
1986.04.26
With the development of the times,many countries along the "One Belt One Road" have built their own nuclear power plants. The nuclear accidents in the past few decades have reminded us that for the well-being of all mankind,we must pay attention to nuclear safety and avoid these nuclear accidents from happening again.
随着时代的发展，“一带一路”沿线的许多国家都建设了自己的核电站。几十年来的核事故提醒我们，为了全人类的福祉，必须注意核安全，避免这些核事故再次发生。

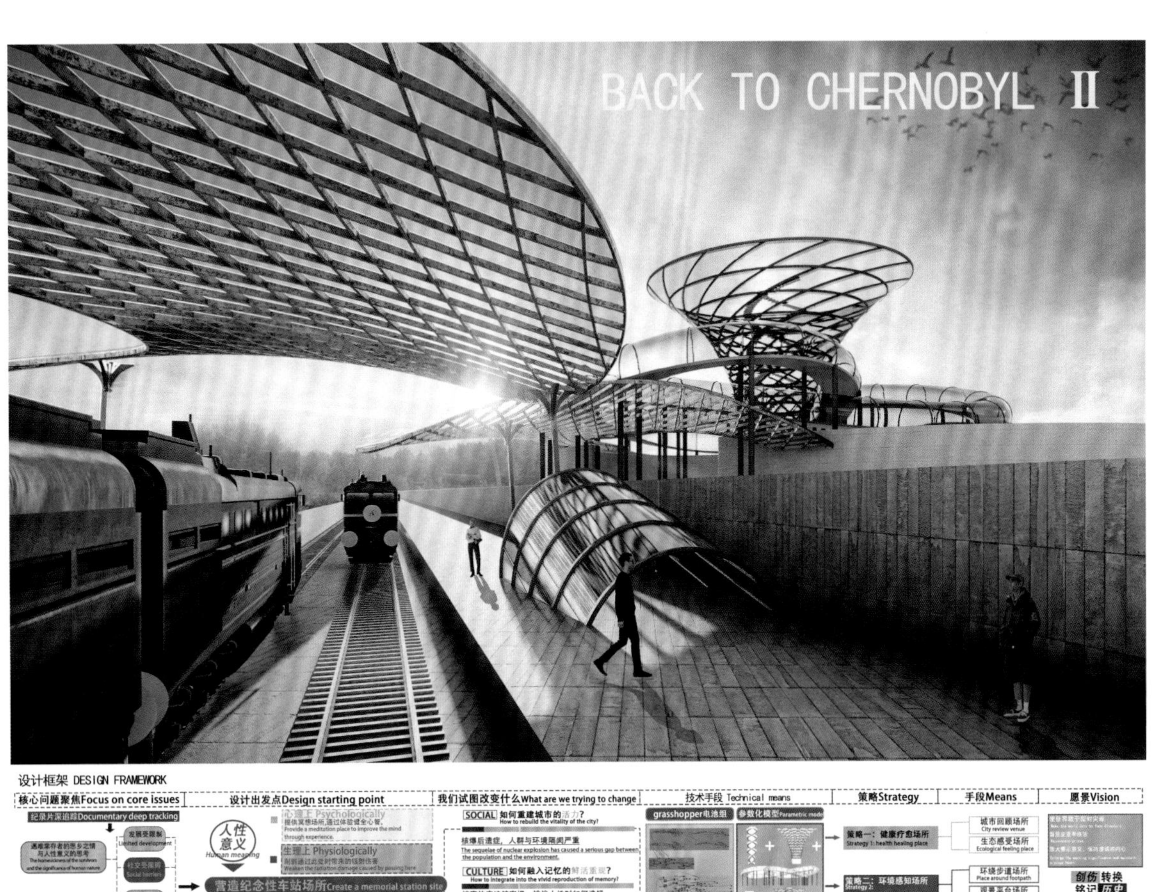
BACK TO CHERNOBYL Ⅱ

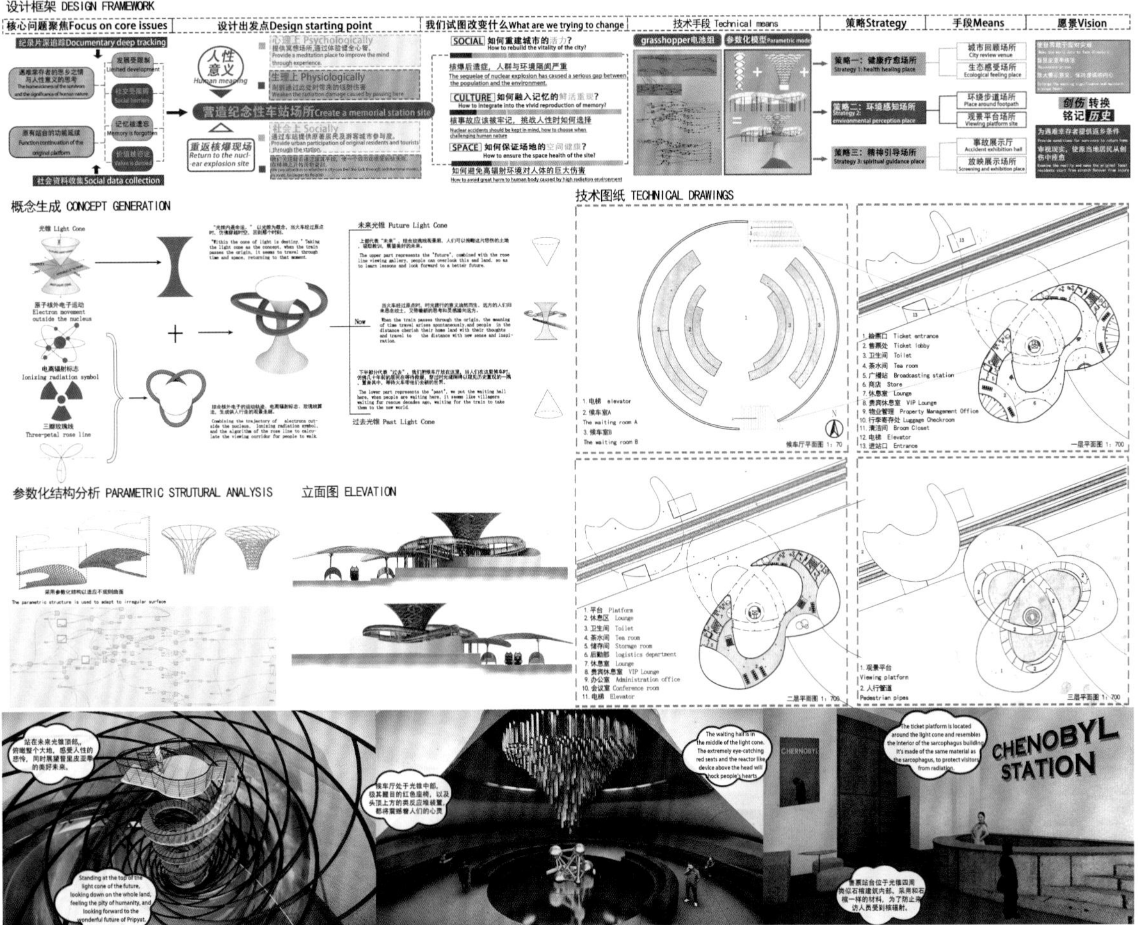
设计框架 DESIGN FRAMEWORK
概念生成 CONCEPT GENERATION
技术图纸 TECHNICAL DRAWINGS
参数化结构分析 PARAMETRIC STRUTURAL ANALYSIS
立面图 ELEVATION
CHENOBYL STATION

铁路集线器——偶然发生的相遇

Incidental Hub

参赛学校 University/College
天津城建大学
Tianjin Chengjian University, China

指导教师 Supervisor(s)
万　达 WAN Da
谭立峰 TAN Lifeng

参赛学生 Participant(s)
黄　琳 HUANG Lin
胡雅琪 HU Yaqi
张如意 ZHANG Ruyi
丁　楠 DING Nan
汪雨婕 WANG Yujie

简介 Description

建筑是城市环境的一部分，本次的设计方案以 incidental hub 为主题，旨在以客运站为载体，建立一个城市发生空间，约旦人民重视社交礼仪，设计通过提供更多偶然性的空间，来加强人与人之间的交流与联系。首先设计提取了约旦建筑风格中的拱顶元素，再对拱元素进行一系列的变化，运用到建筑立面和平面上，形成拾级而上的空间形式。通过这次设计，希望提升在快节奏的生活中人们的社交幸福感，促进城市发展。

As part of the urban environment, the design of Incidental Hub aims at presenting a railway station to create an urban space. Jordanian people value social etiquette, so the design provides more incidental space to increase communication and tighten the ties between people. The arch element in Jordanian architectural style was extracted in the first place, based on which a series of changes were allowed on the facade and plane so that the spatial form became ascending stairs. Through this design, we hope to enhance the social happiness of people living a fast-paced life and promote urban development.

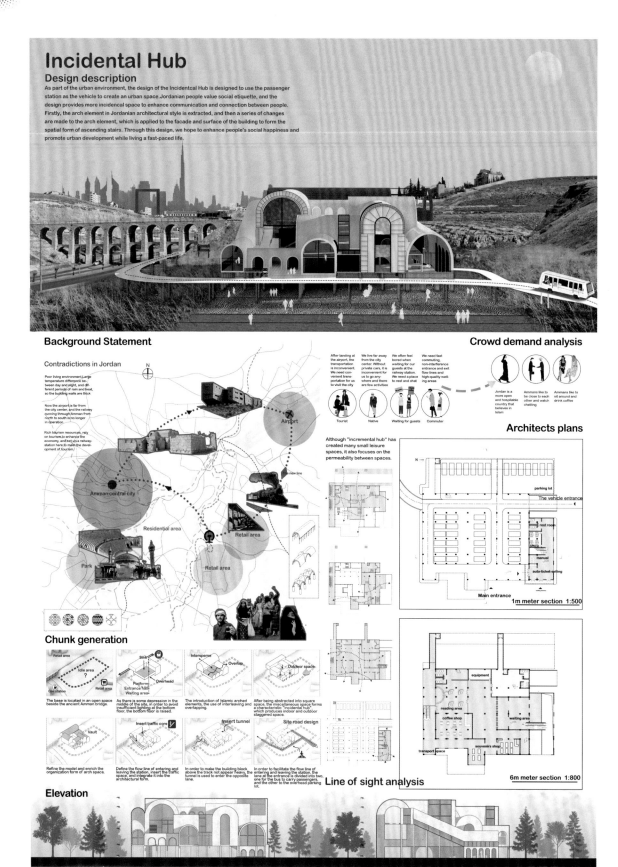

Incidental Hub

Design description

As part of the urban environment, the design of the Incidentcal Hub is designed to use the passenger station as the vehicle to create an urban space.Jordanian people value social etiquette, and the design provides more incidencal space to enhance communication and connection between people. Firstly, the arch element in Jordanian architectural style is extracted, and then a series of changes are made to the arch element, which is applied to the facade and surface of the building to form the spatial form of ascending stairs. Through this design, we hope to enhance people's social happiness and promote urban development while living a fast-paced life.

Background Statement

Contradictions in Jordan

Poor living environment.Large temperature difference between day and night, and different periods of rain and heat, so the building walls are thick

Now the airport is far from the city center, and the railway running through Amman from north to south is no longer in operation

Rich tourism resources, rely on tourism to enhance the economy, and set up a railway station here to meet the development of tourism

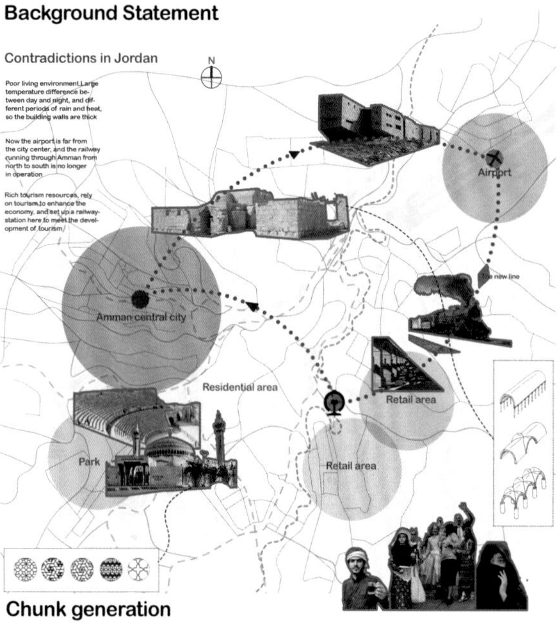

N
Airport
Amman central city
Residential area
Retail area
Retail area
Park

Crowd demand analysis

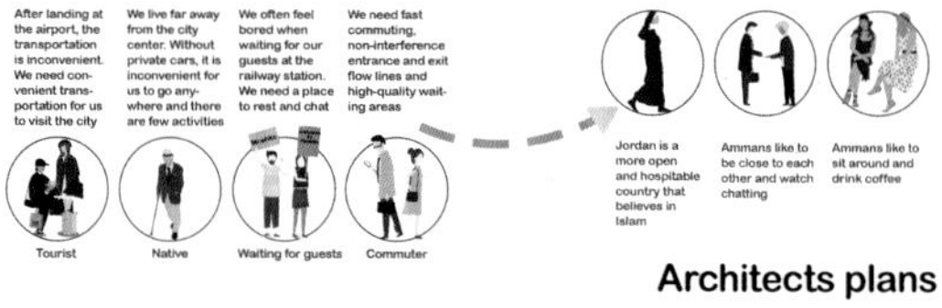

After landing at the airport, the transportation is inconvenient. We need convenient transportation for us to visit the city — Tourist

We live far away from the city center. Without private cars, it is inconvenient for us to go anywhere and there are few activities — Native

We often feel bored when waiting for our guests at the railway station. We need a place to rest and chat — Waiting for guests

We need fast commuting, non-interference entrance and exit flow lines and high-quality waiting areas — Commuter

Jordan is a more open and hospitable country that believes in Islam

Ammans like to be close to each other and watch chatting

Ammans like to sit around and drink coffee

Architects plans

Although "incremental hub" has created many small leisure spaces, it also focuses on the permeability between spaces.

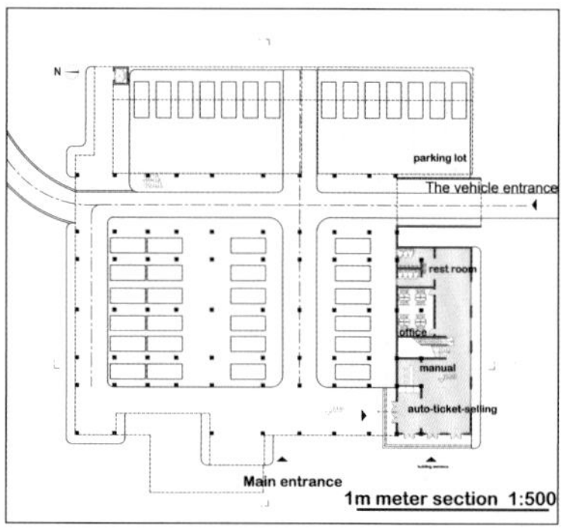

N
parking lot
The vehicle entrance
rest room
manual
auto-ticket-selling
Main entrance
1m meter section 1:500

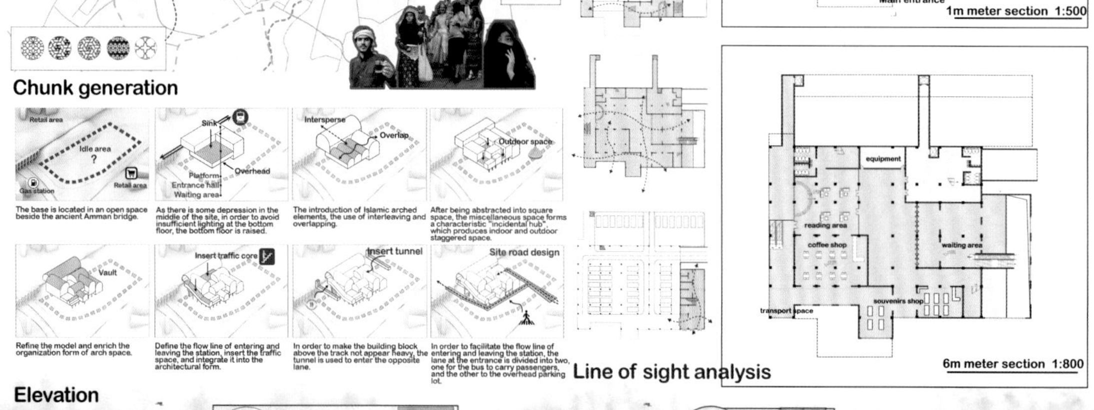

Chunk generation
Retail area
Idle area
?
Gas station
Retail area
The base is located in an open space beside the ancient Amman bridge.
Sink
Overhead
Platform
Entrance hall
Waiting area
As there is some depression in the middle of the site, in order to avoid insufficient lighting at the bottom floor, the bottom floor is raised.
Intersperse
Overlap
The introduction of Islamic arched elements, the use of interleaving and overlapping.
Outdoor space
After being abstracted into square space, the miscellaneous space forms a characteristic "incidental hub", which produces indoor and outdoor staggered space.
Vault
Refine the model and enrich the organization form of arch space.
Insert traffic core
Define the flow line of entering and leaving the station, insert the traffic space, and integrate it into the architectural form.
Insert tunnel
In order to make the building block above the track not appear heavy, the tunnel is used to enter the opposite lane.
Site road design
In order to facilitate the flow line of entering and leaving the station, the lane at the entrance is divided into two, one for the bus to carry passengers, and the other to the overhead parking lot.
equipment
reading area
coffee shop
waiting area
souvenirs shop
transport space
Line of sight analysis
6m meter section 1:800

Elevation

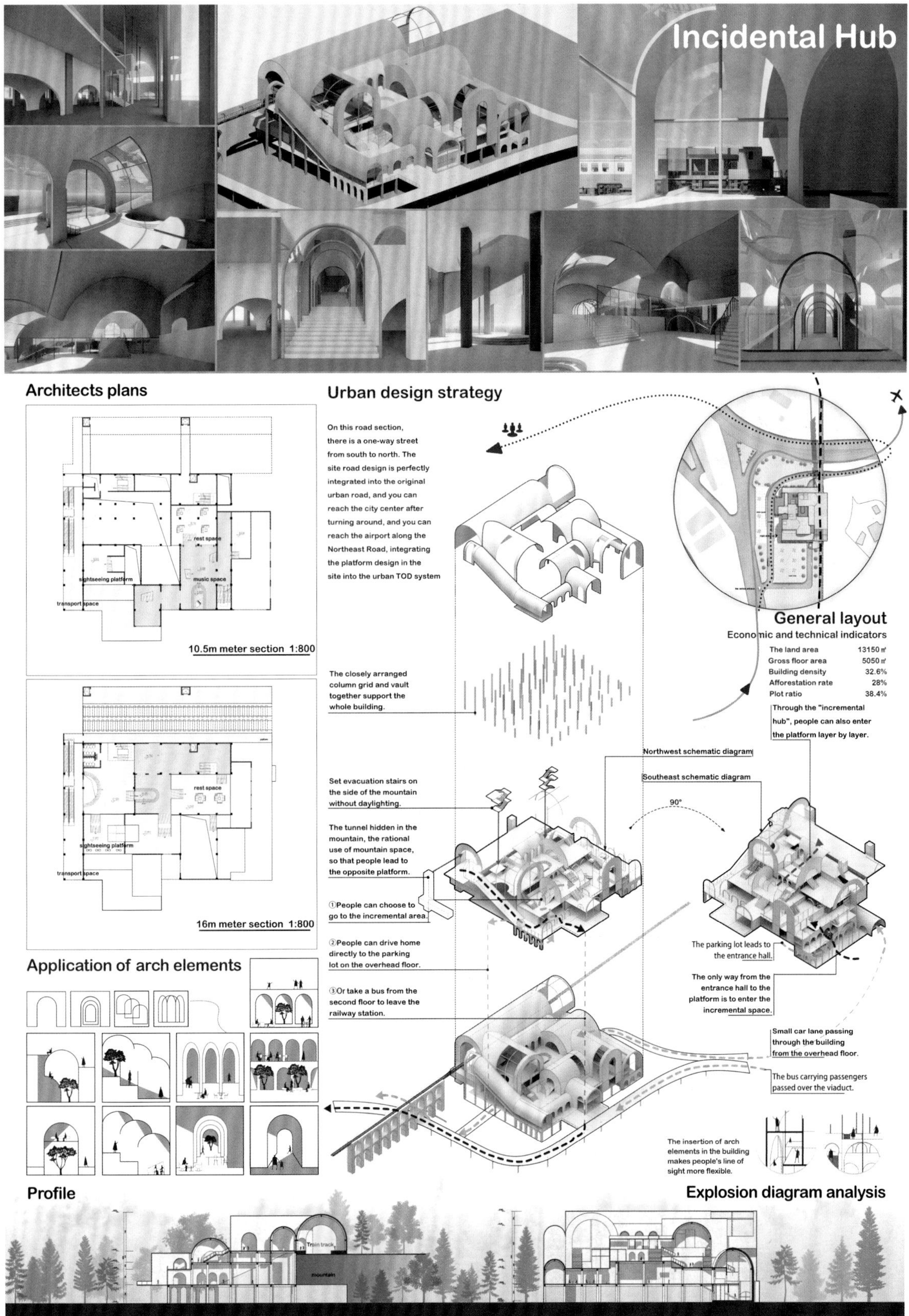
Incidental Hub
Architects plans
rest space
sightseeing platform
music space
transport space
10.5m meter section 1:800
16m meter section 1:800
Urban design strategy
On this road section, there is a one-way street from south to north. The site road design is perfectly integrated into the original urban road, and you can reach the city center after turning around, and you can reach the airport along the Northeast Road, integrating the platform design in the site into the urban TOD system
The closely arranged column grid and vault together support the whole building.
Set evacuation stairs on the side of the mountain without daylighting.
The tunnel hidden in the mountain, the rational use of mountain space, so that people lead to the opposite platform.
①People can choose to go to the incremental area.
②People can drive home directly to the parking lot on the overhead floor.
③Or take a bus from the second floor to leave the railway station.
General layout
Economic and technical indicators
The land area 13150㎡
Gross floor area 5050㎡
Building density 32.6%
Afforestation rate 28%
Plot ratio 38.4%
Through the "incremental hub", people can also enter the platform layer by layer.
Northwest schematic diagram
Southeast schematic diagram
90°
The parking lot leads to the entrance hall.
The only way from the entrance hall to the platform is to enter the incremental space.
Small car lane passing through the building from the overhead floor.
The bus carrying passengers passed over the viaduct.
The insertion of arch elements in the building makes people's line of sight more flexible.
Application of arch elements
Profile
Explosion diagram analysis
Train track
mountain

科尔维尔车站的原木模块化设计

Modular Timber Coalville Station

参赛学校 University/College

英国东伦敦大学

University of East London, the UK

指导教师 Supervisor(s)

Fulvio Wirz

参赛学生 Participant(s)

Manali Siddhesh Walkar

Riyad Hossain

Gunjankumar Jaysinh Barot

Mourtada Baboukari

简介 Description

这座建筑背后的设计理念是模块化设计、迅速施工和可持续性建筑。建筑材料包括木材、亚克力板、玻璃和面板。火车不断地进出往来于车站，因此在设计中采用这种流动的形式，将其应用于建筑形状，使其具有流体结构特点。这种流体结构形状有助于分散负载，保障结构稳定性。

车站周围景观也采用车站建筑的模块化设计，并将其扩展至对休息区和通道的设计。休息区采用了绿色空间设计，能为附近地区提供部分的凉爽环境。

The entire concept behind this building is modular, rapid and sustainable construction.

Building materials include wood, acrylic sheets, glass and panels. Railways are constantly arriving and departing and adopting that flow of trains has been applied on the shape of the building to make it fluid. This fluid shape helps to distribute load and provide structural stability.

The surrounding landscape of the station also adopted the module of the building and repeated that into sitting areas and pathways. Those sitting areas are designed with green space to createa temporary climate condition which provides a cool environment to the nearby areas.

Masterplan

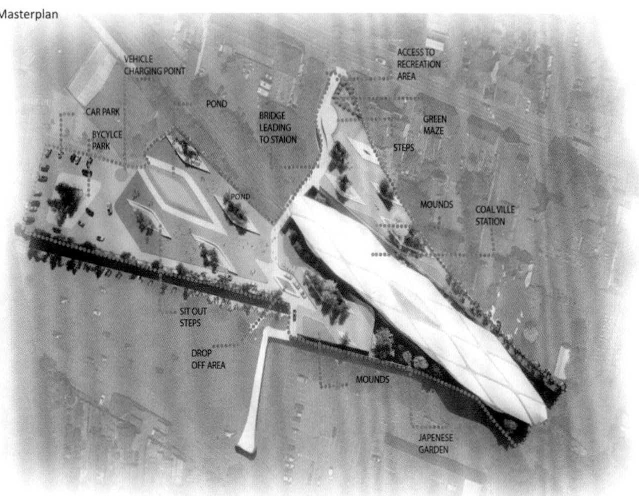

Coalville is a town in the Leicestershire county of England. It is approximately 100 miles to the northwest of London. The town serves a population of 31900 persons, covering around 108 sq. miles of area. As the name suggests, Coalville was best known for coal – mining. There were many quarries and industries started back in the 1800s. This is what led to the rapid industrialization of the town.

Although it was never a tourist center, there have been many interesting developments lately to cater to the touristy need. Inspite of all the developments there has been a major infrastructure lag. The railway line passing by the coalville town is the East Middle Lands railway line and has stations in Leicester and Long Eaton. Coalville lies between the 2 stations. Inspite of the historic importance of the town which is now further developed as a touristic centre, it still does not have a main line railways station. In order to improve the visitor experience, and travel ease to the habitants, we propose the Coalville Railway Station.

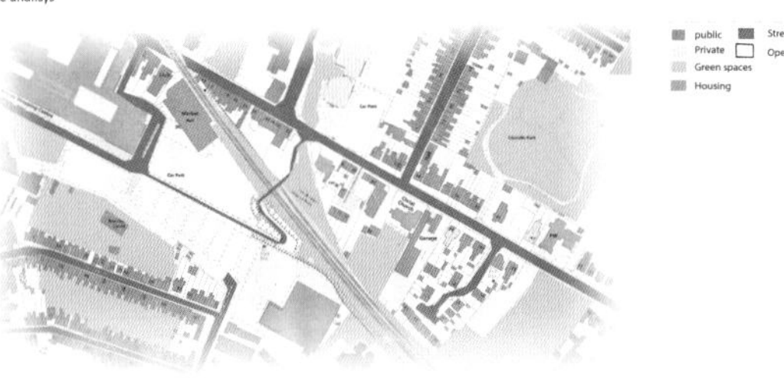

Site analysis

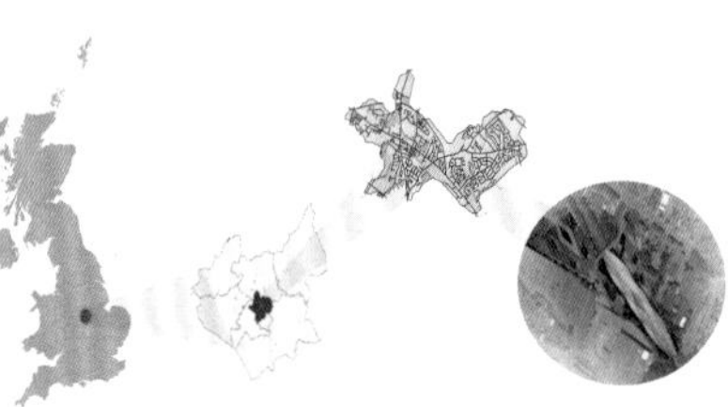

Important points of interest around Coalvile and surrounding cities

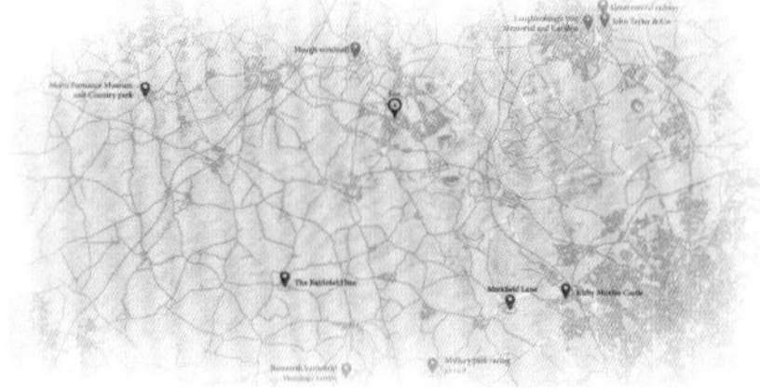

Mobility and transport

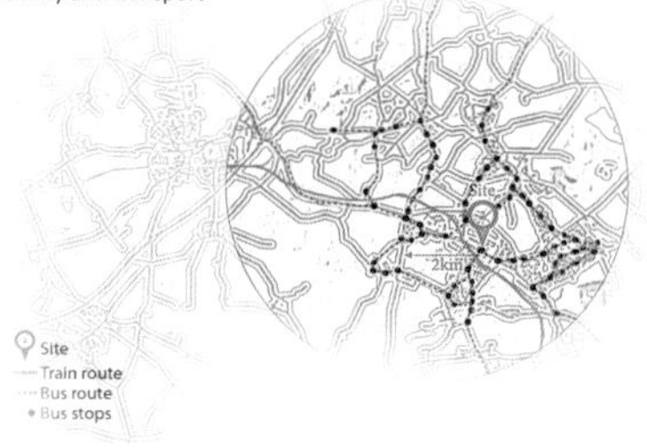

Design Goals:
The station aims to provide a first-class travel experience to the users.
The project also aims to make a significant rise in visitors.
This will help in generating revenue for the town which can be used for further developments.
We want to provide a break-out space surrounding the station building, along with the right circulation for private and public vehicles and drop-off areas.
We have also catered to the path of sustainability by using the right building materials, controlling the carbon footprint, and introducing sustainable travel programs.

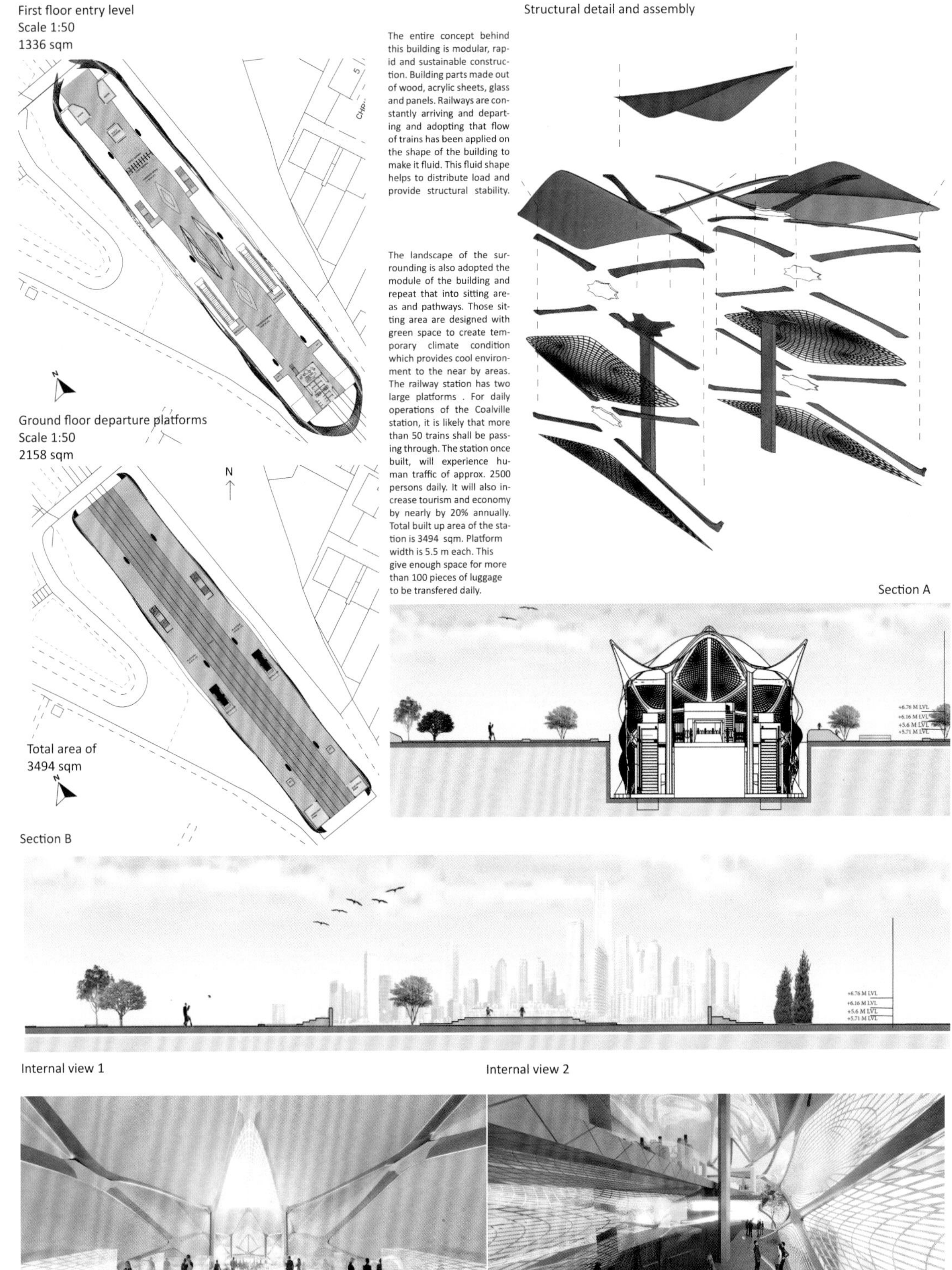

The entire concept behind this building is modular, rapid and sustainable construction. Building parts made out of wood, acrylic sheets, glass and panels. Railways are constantly arriving and departing and adopting that flow of trains has been applied on the shape of the building to make it fluid. This fluid shape helps to distribute load and provide structural stability.

The landscape of the surrounding is also adopted the module of the building and repeat that into sitting areas and pathways. Those sitting area are designed with green space to create temporary climate condition which provides cool environment to the near by areas. The railway station has two large platforms . For daily operations of the Coalville station, it is likely that more than 50 trains shall be passing through. The station once built, will experience human traffic of approx. 2500 persons daily. It will also increase tourism and economy by nearly by 20% annually. Total built up area of the station is 3494 sqm. Platform width is 5.5 m each. This give enough space for more than 100 pieces of luggage to be transfered daily.

数字化结构设计

Category B: Structural Design

跨海峡两用立式浮桥

Vertical Channel-crossing Pontoon Bridge

参赛学校 University/College

安徽建筑大学

Anhui Jianzhu University, China

指导教师 Supervisor(s)

周　宇　ZHOU Yu

马　巍　MA Wei

参赛学生 Participant(s)

吴飞羊　WU Feiyang

叶俊瑶　YE Junyao

王迅周　WANG Xunzhou

李彦锴　LI Yankai

孙　文　SUN Wen

简介 Description

本作品以悬浮轨道为灵感，将桥隧上下双层交通与动力舟相结合，创新出下挂浮管隧道的装配式动力舟桥，并在其单元体的基础上进行优化设计，实现桥隧互通、桥隧合并、桥面通车、隧道通轨等多种形式的桥梁结构以满足各类交通需要。该桥能为我国海峡地区经济文化交流提供交通基础设施支撑，有着极为重要的战略意义和十分广阔的应用前景。

This work draws inspiration from floating tracks and combines a multi-level bridge or tunnel with a power-driven vessel. The innovative product is a modular power-driven bridge vessel with floating tube tunnels installed on its top and bottom. Each unit of the vessel is optimized to achieve various forms of bridge structure like bridge-tunnel connection, bridge-tunnel integration, traffic on the bridge deck and railways laid through the tunnel to meet different traffic needs. This bridge serves as an infrastructure supporting the economic and cultural exchanges near and across the strait, which has great strategic significance and wide prospect for application.

二等奖 >> Second Prize

荔樟丝梦

Libo Dream-Silk Bridge

参赛学校 University/College
北京建筑大学
Beijing University of Civil Engineering and Architecture, China

指导教师 Supervisor(s)
胡梦涵 HU Menghan
龙佩恒 LONG Peiheng

参赛学生 Participant(s)
孟斯超 MENG Sichao
万　幸 WAN Xing
马仕杰 MA Shijie
吴世宇 WU Shiyu
郭明暄 GUO Mingxuan

简介 Description

荔樟丝梦特大桥，一座拟建于贵州省荔波县樟江之上的铁路斜拉桥。作为西部陆海新通道沿线——贵南高铁沿线的重要节点，它为西部大开发战略及乡村振兴战略贡献力量。本桥以丝绸之路、丝带、梦想作为意象，设计了“人字形”“丝带状”双主塔，一高一矮，体现人们携手共进的人类命运共同体理念。同时，应用 BIM 等新技术，打造数字化、智能化桥梁设计，突出安全、耐久、环保、美观、可施工、可管养的设计理念。

Lizhang Dream-Silk Bridge is a cable-stayed railway bridge proposed to be built over the Zhangjiang River in Libo County, Guizhou Province in China. As an important node along the Guinan high-speed railway along the new western land-sea corridor, it contributes to the strategy of developing western regions and the rural revitalization strategy. Based on the images of the Silk Road, ribbon and the dream, the bridge has designed double main towers in the shape of "herringbone" and "ribbon", one of which is high and the other is low, jointly reflecting the concept of a community with a shared future for mankind. New technologies such as BIM were applied to achieve a digital and intelligent bridge design, highlighting the design concepts of safety, durability, environmental protection, beauty and allowing for modification and maintenance.

丝绸之桥

Bridge of Silk

参赛学校 University/College

苏州科技大学

Suzhou University of Science and Technology, China

指导教师 Supervisor(s)

冯　进　FENG Jin

董志国　DONG Zhiguo

参赛学生 Participant(s)

冯卓远　FENG Zhuoyuan

吴亦风　WU Yifeng

陆毅涵　LU Yihan

陈志同　CHEN Zhitong

刘启航　LIU Qihang

简介 Description

设计尝试探索双向渐进结构拓扑优化算法（BESO）和湿网格算法的结合，并设计出承载展览空间、高铁轨道的创新型结构的桥梁。设计的功能框架定位为展览空间并以承载丝绸之路历史文化、“一带一路”为主题。利用湿网格可以仿刻拓扑结构的特点，设计将 BESO 计算得到的拓扑结构节点和功能吸引点结合在一起，并按照结构和功能两者的权重比例干扰湿网格，最终生成得到兼具结构拓扑优化和功能路径拓扑优化的桥梁设计，使得功能与形式统一。

This design attempts to explore the combination of Bi-directional Evolutionary Structural Optimization (BESO) and the Wool-Thread Algorithm to build a bridge with an innovative structure that includes an exhibition space and railway tracks. The major function designed for the bridge is a space for exhibitions about the history and culture of the Silk Road and the “Belt and Road” Initiative. Wool-Thread can imitate the features of the topological structures generated by the calculations by BESO and combine them with functional attraction points. On top of that, the Wool-thread would be interfered with according to the weight ratio of structure and function to coordinate optimized topologies of structure and function and achieve integration of form and utility.

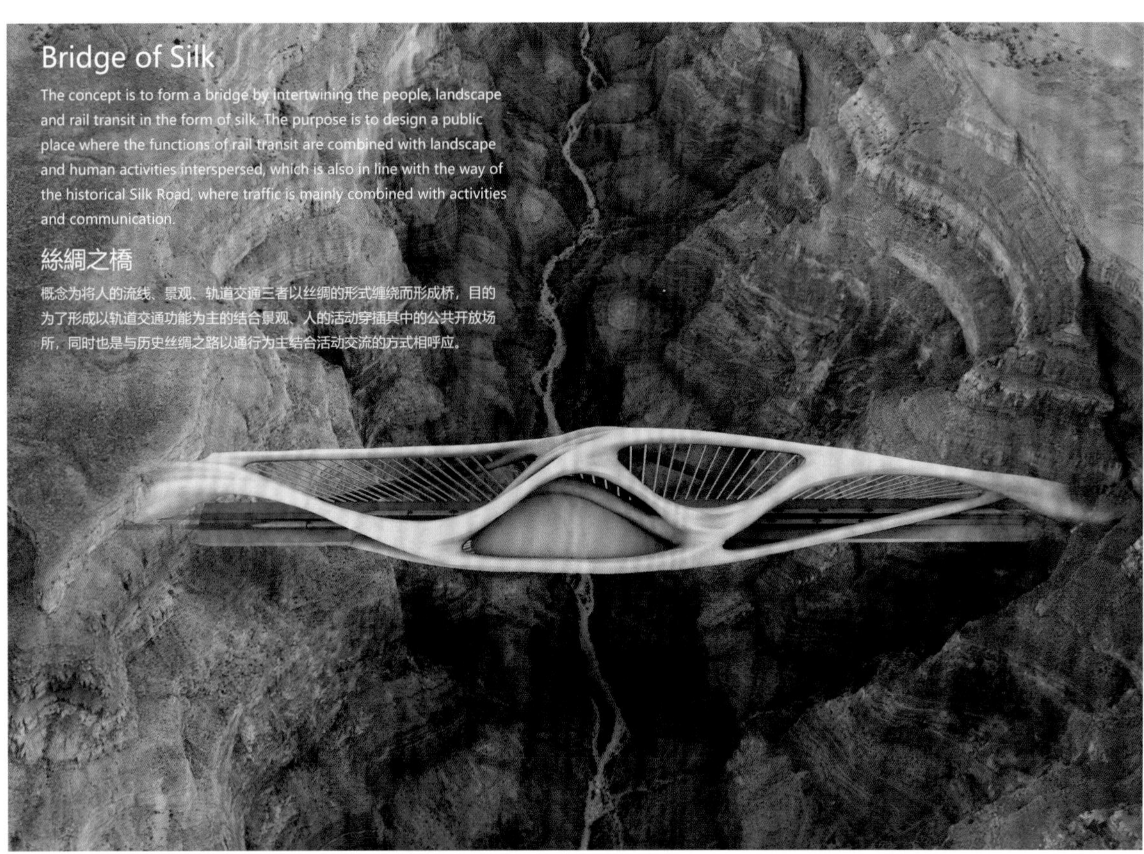
Bridge of Silk
The concept is to form a bridge by intertwining the people, landscape and rail transit in the form of silk. The purpose is to design a public place where the functions of rail transit are combined with landscape and human activities interspersed, which is also in line with the way of the historical Silk Road, where traffic is mainly combined with activities and communication.
絲綢之橋
概念为将人的流线、景观、轨道交通三者以丝绸的形式缠绕而形成桥，目的为了形成以轨道交通功能为主的结合景观、人的活动穿插其中的公共开放场所，同时也是与历史丝绸之路以通行为主结合活动交流的方式相呼应。

Innovative Structure Design based on the Combination of Topological Optimization and Wool-Thread Algorithm
This paper attempts to explore the combination of Bi-directional Evolutionary Structural Optimization (BESO) and the Wool-Thread Algorithm, which helped to design an innovative structure bridge with the function of exhibition and the high-speed rail track. The content mainly involves two different processes, which are using BESO to generate reasonable topological structure nodes of the bridge and combining architecture functional attraction points to generate optimal topological paths with wool-thread algorithm. The characteristics of topological structure can be imitated by wool-thread, so that the design combines the topological structure nodes and functional attraction points calculated together. The wool-thread is disturbed by the weight ratio of both structure and function, and finally a bridge design with both structural topology and functional path topology is generated.
基于双向渐进结构拓扑优化和湿网格算法的创新型结构桥梁设计
设计尝试探索双向渐进结构拓扑优化算法和湿网格算法的结合，并设计出承载展览空间、高铁轨道的创新型结构的桥梁。文章主要从两个方向阐述设计过程，利用BESO生成桥梁合理的拓扑结构节点；并结合建筑功能吸引点用湿网格生成最优拓扑路径。利用湿网格可以仿刻拓扑结构的特点，设计将BESO计算得到的拓扑结构节点和功能吸引点结合在一起。并按照结构和功能两者的权重比例干扰湿网格，最终生成得到兼具结构拓扑和功能路径拓扑的桥梁设计

三等奖 >> Third Prize

冰雪之缘

Frozen

参赛学校 University/College

河北建筑工程学院

Hebei University of Architecture, China

指导教师 Supervisor(s)

李玉忠 LI Yuzhong

胡建林 HU Jianlin

参赛学生 Participant(s)

贾　琦 JIA Qi

李子晗 LI Zihan

吴依璇 WU Yixuan

胡　馨 HU Xin

蔡庆文 CAI Qingwen

简介 Description

本桥以“一带一路”和冬奥会为背景进行设计。桥址位于张家口市赤城县以北。本着安全、耐久、适用、经济和美观的原则，设计为中承式拱桥。铁路等级 I 级，双线，时速 250 千米，中活载。桥梁全长 156 米，拱肋采用劲性骨架钢管混凝土拱，拱轴线为悬链线，材料为 C55 混凝土；拱上立柱为双斜式柱，材料为 C40 混凝土；吊杆为 PES5-121 丝镀锌高强平行钢丝；纵梁为钢筋混凝土箱形梁，材料为 C45 混凝土。拱肋骨架采用悬臂法进行施工，拱肋混凝土采用分层分环法施工。

The bridge was designed in the context of the "Belt and Road" Initiative and the Winter Olympics. Located in the north of the Chicheng County, Zhangjiakou City in China, the mid-support arch bridge is compliant with the principles of safety, durability, applicability, economy and beauty. The railway is of double-track and classified as class I, with a designed speed of 250km/h and a medium live load. The bridge is 156m long and the arch rib adopts a concrete-filled steel tube skeleton arch. The arch axis is the catenary made from C55 concrete. The upright column on the arch is a double-inclined one made from C40 concrete. The boom uses PES5-121 galvanized high-strength parallel steel wire, while the beams adopt reinforced concrete box made from C45 concrete. The arch rib skeleton is constructed by the cantilever method and the arch rib concrete by the sub-section method.

STS 大桥

STS-Bridge

参赛学校 University/College
河南城建学院
Henan University of Urban Construction, China

指导教师 Supervisor(s)
屈讼昭 QU Songzhao
王　仪 WANG Yi

参赛学生 Participant(s)
秦承成 QIN Chengcheng
李泰达 LI Taida
耿传绪 GENG Chuanxu
李秀秀 LI Xiuxiu
张耀鹏 ZHANG Yaopeng

简介 Description

六三六大桥是基于“一带一路”的支援建设，为推进成员国城市化建设，设计应用于城市空间下的公轨的两用桥梁，主梁为曲线上加劲连续钢梢梁结构，双层桥面布置，上层为八车道公路，下层为双线高铁；下层桥面外侧挑臂设置有维修栈道。

六三取自2063年“一带一路”美好愿景愿各国友谊长存，六取自今年是“一带一路”倡议第六年。六和三也有“六六大顺”“三阳开泰”之意，，六三六大桥拟为连接中国与其他各国之间的纽带，寓意各国联动共同推进“一带一路”的建设。

The 636 Bridge is a double-use public railway bridge designed for roads in urban space to support the building of the “Belt and Road” Initiative and promote the urbanization of member states. The structure of the main beam is a curved reinforced continuous steel tip beam. The bridge deck is double-layered with an eight-lane highway on the upper level, with a double-line high-speed rail on the lower level; the pick arm outside the lower deck is equipped with a maintenance plank.

“63” refers to the year 2063 when the friendship of countries along the “Belt and Road” can last for long according to our aspiration, while “6” means that it is the sixth year since the “Belt and Road” Initiative was launched. In Chinese culture, “6” signifies good fortune and “3” represents peace and smoothness. We hope that the 636 Bridge could serve as the bond between China and other “Belt and Road” countries that join hands to fuel the construction of the “Belt and Road”.

中俄首座“国家团结—人民友谊”客运铁路桥

The First Passenger Railway Bridge Russia-China “Unity of Countries-Friendship of People”

参赛学校 University/College

俄罗斯莫斯科国立建筑大学

Moscow State University of Civil Engineering, Russia

指导教师 Supervisor(s)

Saltykov Ivan

参赛学生 Participant(s)

Zenkina Alisa

Kavinina Elizaveta

Papikian Karine

Zhuravleva Daria

简介 Description

我们设计了一座横跨黑龙江的铁路桥。黑龙江是俄罗斯联邦和中华人民共和国的界河。历史上，两个边境城镇（俄罗斯布拉戈维申斯克和中国黑河）位于黑龙江两岸。

铁路桥的主要设计理念是两种不同的文化的融合，体现这两种文化的相互影响，同时尽可能保留各自独特性。

在分析两种文化，寻找设计灵感时，我们决定使用两种具有象征意义的动物分别代表两个国家：龙代表中国，熊则代表俄罗斯。

从俄罗斯的文化视角来看，铁路桥需要营造出大规模的效果。因此，我们开始衡量使用胶合木材的可能性。在木工行业发达的今天，这种材料强调自然元素，也非常易得。建筑的主要特点是半拱形设计，我们给它设计了带支撑的结构。从中国的文化视角来看，设计既要实现建筑效果又要达到化繁为简的效果，因此我们选择了钢筋混凝土，并应用钢缆作为主要元素。

We designed railway bridge across Amur River. The Amur River is the water border between the Russian Federation and the People's Republic of China. Historically, two border towns (Russian-Blagoveshchensk; Chinese-Heihe) have been located in one region on opposite banks of the river.

The main idea of the bridge is the idea of merging two different cultures, how they influence each other while remaining as original as possible.

When analyzing two cultures and searching for images, we decided to take two symbolic animals, each of which would be associated with a separate country. So the bear became an allegory for Russia, and the dragon became China.

On the part of Russia, it was necessary to create the effect of massiveness. So we turned our eyes to the possibilities of glued wood, which is very easy to get in view of the developed woodworking industry and which emphasizes with nature. The main point in the construction of the semi-arch was that we set it as a farm with braces. On the part of China, in turn, the design should meet the architectural effect and achieve the effect of simplifying the complexity. Reinforced concrete was also chosen as the material, and the main elements of the structure were cables.

丝路红环

Red Ring Bridge

参赛学校 University/College

沈阳建筑大学

Shenyang Jianzhu University, China

指导教师 Supervisor(s)

王子一 WANG Ziyi

周　烨 ZHOU Ye

参赛学生 Participant(s)

李大超 LI Dachao

沈欣元 SHEN Xinyuan

宣景庆 XUAN Jingqing

赵洛愉 ZHAO Luoyu

KULIKOVA ALINA

简介 Description

“丝路红环”铁路桥址选在古丝绸之路的起点西安。其结构类型为双塔双索面预应力混凝土斜拉桥。“丝路红环”的两个索塔的外观为椭圆环状，两个椭圆环的中心在一条水平线上，中心索塔颜色主调为中国红，寓意着中国愿与“一带一路”沿线国家和地区齐心协力地打破发展瓶颈，缩小发展差距，共享发展成果，打造甘苦与共、命运相连的发展共同体。

The “Red Ring” railway bridge is located in Xi’an, China, the starting point of the ancient Silk Road. Its structure is a prestressed concrete cable-stayed bridge with two towers and two cable planes. The two towers of the “Red Ring Bridge” look like an elliptical ring, the centers of which are on a horizontal line. The central tower is in red, symbolizing China. We use this to express our hope that China is willing to work together with countries and regions along the “Belt and Road” to eliminate bottlenecks, narrow gaps and share the fruits of development to create a community facing common difficulties with a shared destiny.

丝路之脉

The Vein of the Silk Road

参赛学校 University/College

苏州科技大学

Suzhou University of Science and Technology, China

指导教师 Supervisor(s)

王大鹏 WANG Dapeng

蔡新江 CAI Xinjiang

参赛学生 Participant(s)

钱　鹏 QIAN Peng

朱　敏 ZHU Min

赖　一 LAI Yi

刘玉悦 LIU Yuyue

钟　鹏 ZHONG Peng

简介 Description

拟建“丝路之脉”大桥位于中巴铁路线路旁的一段，设定该线路为高速铁路路线，桥址位于铁路墨玉站附近，考虑为进出站线路，有多条线路已开始交汇。

桥上部结构形式采用三跨预应力混凝土连续钢构，主跨跨度 115 米。主桥跨径布置为，总长 244 米。选用箱型截面，主桥纵桥向结构布置形式采用不等跨变截面。桥墩选用竖直双肢薄壁墩。基础采用预制装配式桩基础。

主桥两侧安装不锈钢镜面吸引复合板，组成丝绸状飘带形式，融合环境，契合主题。

The proposed bridge named "The Vein of the Silk Road" is a high-speed railway bridge next to the China-Pakistan railway line near the Moyu Station. It has both entry and exit lines that many of which have begun to intersect.

The upper structure of the bridge is a three-span prestressed concrete continuous rigid frame with a main span of 115m. The span of the bridge body is 244m. A box section is selected and the longitudinal structure of the main bridge is unequal span and variable cross section. The bridge piers are vertical double-limb thin-walled piers. The foundation adopts prefabricated piles.

Stainless steel mirrors are to be installed on both sides of the main bridge to attach composite panels, composing a structure like silk, consistent with the environment and highlighting the theme.

伦敦格罗夫纳新铁路桥

New Grosvenor Railway Bridge London

参赛学校 University/College

英国东伦敦大学

University of East London, the UK

指导教师 Supervisor(s)

Ali Abbas

Arya Langroudi

参赛学生 Participant(s)

Mathura Mahadevan

Jaykumar Lakhani

Saru Prajapati

Waleed Anwar

Briston Joseph

简介 Description

我们团队最终确定了本次铁路桥建筑比赛的钢缆拱桥设计。铁路桥有两个拱形结构，在桥的中心点汇合，钢缆将桥面和拱顶连接在一起。拱顶由混凝土建成，因此，铁路桥将同时承受拱顶的压力和拉索的拉力。这种设计不会造成铁路桥扭转效应。此外，桥面由预制混凝土制成，预制混凝土在施工中具有一定的经济效益，能改善施工现场的安全和健康条件，减少建筑废料，具有抗火性，能抵御极端天气，节约总体建筑成本。我们团队融合了不同的设计理念，包括创新、美学、易于建造和施工，以及可持续发展，也兼顾安全性、适用性、环境可持续性、低碳发展和可建造性。

The team has finalized the cable arch bridge for this railway bridge competition. The bridge has two arches which will meet at a centre point. Also, the cables are connecting the deck of the bridge and the arches together. The arches are made with concrete. So the bridge will carry the compression by the arch and tension of the cables. Also, this design will not allow the torsion effect to the bridge. Further the decks are made with precast concrete. The precast concrete method will provide economy on site, improve health and safety on site, reducte waste on site, improve fire resistance, resist to weather and whole building savings. Also, the team is mainly considered the design concept for railway bridge such as an innovative, aesthetics, easy to build and implement, and sustainable solution. At the same time, the team considered other design concepts such as safety, applicability, environmental sustainability, low carbon consideration, and constructability.

绸安桥

Chouan Bridge

参赛学校 University/College
西安建筑科技大学
Xi'an University of Architecture and Technology, China

指导教师 Supervisor(s)
孙建鹏 SUN Jianpeng
门进杰 MEN Jinjie

参赛学生 Participant(s)
邢重阳 XING Chongyang
黄瑞祺 HUANG Ruiqi
郑仕豪 ZHEGN Shihao
徐伟超 XU Weichao
尹 鹏 YIN Peng

简介 Description

本设计桥梁为上承式圬工拱桥，为铁路桥梁。桥名为绸安桥，位于西安浐河之上。

气象资料：

西安市平原地区属暖温带半湿润大陆性季风气候，年平均气温 13.0~13.7℃，最冷 1 月份平均气温 -1.2~0℃，最热 7 月份平均气温 26.3~26.6℃。

基本设计资料：

桥面宽度：13.2 米；

铁路等级：高速铁路；

设计行车速度：160 千米 / 小时；

设计荷载：ZC 荷载；

线路情况：CRTSI 型双块板式无砟轨道，双线，线间距 4.6 米；

设计使用年限：正常使用条件下主体结构设计使用年限为 100 年。

The designed bridge called Chou'an (Silent Silk) Bridge is a top-loaded masonry arch bridge, which is a railroad bridge. It is on the Chanhe River in Xi'an, China.

Meteorological data:

The plain area that Xi' an belongs to has a warm temperate semi-humid continental monsoon climate. Its annual average temperature is 13.0~13.7℃. The lowest average temperature in January is -1.2~0℃ . The highest average temperature in July is 26.3~26.6℃ .

Basic information about the design:

Width of the bridge deck: 13.2m.

Class of the railway: High-speed railway.

Designed speed: 160km/h.

Designed load: ZC load.

Track situation: CRTSI, double slab, ballastless track, double line, central lines of track 4.6m.

Designed service life: 100 years for the main structure under normal conditions.

尊严之桥

Dignity Bridge

参赛学校 University/College
长安大学
Chang'an University, China

指导教师 Supervisor(s)
罗晓瑜 LUO Xiaoyu

参赛学生 Participant(s)
法瓦兹 Matanmi fawas

简介 Description

尼日利亚是一个拥有 2 亿多人口的多文化、多民族和多宗教国家。尼日利亚的前首都拉各斯州是我的家乡，同时也是非洲人口最多的城市。拉各斯非常拥挤，每天频繁的贸易往来、城市中的众多人口都增加了城市内人员和货物的交通运输需求。因此，我计划在我就读的中学附近某处设计一座铁路桥，铁路轨道垂直于一条四车道的道路。这片区域有众多的企业和行人，每当火车经过，道路就会堵塞至少十五分钟，导致学校、医院和公司等繁忙地段形成严重的交通堵塞，我们上学也经常因此迟到。修建铁路桥将大大缓解该地区的交通堵塞，确保医院救护车、校车等车辆能及时、正常通行。

第一幅图显示人们如何在铁路旁工作，他们通常在火车快要经过的时候装载货物，这对他们自己和货物都有巨大风险。铁路桥下面就是一条与铁轨垂直的四车道公路。

同时，我还计划设计一个斜拉桥，桥塔的设计灵感来自尼日利亚盾徽，特别是盾徽中两匹马所代表的“尊严”和“为国家服务”的理念。我称这座桥为“尊严之桥”。尊严桥为尼日利亚人民带来荣誉、尊重和尊严，也促进了尼日利亚的绿色文化和传统文化的发展。

Nigeria has a population of over 200 million people with different cultures, ethnicities and religions. It's former capital was "Lagos state" which happens to be my hometown and the most populated city in Africa. Lagos is a very crowded place with a lot of businesses, day in and day out. The large numbers of people in the city have increased the transportation demand of both the people and goods transported within city. I decided to design a railway bridge to somewhere close to my secondary school, where there is a railway track perpendicular a road of 4 car lanes with both small enterprises and pedestrians in both sides of the road. Whenever the train approaches, the road will be blocked for at least 15 minutes causing huge traffic jam in a busy area full of schools, hospitals, companies and so on. This has several occasions made some of my classmates come late to late in numerous times. Constructing a railway bridge will drastically reduce the traffic jam in this area and make the hospital ambulances, school buses etc. operate normally without delay.

The first picture shows how the people operate beside the railway track , they usually arrange their things whenever the train approaches, which is risky to them and their goods. Under this railway track there is a perpendicular 4 car lanes to the railway track.

Moreover, I have decided to design a cable stayed bridge, the bridge tower is designed from a philosophy behind the Nigerian coat of arm, particularly from the two horses in the coat of arm, which represent dignity and readiness to serve the country. I called the bridge "DIGNITY BRIDGE". The dignity bridge brings honor, respect and of course dignity to the people of the country, also promoting the green culture and traditional cultwre of Nigeria.

邕宁大桥

Yongning Bridge

参赛学校 University/College
长春建筑学院
Changchun University of Architecture and Civil Engineering, China

指导教师 Supervisor(s)
张广萍 ZHANG Guangping
李厚萱 LI Houxuan

参赛学生 Participant(s)
刘昊坤 LIU Haokun
陈世纯 CHEN Shichun
吴文杰 WU Wenjie
张纯贵 ZHANG Chungui
李冬雪 LI Dongxue

简介 Description

邕宁大桥坐落在广西南宁邕江之上，是衔接“一带一路”陆路南北新通道的重要组成部分。此桥为双塔无背索斜拉桥，桥梁总长 512 米，跨径组合为 80 米 +352 米 +80 米，本桥两座主塔对称分布，索塔桥面以上高度 72 米，桥面以下高度 12 米。两座主塔以丝绸之路传入的乐器箜篌为景观设计，箜篌象征着世界性的凤凰文化，且独具东方民族之美。民族的就是世界的，民族有梦想，国家有梦想，世界有梦想。有梦想，有追求，有奋斗，这将会给世界带来无限生机和美好前景。

The Yongning Bridge is on the Yongning River in the city of Nanning in Guangxi of China. It is an important part of the new north-south road connecting the "Belt and Road" overland route. The bridge is a double-pylon cable-stayed bridge with a total length of 512m and a span combination of 80+352+80m. Its two main towers are at symmetric positions, with a height of 72m above and 12m below the bridge surface. The two main towers are designed with Konghou, an instrument spread to China by the Silk Road, which is a symbol of phoenix with unique oriental beauty that can resonate with the rest of the world. National culture is an asset of the world. If nations all have dreams, a country will have dreams and the world will have dreams. Dreams, pursuits and efforts will surely bring to the world infinite vitality and a bright future.

时律桥

Infinite

参赛学校 University/College
重庆交通大学
Chongqing Jiaotong University, China

指导教师 Supervisor(s)
董莉莉 DONG Lili
徐略勤 XU Lueqin

参赛学生 Participant(s)
何雨峰 HE Yufeng
邓 琼 DENG Qiong
王 天 WANG Tian
董美馨 DONG Meixin
刘 祥 LIU Xiang

简介 Description

以人为本是本桥设计理念的核心内容，意在给居民与游客提供更快的通勤方式和更好的观景体验。绿色低碳理念贯穿设计始末，设计阶段时纳入装配式设计思路；建筑运行阶段采用摩擦式纳米发电机主动设备提供电能。时空概念也是重要的设计理念，INFINITE——时律桥两翼以流形的空间构造阐述了广义相对论的时空观。钢桁架的蓝灰桥身保证了铁路与公路并存，并延续了老桥的造型与使命，保持了桥群视觉效果的连续性，一幅“过去—现在—未来”的桥景映入眼帘。

The core of the bridge's design philosophy is a people-oriented approach, aiming at faster commuting and a better viewing experience for residents and visitors. Green, low-carbon concepts and a modular method were incorporated throughout the design. The functioning of the building uses a triboelectric nanogenerator to provide electricity. The concept of space-time is also a key, so the two wings of the INFINITE-Time Bridge illustrate the general relativistic view of space-time in manifold configuration. The blue-gray body of the steel truss ensures the coexistence of rail and road, preserves the look and mission of the old bridge and maintains a visual continuity of the bridge complex. What is in front of our eyes is a view of the “past-present-future” of the bridge.

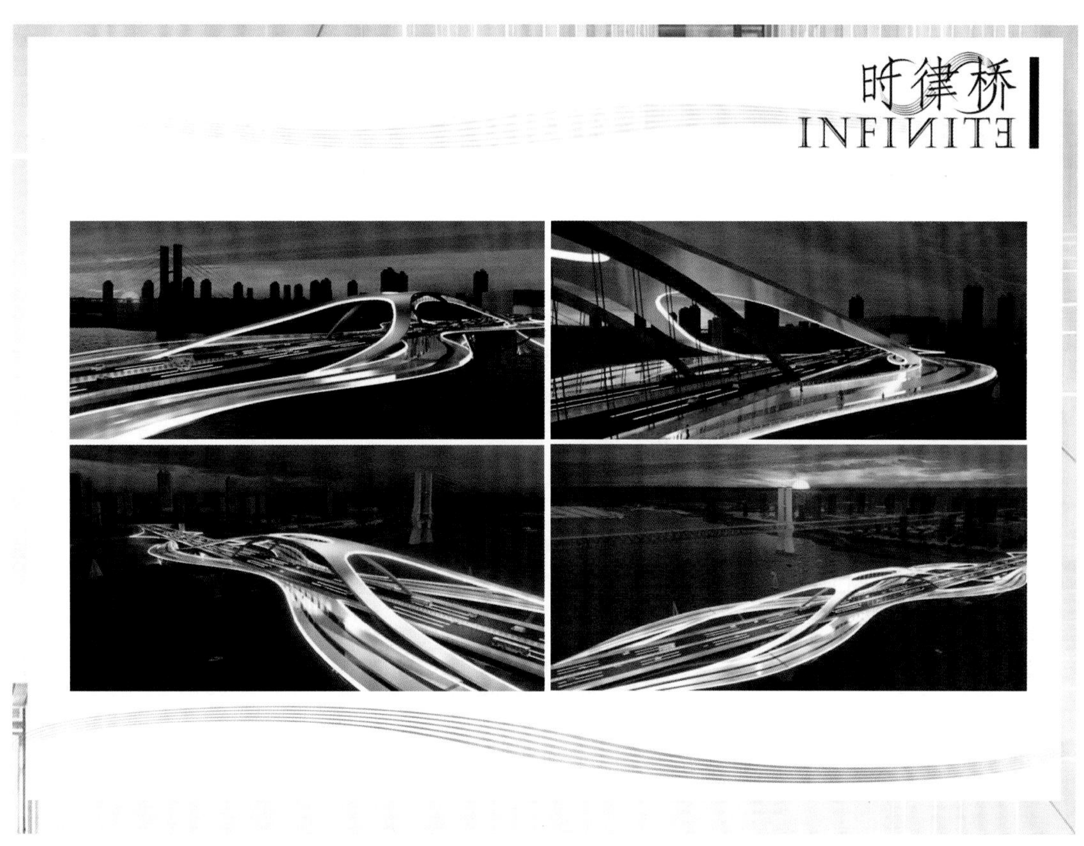
时律桥
INFINITE

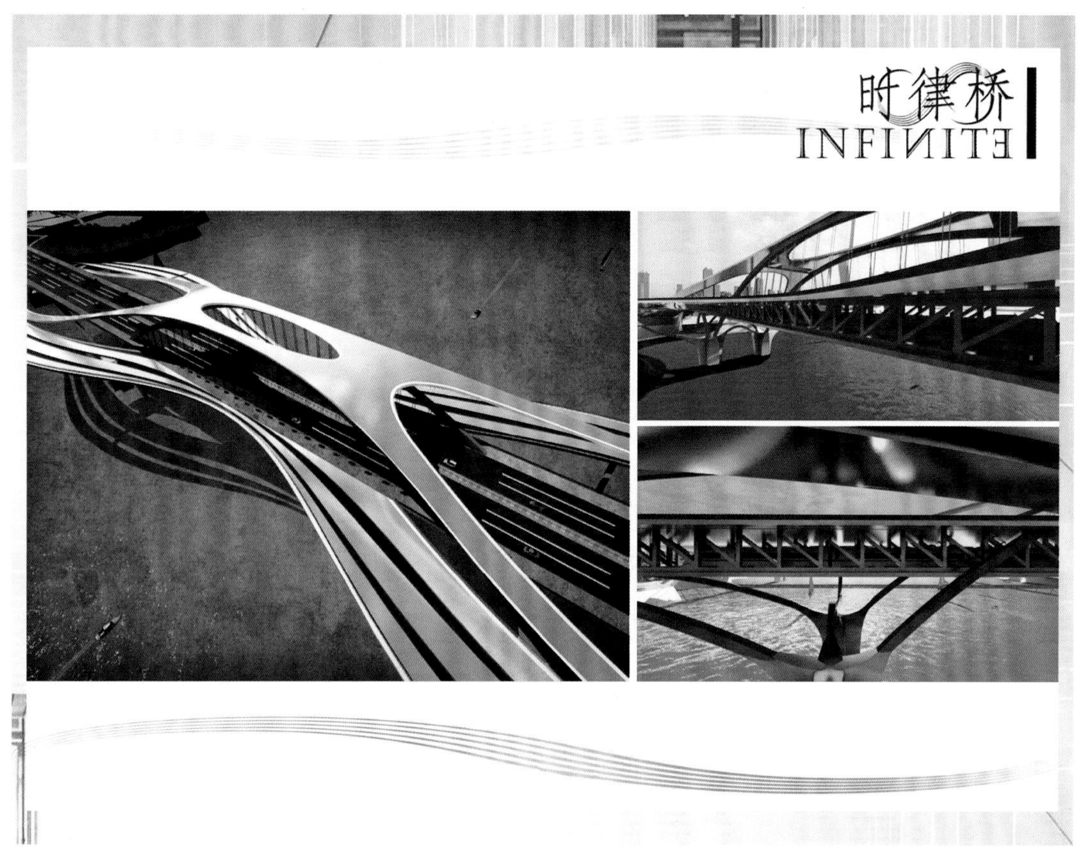
时律桥
INFINITE

清江络鸿桥

Wild Gooses Bridge

参赛学校 University/College
重庆交通大学
Chongqing Jiaotong University, China

指导教师 Supervisor(s)
董莉莉 DONG Lili
徐略勤 XU Lueqin

参赛学生 Participant(s)
伍慧萌 WU Huimeng
陈惠历 CHEN Huili
贺洪滔 HE Hongtao
夏瑞洁 XIA Ruijie
曾 沁 ZENG Qin

简介 Description

清江络鸿桥打破传统铁路桥仅服务于列车通行的主要功能，结合周边情况，尝试在满足列车通行的同时满足居民通行及游客游览，并为他们提供更安全舒适，妙趣横生的过桥体验。本次设计采用拱式桥梁结构，两根主拱，四根副拱，主拱向内倾斜，副拱向外倾斜，并由人行自重、与主梁连接的横向拉索进行横向力平衡。大小不一，方向各异的拱架起主梁，犹如群雁衔起丝丝飘带，解决货运通车、人行游览需求，打通当地的幸福之路、乡村振兴之路。

Wild Gooses Bridge does not follow the tradition of railway bridges serving merely for transportation by train. Considering the surrounding environment, this design tries to keep the transportation function and also allow residents to pass and tourists to visit, providing them with a safer, more comfortable and surprisingly good experience of bridge crossing. The bridge uses an arch structure with two main arches and four secondary arches. The main arches tilt inwards and the secondary ones outwards to balance the lateral force caused by the weight of the pedestrians and the transverse ties connected to the main girders. The arches of different sizes and directions frame the main girders, just like a flock of geese holding a silk ribbon, solving the demand for freight transportation and pedestrian excursions, opening up the road leading to happiness and rural revitalization for the local people.

红鹰大桥

Hongying Bridge

参赛学校 University/College
河南城建学院
Henan University of Urban Construction, China

指导教师 Supervisor(s)
王　仪　WANG Yi
屈讼昭　QU Songzhao

参赛学生 Participant(s)
王自立　WANG Zili
张时可　ZHANG Shike
田华鑫　TIAN Huaxin
曾兴涛　ZENG Xingtao
宋　毅　SONG Yi

简介 Description

设计桥梁位于甘肃省兰州市七里河区，地跨黄河，是负责联通七里河区和安宁区经济贸易的交通要道。兰州古称金城，是“控扼冲要、道通西域”的丝路要地，也是中华民族多元一体进程的见证地，历史文化厚重，文物遗存丰富。新桥名为红鹰大桥，采用了斜拉桥的现代化建桥方案，桥跨方向由南至北，桥面宽 42 米，桥梁行车运营方案按照两侧公路，中间铁路方案布局行驶，且桥面宽度按 11 米 +12 米 +11 米进行布置，两侧各设 3 米路缘带，总桥长 1000 米。

The designed bridge is located in the Qilihe District, Lanzhou City, Gansu Province in China. It is a significant channel across the Yellow River, connecting the Qilihe District and the Anning District for the economic and trade exchanges between the two. Lanzhou, known as Jincheng in ancient times, is an important town on the Silk Road that “controls the key bastions and leads to the western regions”. It is also a witness of the integration of the multiple ethnic groups in China boasting profound history, culture and heritages. The new bridge, named the Hongying Bridge, was designed as a cable-stayed modern bridge. From south to north, the bridge span is 42m. The traffic planned for the bridge is highways on both sides and a railway in the middle. The width was arranged as 11m+12m+11m, leaving 3m on both sides as curb belts. The total length of the bridge is 1000m.

全梁式铁路桥

Only Beams

参赛学校 University/College

俄罗斯莫斯科国立建筑大学

Moscow State University of Civil Engineering, Russia

指导教师 Supervisor(s)

Evgeny Leonidovich Bezborodov

Garanzha Igor Mikhailovich

参赛学生 Participant(s)

Savitskii Maxim Alexandrovich

Sekachev Egor Andreevich

Golubeva Anna Andreevna

Sergeeva Inna Alekseevna

简介 Description

我们的建筑项目名为全梁式铁路桥（OnlyBeams）。

在设计时，团队考虑了两种铁路桥结构安排：桁架结构和梁式结构考虑到方案的可行性，最终采用了梁式结构设计。长 6 米的梁可以通过铁路轨道运输，而桁架需要在施工现场制成大跨度结构，提高建筑质量的同时会延长工期，增加建筑成本。桁架桥（高架桥）的维护存在一些死角，需要定期维护，此外，货运列车通常使用铰链边，可以根据需要在铁路栈桥上的任意地点装卸货物，而如果采用桁架结构，则难以实现。桁架结构也会涉及更复杂的计算，更容易产生各种变形。因此，经过计算，我们最终选择了梁式结构。

Our project is called OnlyBeams.

Our team considered two structural arrangements for the bridge: truss and girder. In the end, it was decided to use the girder design. It was primarily due to the manufacturability of this solution, the beams with a length of 6m can be transported by railroad tracks, the truss construction is enlarged at the construction site due to its dimensions, which increases the construction period and the quality of the construction, and makes it economically unreasonable. Also truss bridge (trestle) has hard-to-reach places for maintenance, which should be carried out regularly. In addition, due to the nature of production, freight trains are often used with hinged sides, which allows unloading at any required location along the railroad trestle; the use of a truss will limit these possibilities. Designing trusses will lead to more complex calculations, and such structures are more sensitive to deformations of various kinds. Also we did the calculations to select the main beam.

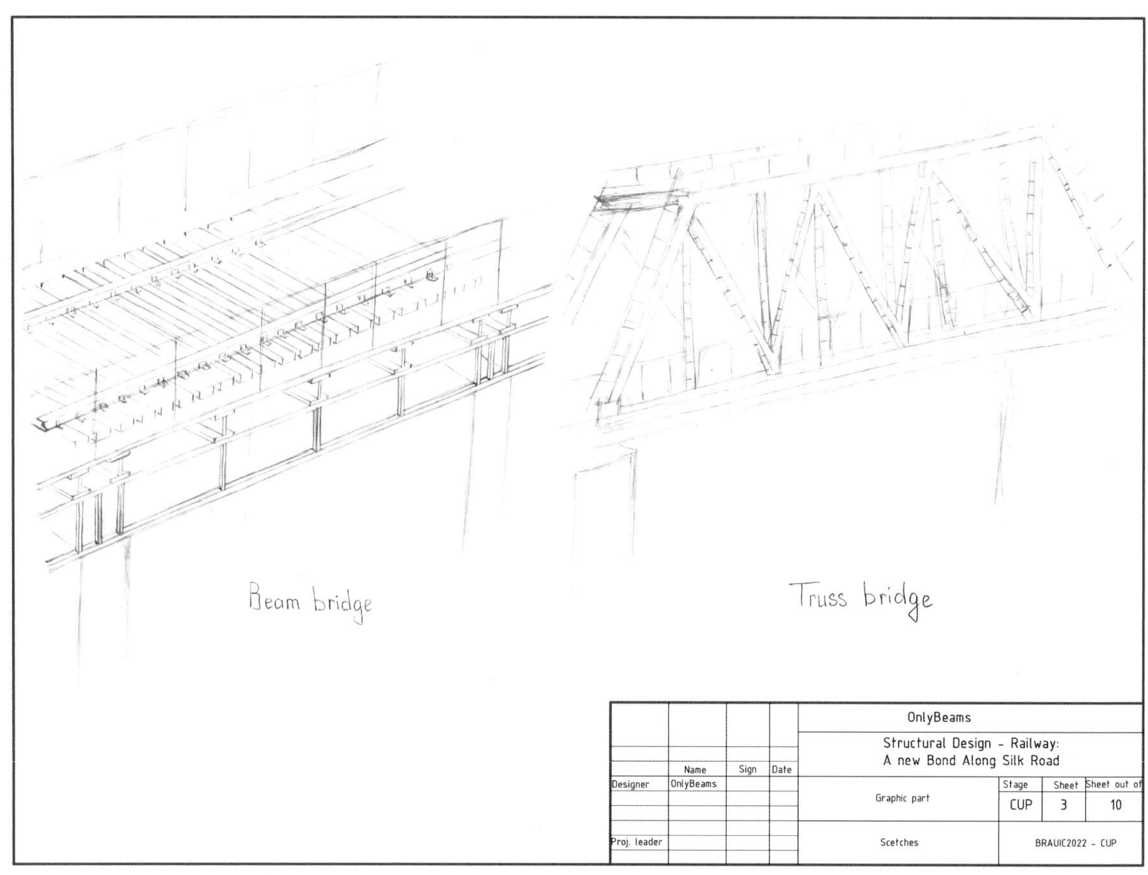
Beam bridge
Truss bridge
OnlyBeams
Structural Design - Railway:
A new Bond Along Silk Road
Name
Sign
Date
Designer
OnlyBeams
Graphic part
Stage
Sheet
Sheet out of
CUP
3
10
Proj. leader
Scetches
BRAUIC2022 - CUP

埃里温（Yerevan）—久姆里（Gyumri）—瓦纳佐尔（Vanadzor）—艾鲁姆（Ayrum）—格鲁吉亚边境铁路线上 Fialetovo 市—瓦纳佐尔（Vanadzor）段 KM20+300 铁路桥设计

Design of the Bridge on the KM20+300 of the Section Fialetovo-Vanadzor of Yerevan-Ayrum-Georgian Border Railway

参赛学校 University/College

亚美尼亚国立建筑大学

National University of Architecture and Construction of Armenia

指导教师 Supervisor(s)

Artashes Sargsyan

Aren Soghoyan

参赛学生 Participant(s)

Taron Baghdasaryan

Sergey Dadoyan

Hayk Dadayan

Khachik Chkolyan

简介 Description

本项目计划在 Fialetovo 市瓦纳佐尔（Vanadzor）段铁路线上修建铁路桥，该铁路桥可以连接亚美尼亚现有的两条铁路：一条经埃里温（Yerevan）、久姆里（Gyumri）、瓦纳佐尔（Vanadzor）、艾鲁姆（Ayrum）直到格鲁吉亚边境，另一条经埃里温（Yerevan）、赫拉兹丹（Hrazdan）、伊杰万（Idjevan）直到阿塞拜疆边境。这个铁路段将成为从阿塞拜疆到土耳其途径哈萨克斯坦、伊杰万（Idjevan）、迪利然（Dilijan）、Fialetovo 市、瓦纳佐尔（Vanadzor）、久姆里（Gyumri）和卡尔斯（Kars）最近的铁路线路，并将成为丝绸之路经济带的一部分。

该铁路将以位于瓦纳佐尔市南部边界的 RS59+00 到 RS62+40 路段的铁路桥为基础。铁路桥的水平曲度半径为 R=400 米，纵向坡度为 5‰。

铁路桥有 7 个桥跨结构，上部结构设计为组合梁 L=34.2 米，由 2 根钢制的双 T 梁和顶部的混凝土桥面板组成。主梁之间的间距为 2.0 米。在梁之间预留了两个技术人行道和技术通道，用于后期的桥梁维护。铁路桥上的轨道交通计划采用电力牵引，预计将在桥墩顶部的悬臂上安装柱子。为减少列车噪声对环境的影响，可考虑安装隔声屏障。桥墩高 11~24 米，桥台采用现浇钢筋混凝土设计。

The railway bridge designed on the proposed Fialetovo-Vanadzor Railway Line. Proposed railway can connect two existing railways in Armenia: Yerevan-Gyumri-Vanadzor-Ayrum-Georgian Border and Yerevan-Hrazdan-Idjevan-Azerbaijan Border. Poposed line will become the nearest railway route from Azerbaijan to Turkey through Kazakh-Ijevan-Dilijan-Fialetovo-Vanadzor-Gyumri–Kars and can became a part of the Silk Road Economic Belt.

Railway alignment assumes the construction of bridge on the section RS59+00 to RS62+40 which runs at the south border of Vanadzor City. The bridge placed on the horizontal curve with a radius R=400m and 5‰ longitudinal slope.

Bridge has 7 spans and super structure designed as composite beam L=34.2m, what consists of 2 steel double-T beams with concrete deck slab on the top. The space between the main girders is 2.0m. Two technical sidewalks and technical passage between the beams were foreseen for the bridge maintenance. The rail traffic over the bridge will run with electric traction and it is designed to install posts on cantilevers placed on the top of the piers. For the decreasing of the trains noise influence on the environment, it is considered to install the noise fences. Piers 11 to 24 metres height and abutments designed from in-situ reinforced concrete.

允南桥

Yunnan Bridge

参赛学校 | University/College
沈阳建筑大学
Shenyang Jianzhu University, China

指导教师 | Supervisor(s)
张素杰 ZHANG Sujie
张可心 ZHANG Kexin

参赛学生 | Participant(s)
齐天玉 QI Tianyu
朱笑莹 ZHU Xiaoying
刘继诚 LIU Jicheng
范亦珍 FAN Yizhen
王文诚 WANG Wencheng

简介 Description

“允南桥”即“允恭克让，明烛天南”，意思是诚实、恭敬又能够谦让，光照亮了南面的天空。从天空鸟瞰全局，桥的主体像天鹅翱翔在天宇之间，给人以和平的视觉冲击。从河对岸正目而视，整座大桥犹如一把竖琴，典雅美观，散发着高雅的神韵。允南桥既有着对人恭敬谦和的品格，又有崇尚和平的美好精神追求，把对未来生活的美好憧憬都融于桥的寓意中，象征的是心灵之桥，人与人之间需要理解和宽容。

Yunnan, the name of the bridge, is extracted from "Yun Gong Ke Rang, Ming Zhu Tian Nan", which means that honesty, respect and humility form the light that can illuminate the sky in the south. From a bird's eye view, the main body of the bridge looks like a swan flying in the sky, showing people a peaceful scene. Looking straight from the other side of the river, the entire bridge is like a harp, gorgeous, exuding an elegant charm. The Yunnan Bridge both expresses the ethics of respect and humbleness and contains our pursuit of peace-keeping. People's aspirations for their future life are all infused into the name of "Yunnan". The bridge is expected to connect people's hearts and bring the understanding and tolerance that we all need.

Amtrak

甸白铁路桥梁设计

Dianbai Railway Bridge Design

参赛学校 University/College
天津城建大学
Tianjin Chengjian University, China

指导教师 Supervisor(s)
薛　江　XUE Jiang
刘进军　LIU Jinjun

参赛学生 Participant(s)
阮观浩　RUAN Guanhao
吴飞龙　WU Feilong
吴　昊　WU Hao
段敏坤　DUAN Minkun
何玥柯　HE Yueke

简介 Description

甸白路桥位于甘肃省天水市麦积区渭南镇、石佛镇及中滩镇三镇的交汇地带（三阳川新区），下穿宝兰客专后和现状石佛支线，横跨甸白路，桥梁全长 120 米，为三跨钢管桁架拱桥（30+60+30）米，桥面宽 13 米，采用单箱单室等截面。拱式组合体系桥能将梁和拱两种基本结构组合起来，充分发挥了主梁承重，吊杆进行荷载传递等结构特性及组合作用。除了体现安全、适用、环保、节能等理念外，与普通桥梁相比，不仅更经济而且在景观方面也是独树一帜。

The Dianbai Highway Bridge is at the intersection of Weinan Town, Shifo Town and Zhongtan Town (the Sanyangchuan New District) in the Maiji District, Tianshui City, Gansu Province in China. Stretching beneath the Baoji-Lanzhou High-speed Railway, the bridge joins with the Shifo Branch Line and crosses the Dianbai Road. The bridge is a 120m long and three-span steel pipe truss arch bridge (30+60+30)m. With a width of 13m, the bridge adopts a single-box, single-cell cross section. The arch-type composite system bridge can combine the two basic structures of beam and arch and give full play to the structural characteristics as well as the combined effects of the main beam bearing load and the suspension rod for load transmission. In addition to embodying the concepts of safety, suitability, environmental protection and energy saving, this bridge is more economical and special in landscape than an average bridge.

船路皆通可伸缩桥

Telescopic Bridge Connecting Ship and Road

参赛学校 University/College

中原工学院

Zhongyuan University of Technology, China

指导教师 Supervisor(s)

纠永志 JIU Yongzhi

赵 毅 ZHAO Yi

参赛学生 Participant(s)

董凯钧 DONG Kaijun

谷凌云 GU Lingyun

司卓卓 SI Zhuozhuo

刘幸义 LIU Xingyi

李子山 LI Zishan

简介 Description

目前，规划“西合高铁”过境唐县，而跨唐河之上修建铁路桥梁成为必然趋势，我们所设计的桥梁不仅满足铁路运输，而且还可通行船只，达到了一桥两用的目的。

我们所设计的桥梁总长 150 米，总宽度 50 米，为中小型桥梁。桥梁类型为悬索钢架桥，该桥梁打破了直通路或只通船的旧观念，成为新型桥梁的代表，该桥梁的设计理念为“美、创新、环保、安全。”

According to the current plan, the Xihe high-speed railway will pass Tangxian County, so it is an inevitable trend to build a railway bridge over the Tanghe River. The bridge that we designed not only meets the need of railway transportation, but also allows ships to pass, achieving the goal of one bridge for two purposes.

The bridge we designed has a total length of 150m and a total width of 50m. It is a medium and small-sized bridge classified as a suspension steel frame bridge. This bridge denies the outdated concept of road-only or ship-only and has become an example of a new bridge. The design concept of this bridge is "beauty, innovation, environmentally friendly and safety".

Category C: Scene Modeling and Visualization

中老铁路上的“雀舞之翎”

Dancing Feather on China-Laos Railway—Xishuangbanna Station

参赛学校 University/College
北京建筑大学
Beijing University of Civil Engineering and Architecture, China

指导教师 Supervisor(s)
王国利 WANG Guoli
郭　明 GUO Ming

参赛学生 Participant(s)
付泽昕 FU Zexin
梁轩硕 LIANG Xuanshuo
付思源 FU Siyuan
金　璐 JIN Lu
黄圣佳 HUANG Shengjia

简介 Description

为响应 2022“一带一路”国际大学生数字建筑设计竞赛号召，特依据个人专业背景及特长组建本团队，根据竞赛要求并综合考虑“一带一路”倡议、地域及数据源等因素，选择云南省的西双版纳火车站作为地理场景建模区域。采用 SketchUp 软件，基于图纸数据及影像进行三维实体建模；最后利用 Lumion 软件和 Enscape 软件，渲染出质感以及光影效果，丰富场景，最终得到贴合实际的精细模型。

In response to the call for the 2022 Belt and Road International Student Competition on Digital Architectural Design, we formed this team based on our professional background and individual strengths. According to the requirements of the competition and taking into account the “Belt and Road” Intiatrke, geography, data sources, etc., the Xishuangbanna Railway Station in China’s Yunnan Province was chosen as the area for geographic scene modeling. We used SketchUp software for 3D solid modeling based on the drawing data and images; the final step is the rendering of the texture as well as light and shadow effects using the software of Lumion and Enscape to obtain a detailed and realistic model.

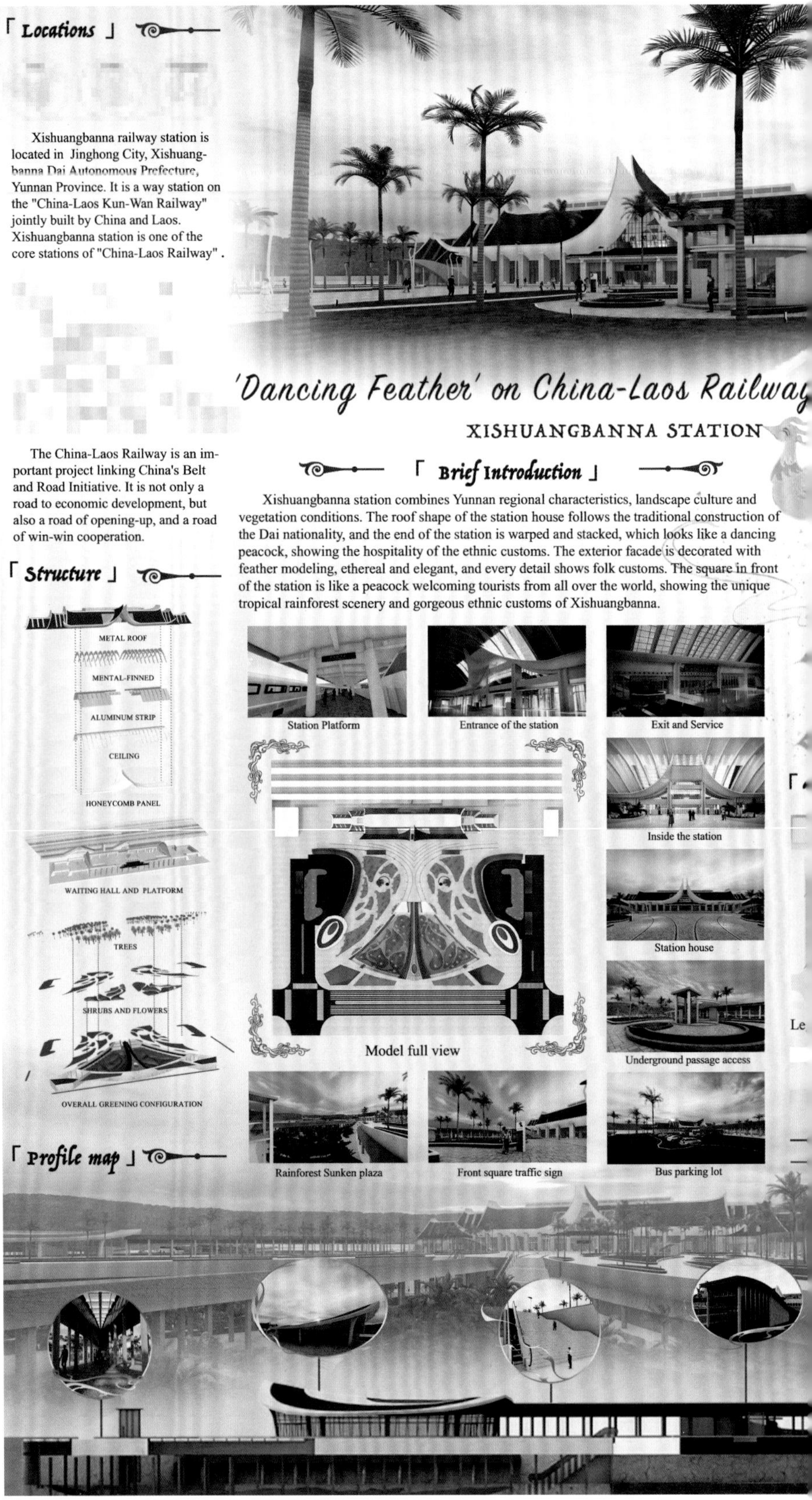

「Locations」

Xishuangbanna railway station is located in Jinghong City, Xishuangbanna Dai Autonomous Prefecture, Yunnan Province. It is a way station on the "China-Laos Kun-Wan Railway" jointly built by China and Laos. Xishuangbanna station is one of the core stations of "China-Laos Railway".

The China-Laos Railway is an important project linking China's Belt and Road Initiative. It is not only a road to economic development, but also a road of opening-up, and a road of win-win cooperation.

'Dancing Feather' on China-Laos Railway

XISHUANGBANNA STATION

「Brief Introduction」

Xishuangbanna station combines Yunnan regional characteristics, landscape culture and vegetation conditions. The roof shape of the station house follows the traditional construction of the Dai nationality, and the end of the station is warped and stacked, which looks like a dancing peacock, showing the hospitality of the ethnic customs. The exterior facade is decorated with feather modeling, ethereal and elegant, and every detail shows folk customs. The square in front of the station is like a peacock welcoming tourists from all over the world, showing the unique tropical rainforest scenery and gorgeous ethnic customs of Xishuangbanna.

「Structure」

METAL ROOF
MENTAL-FINNED
ALUMINUM STRIP
CEILING
HONEYCOMB PANEL
WAITING HALL AND PLATFORM
TREES
SHRUBS AND FLOWERS
OVERALL GREENING CONFIGURATION

Station Platform

Entrance of the station

Exit and Service

Inside the station

Station house

Model full view

Underground passage access

Rainforest Sunken plaza

Front square traffic sign

Bus parking lot

「Profile map」

「区位分析」
西双版纳站位于云南省西双版纳傣族自治州景洪市嘎洒镇曼暖龙村是中国和老挝两国两党共同兴建的"中老昆万铁路"上的途径站。西双版纳站是"中老铁路"的核心站点之一，是紧密联系"中老"边境的重要结点。
于2021年正式开通的中老铁路是中国"一带一路"倡议与老挝"变陆锁国为陆联国"战略对接的重要项目，它既是一条经济发展之路，也是对外开放之路，更是一条合作共赢的友谊之路。
中老铁路上的"雀舞之翎"
——西双版纳站
「作品简介」
西双版纳站以"雀舞彩云，灵动版纳"为设计理念，结合云南地域特色、景观文化和植被状况等开展绿色通道专项设计。站房屋顶造型沿用傣族传统建筑，端部起翘层叠挑出，如展翅的孔雀翩翩起舞，彰显热情好客的民族风情。外立面采用花翎造型进行点缀，空灵飘逸，每处细节都彰显出民俗风情。站前广场与站房相辉映，又似展屏的孔雀喜迎八方游客，展示出西双版纳独特的热带雨林风光和绚丽的民族风情。
「结构分解」
建筑金属屋面
金属肋
铝条板
天花吊顶
金属蜂窝板
候车大厅及火车站台
树木
灌木&花卉
整体绿化配置
火车站台
站房入口
站房出口及服务台
站房内部
站房及地下通道出入口
地下通道直梯出入口
模型全貌展示
雨林下沉广场
站前广场交通指示牌
大巴及公交停车场
「功能分区」
图例
人行道路
车行道路
下沉广场
绿化区域
地下通道口
服务中心及通道
站房及周边建筑
公交大巴停车场
消防车流线
公交大巴车流线
社会车辆车流线
「剖面分解」

二等奖 » Second Prize

“色、质、形、名”——百年建筑光影漫步

“Color, Quality, Shape, Name”—Walk Through A Century of Architectural Light and Shadow

参赛学校 University/College

安徽建筑大学

Anhui Jianzhu University, China

指导教师 Supervisor(s)

王 艳 WANG Yan

邢 瑜 XING Yu

参赛学生 Participant(s)

殷晓雨 YIN Xiaoyu

陈 泽 CHEN Ze

王艺璇 WANG Yixuan

张 墨 ZHANG Mo

毛 蕾 MAO Lei

简介 Description

门台子烤烟厂与津浦铁路相接，该厂完整保留了 20 世纪早期西方列强在华烟草加工厂的整体格局与建筑形式，工业厂房、办公建筑和高级别墅体现了中西建筑风格、技术和文化的交融，是见证并记录中国烤烟工业发展的活态博物馆。通过打造实景

三维与数字场景相结合的体验空间，充分调动和增强沉浸体验与到场感知力，激发并引导受众对历史建筑价值的求知欲和保护欲，使其历史、科学、艺术、社会和文化价值得到深层理解与传承。

The Mentaizi Flue-cured Tobacco Factory is connected with the Jinpu Railway. The factory completely retains the overall pattern and architectural form of the tobacco processing factories of Western powers left in China in the early 20th century. The industrial plants, office buildings and high-end villas reflect the integration of the architectural style, technology and culture of China and the West. This is why the factory can even be regarded as a living museum that witnesses and records the development of China's flue-cured tobacco industry. By creating an experience space that combines real three-dimensional and digital scenes and fully mobilizing and enhancing the immersive experiences and perceptions gained by people at present, the building could stimulate their desire for knowledge and raise their awareness of protecting the values of these historical buildings with the aim of deeply understanding and inheriting the history, science, art, society and cultural values.

门台子烟叶复烤厂

“凭”江“固”苏

Traffic Consolidation Through Rivers in Suzhou

参赛学校 University/College
长春建筑学院
Changchun University of Architecture and Civil Engineering, China

指导教师 Supervisor(s)
尹鹏程 YIN Pengcheng

参赛学生 Participant(s)
高朋博 GAO Pengbo
彭　晶 PENG Jing
王冰旭 WANG Bingxu
帅琦旭 SHUAI Qixu
董　震 DONG Zhen

简介 Description

《凭江固苏》名字的灵感：苏州古称平江又名姑苏。“江”代表交通的意思，“固”是巩固的意思。取谐音和文字意义，凭借交通巩固了苏州。体现“一带一路”倡议所带来的经济发展和文化交流给苏州又增添了新活力。苏州站相对于全国的火车站来说，规模上不算特别庞大，但是绝对是一个具有苏州特色的火车站。远看苏州火车站外形中轴对称，车站在线路的南北两侧设置了部分站房，实则正是这样简单的外形却隐藏着浩瀚星海的历史文化。设计者采用 Blender 软件进行建筑主体的建模，使用 SU 软件进行辅助设计，应用 Lumion 软件进行渲染。

The name of the station "Traffic Consolidation through Rivers in Suzhou" draws inspiration from the following: Suzhou was called Pingjiang (the Pingjiang River), also known as Gusu (the same pronunciation as "consolidating Suzhou" in Chinese). Jiang (River) represents traffic and Gu means consolidation. This is why the complete meaning of this homophone is consolidating the traffic in Suzhou with rivers, highlighting the economic development and the increase of cultural exchanges benefited from the "Belt and Road" policy and the new vitality added to Suzhou. Compared with other railway stations in China, the Suzhou Station is not particularly large in scale, but it is the one with the most Suzhou characteristics. Looking from the distance, the Suzhou Station is symmetric on both sides of its axis. Some station facilities are set up on the north and south sides of the railway. Such a simple layout contains profound history and culture. The designer use Blender software to model the mainbody of the building, use sketch Up softwene to assist in the design, and apply Lumion software to render.

苏州站
憑江圖蘇
苏州站

憑江圖蘇
苏州站

三等奖 >> Third Prize

"一带一路"：运城老站

Old Railway Station in Yuncheng along the Belt and Road

参赛学校 University/College
北京建筑大学
Beijing University of Civil Engineering and Architecture, China

指导教师 Supervisor(s)
刘　扬 LIU Yang
李　洋 LI Yang

参赛学生 Participant(s)
张戴新月 ZHANG Daixinyue
贾竞珏 JIA Jingyu
杨　恒 YANG Heng
孙　尹 SUN Yin
王博涵 WANG Bohan

简介 Description

运城市是位于中国山西省的一个地级市，以盐运出名，同时此地也具备着浓厚的历史文化底蕴，关帝庙、后土祠、永乐宫等颇具盛名的历史景点，吸引了大批游客。通过选取运城市内的典型历史景点（解州关帝庙）、曾具有重要战略意义的老车站（解县火车站）、闻名天下的盐业相关景点（盐池广场、盐化工厂物流站），利用无人机采取相关数据，再进行三维建模，得到运城市内具有战略特点和历史特点的模型。

Yuncheng city is a prefecture-level city located in Shanxi Province in China. It is famous for salt transportation but also has much historical and cultural heritage like the Guandi Temple (temple for a general renowned for integrity who was later worshiped by the people as an emperor), the Houtu Temple (temple for Nvwa, the Goddess of land and also the creator of mankind in Chinese legends), the Yongle Palace (in memorial for Lyu Dongbing, a god in Taoism) that attracts a large number of tourists. To obtain models with strategic values and historical characteristics in Yuncheng city, we selected a typical historical scenic spot there (the Guandi Temple in Haizhou Town), an old station that once had strategic significance (the Haixian Railway Station), and two famous scenic spots related to the salt industry (The Salt Pond Square and the Logistics Station of the Salt Chemical Plant), used drones to collect relevant data and finally conducted three-dimensional modeling.

敕勒川

Chilechuan

参赛学校 University/College

长春建筑学院

Changchun University of Architecture and Civil Engineering, China

指导教师 Supervisor(s)

尹鹏程 YIN Pengcheng

参赛学生 Participant(s)

杨泺怡 YANG Luoyi

芦　伟 LU Wei

秦鸿艳 QIN Hongyan

田驭洲 TIAN Yuzhou

尚小力 SHANG Xiaoli

简介 Description

呼和浩特火车东站位于呼和浩特市东部的如意开发区。呼和浩特作为西北部的一个城市，拥有独特的蒙元文化，同时，其作为对外开放的窗口，也作为“一带一路”城市发挥着不可或缺的作用。火车东站总体设计理念突出以人为本，结合地域特色和明代特色，综合考虑功能性、系统性、时效性、文化性和经济性，将既有交通建筑与地域文化特色有机结合。

The East Hohhot Railway Station is located in the Ruyi Development Zone in the east of

Hohhot City in China. As a city in the northwest of the country, Hohhot has a unique Mongolian and Yuan culture. At the same time, as a window opening to the outside world, it also plays an indispensable role as a “Belt and Road” city. The overall design concept of the East Railway Station highlights people orientation, combined with regional characteristics and the features of the Ming Dynasty (1368—1644). Its functionality, systematization, timeliness, cultural values and economy were coordinated so that it becomes a traffic building with local characteristics and the elements of the Ming culture.

“形由力生”——基于计算机力学模拟的成渝地区铁路站场景建模

Scene Modeling of the Chengdu Chongqing Railway Station Based on Computer Mechanical Simulation

参赛学校 University/College
重庆交通大学
Chongqing Jiaotong University, China

指导教师 Supervisor(s)
刘 锐 LIU Rui
雷 怡 LEI Yi

参赛学生 Participant(s)
董乃乾 DONG Naiqian
冯建树 FENG Jianshu
刘意湖 LIU Yihu
柯莉娟 KE Lijuan

简介 Description

对于铁路站的外部环境，我们在车站前广场设置了供人休憩遮荫的拱壳构造来激活这部分空间，旨在呼应建筑形态，保持场地的连续性，使广场空间作为城市的一部分成为能为人们所利用的公共空间。高铁站站楼采用拱壳造型，为了使候车厅能够获取大跨度的无柱空间，拱壳找形时从平面流线出发，确定开口和支座位置，并利用参数化手段进行力学找形。

For the external environment of the railway station, we set up an arch shell structure in the square in the front for people to rest and shade, activate this part of the space and fulfill its function of highlighting the architectural form of the station, maintaining the continuity of the site and make the square a public space of the city where people have access to everyday. The station building is in the shape of an arch shell. To offer the waiting hall a column-free space with a large span, the openings and supporting points of the shell are determined based on the plane streamline, and parametric means are resorted to for its mechanical shape finding.

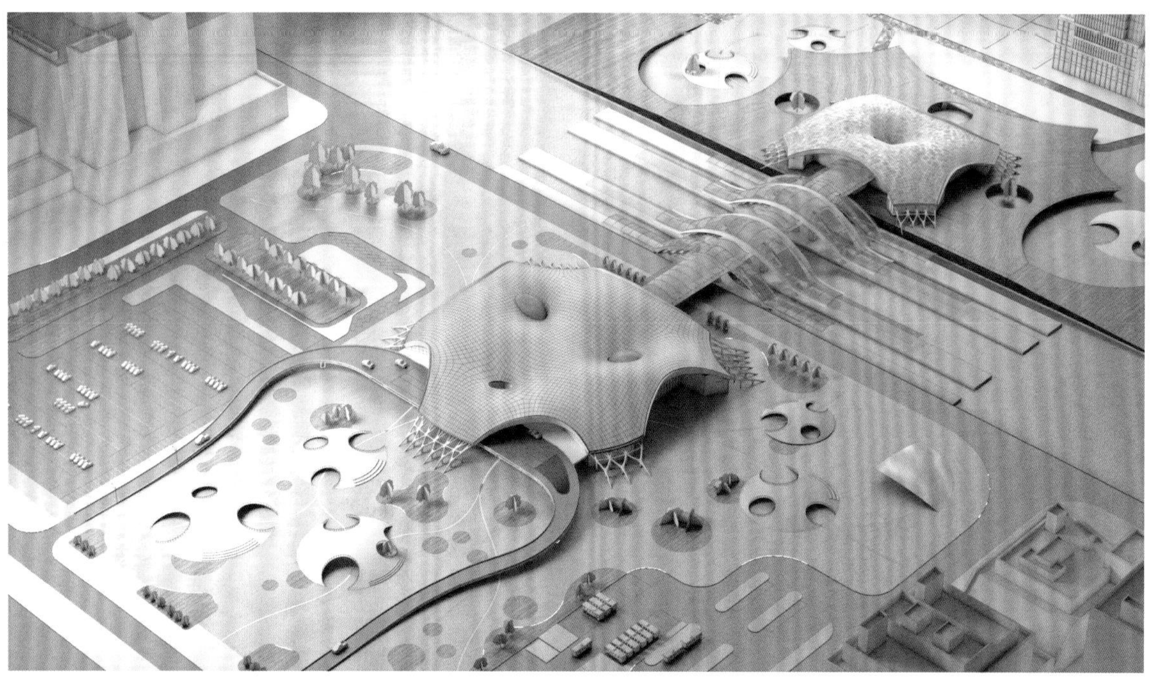

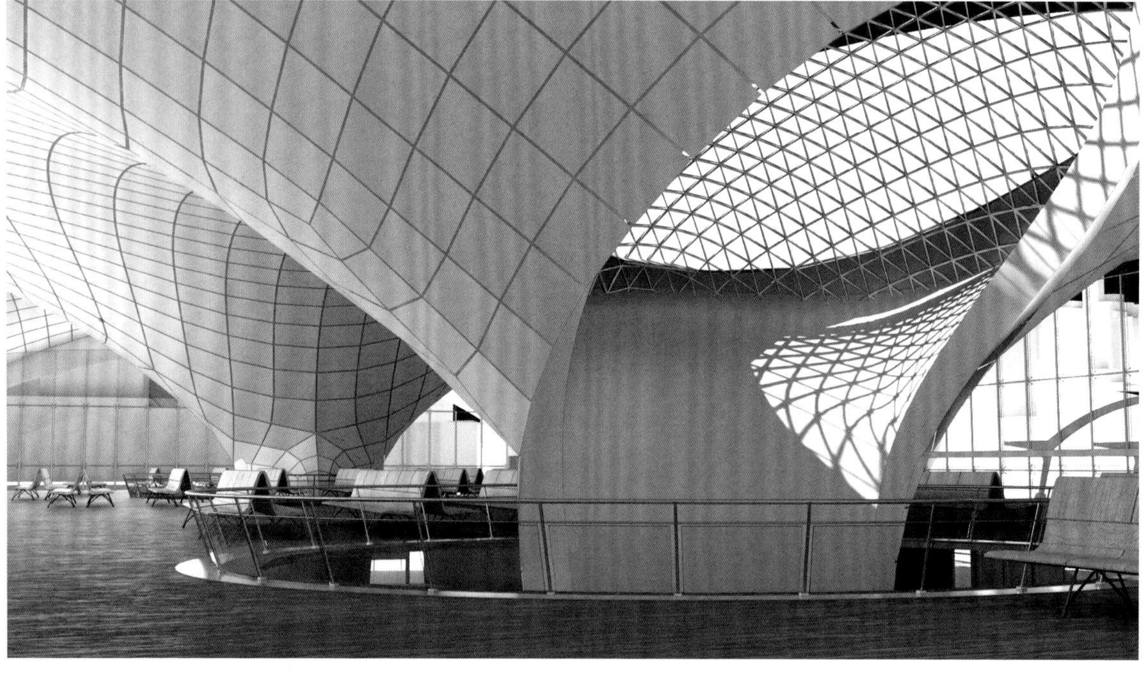

“铁路”回归，建设文明新载体

The Return of Railway as a New Carrier of Civilization

参赛学校 University/College
重庆交通大学
Chongqing Jiaotong University, China

指导教师 Supervisor(s)
李华蓉 LI Huarong

参赛学生 Participant(s)
门月阳 MEN Yueyang
毛宏宇 MAO Hongyu
余 双 YU Shuang

简介 Description

随着高铁及许多铁路线的不断开通，铁路对于“一带一路”沿线国家所产生的经济效益是有目共睹的。与新的铁路线相比，老铁路线退出了运输舞台，面临废弃的处境，但改造后的老铁路线却能重焕生机。以川外小铁路为例，利用隧道和铁轨打造了具有重庆特色的步道，并将红岩文化景点串联起来，延续历史文脉，建设为一个网红打卡景点，进而带动当地旅游业。因此利用 ArcGIS 和 SketchUp 等软件对川外小铁路红梅林站的景观要素进行场景建模。

With the opening of more high-speed and ordinary railway lines, the economic benefits of railways for countries along the “Belt and Road” are widely recognized. Unlike new railway lines, the old ones are phased out as means of transportation, at the brink of being dilapidated. However, these old railway lines can be revitalized once experienced renovation. Taking the Chuanwai Small Railway as an example, tunnels and railway tracks are used to create a pedestrian trail with Chongqing characteristics that connects the Hongyan cultural scenic spots to inherit the historical heritage. The trail has become a super popular online spot, driving the local tourism industry. Therefore, we used ArcGIS, SketchUp and other software are used to model the Hongmeilin Station on the Chuanwai Small Railway.

梁赞市火车站

Railway Station in the City of Ryazan

参赛学校 University/College
俄罗斯莫斯科理工大学梁赞分校
Ryazan Institute, Moscow Polytechnic University, Russia

指导教师 Supervisor(s)
Karetnikova Svetlana Veniaminovna

参赛学生 Participant(s)
Zharinov Alexey Sergeevich

简介 Description

梁赞是一座古老的城市，也是蓬勃发展的地区中心。许多俄罗斯和外国游客来到这个城市，想要深入了解梁赞这个历史悠久的俄罗斯古城，深入了解该地的民间传统和原始工艺。梁赞的地理位置处于南北和东西交通汇流之地，因此莫斯科铁路线的梁赞方向成为莫斯科铁路枢纽当中客流量最大的支线之一。

参数化设计反映了对未来的现代性设计理念，因此我们采用这种风格设计车站。建筑的主体由带有半透明涂层的双层玻璃构成，配备电热供暖以除去积雪和霜冻，防止水汽凝结，周围还有一个由铰链板制成的天篷。建筑平面图呈椭圆形，由两个椭圆壳体组成。外壳的支撑形成了天篷，在建筑周边形成了几个拱门。建筑外形过渡平滑，为建筑营造出轻盈感。项目在 Rhino Grasshopper 软件中应用参数化建模的方法，并在 ArchiCAD 程序中进一步开发项目。

Ryazan is an ancient city and a developing regional center. Many Russian and foreign tourists come to this city to get acquainted with one of the most ancient Russian territories, as well as learn more about folk traditions and original crafts. The Ryazan direction of the Moscow Railway is one of the busiest in terms of passenger traffic at the Moscow railway junction. Due to its location, north-south and west-east traffic flows converge here.

Parametric design reflects modern ideas about the future, so the station building is made in this style. The composition consists of the main volume with a translucent coating of double-glazed windows with electric heating, thanks to which snow cover and frost are removed, moisture condensation is prevented, and there is a canopy which is made of hinged panels. The building is elliptical in plan and consists of two shells. The supports of the outer shell forming the canopy create several arches around the perimeter of the building. The smoothness of the transition of forms provides the apparent lightness of the structure. The project is carried out by the method of parametric modeling in Rhino Grasshopper, further development of the project is carried out in the ArchiCAD program.

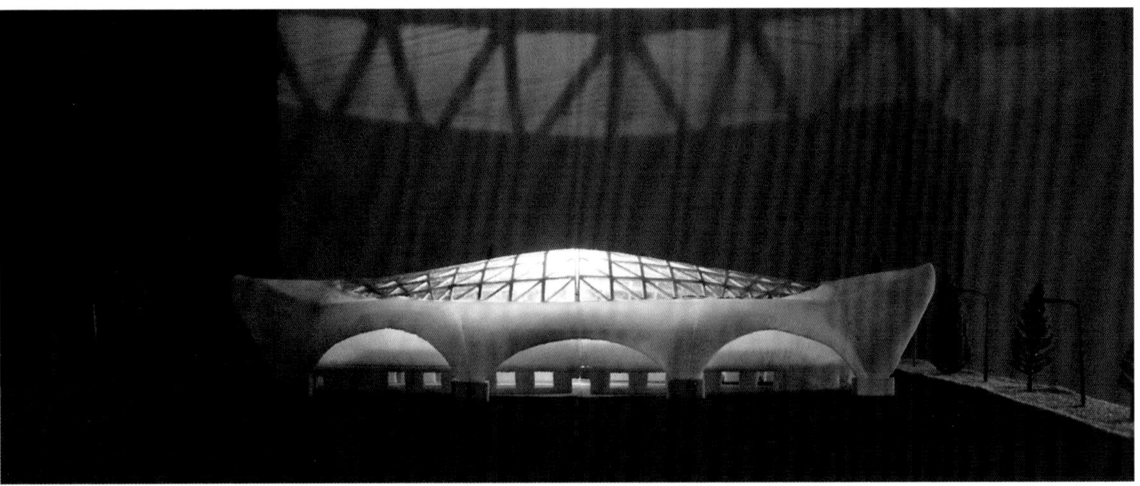

基于数字孪生的铁路小学及其周边场景建模与可视化

Modeling and Visualization of Railway Primary School and Its Surrounding Scenes Based on Digital Twin

参赛学校 University/College

石家庄铁道大学

Shijiazhuang Tiedao University, China

指导教师 Supervisor(s)

聂良涛 NIE Liangtao

参赛学生 Participant(s)

魏世俊 WEI Shijun

梁　雨 LIANG Yu

董雨生 DONG Yusheng

李为民 LI Weimin

简介 Description

铁路小学主体外墙颜色为红色与白色，屋顶为双坡式坡屋顶，且配有幕墙系统；希望小学主体主要分为三部分，西北侧（1 号楼）一幢 8 层建筑；东北侧（3 号楼）一幢 5 层建筑；东南侧（2 号楼）一幢 2 层建筑；三部分由两个连廊连接成一个整体。

模型中利用了 Three.js 技术，轻量化 BIM 模型。随着 BIM 技术的进步、普及，成本逐步降低。因而 BIM 模型的精细程度要求越来越高，体量越来越大，对于一些长大带状的铁路道路地质模型，已有的软件硬件很难流畅运行。

本次竞赛我们建立了一个专业模型，整合后的模型体量过大，运行流畅度低，本团队结合 WebGL 轻量化 BIM 引擎解决此类问题作为创新点。模型最终通过自行编码引擎完成加载。

The main exterior of the railway primary school is in red and white. The roof is in a double-slope form with a curtain wall system. This non-profit primary school has three main parts: an 8-story building (building 1) in the northwest; a 5-story building (Building 3) in the northeast and a 2-story building (Building 2) in the southeast. The three parts are connected by two corridors into a whole.

The Three.js technology and the lightweight BIM model are used in the modeling. As BIM technology progresses and gains popularization, the cost is gradually reduced. This modeling is required to be more precise and large in volume. For the geological models of some long strip railway tracks, it is difficult for the existing software and hardware to run smoothly.

We build a professional model in this competition. The integrated model is too large to run smoothly. The team adopt an innovative approach by combining the WebGL lightweight BIM engine to solve this problem. The model is finally loaded is an engine encoded by ourselves.

广钢旧轨——致敬那个逝去但不曾泯灭的花城工业时代

Old times, Old Guanggang Railways, but Huacheng Industrial Period Is Splendid

参赛学校 University/College
中山大学
Sun Yat-sen University

指导教师 Supervisor(s)
陈定安 CHEN Ding'an

参赛学生 Participant(s)
李名豪 LI Minghao
彭　雄 PENG Xiong
张思敏 ZHANG Simin
丁　莹 DING Ying

简介 Description

广钢铁路是“一带一路”前建设的重要铁路，是广州工业的运输要路，支撑着广州工业几十年的发展。即使广州钢铁厂关闭，广钢铁路仍是一代人的重要记忆。花地河大桥作为重要代表与历史建筑之一，具有深刻的留存价值，运用数字建模技术留存当下历史建筑，是刻不容缓的事情。

The Guanggang Railway, built before the "Belt and Road" Initiative, is an important railway for Guangzhou's industry and has been supporting the development of Guangzhou for decades. These railway tracks remain a memory in the minds of a generation living in Guangzhou even though the Guangzhou Steel Factory was closed forever. The Huadihe Bridge is one of the symbols of the Guanggang Railway and a historical architecture. Given its great value for preservation, there is no time to delay for us to protect it by employing digital modeling technology.

基于风格修复理念的老合肥火车站历史空间场景复原的数字孪生

The Digital Twin of the Restoration of the Historical Space Scene of the Old Hefei Railway Station Based on the Concept of Style Restoration

参赛学校 University/College
安徽建筑大学
Anhui Jianzhu University, China

指导教师 Supervisor(s)
刘存钢 LIU Cungang
桂汪洋 WANG Guiyang

参赛学生 Participant(s)
万智康 WAN Zhikang
张 力 ZHANG Li
周 祥 ZHOU Xiang
武秀玲 WU Xiuling
曹海慧 CAO Haihui

简介 Description

本次建模以老合肥火车站 1966 年时的站房和站台布置为研究和复原对象，基于真实性原则，结合历史卫星照片、历史图像和现状遗迹等进行复原。首先，通过对铁路建设历史背景、车站建设概况、建筑风格及细部特点等进行梳理，明晰了站房基础信息。然后，针对站房图纸遗失、历史信息少等现实因素，从多个角度对站房、站台设计的可能性进行分析，运用三维点云技术进行辅助，最终绘制完成较为完整的图纸，根据图纸完成了此次建模。

This modeling takes the station building and platform of the old Hefei Railway Station in 1966 as the research and restoration object. In line with the principle of authenticity, the restoration was carried out according to past satellite photos, historical images and the status quo of the relics. First, by analyzing the historical background of railway construction, the general situation, the architectural style and detailed characteristics of this specific station, we managed to make clear the basic information of the station building. Then, given the fact that the drawings of this station building were lost and certain historical information was absent, we analyzed different possibilities of the station building and platform design from multiple perspectives, supported by the 3D point cloud technology. Finally, a relatively complete drawing was finished, based on which we accomplished this modeling.

徐州高铁站

Xuzhou High-speed Railway Station

参赛学校 University/College

河南城建学院

Henan University of Urban Construction, China

指导教师 Supervisor(s)

王松威 WANG Songwei

王　仪 WANG Yi

参赛学生 Participant(s)

王子易 WANG Ziyi

李荣耀 LI Rongyao

李泽宇 LI Zeyu

何少康 HE Shaokang

张梓健 ZHANG Zijian

简介 Description

根据地形平坦，道路周围被建筑环绕等情况，我们选择设计搭建高架铁路，以减少对原有交通地形的影响，并且高架铁路还有工程简易、造价低的优势。站台形式则使用较为常见的侧式站台，可以有效地减少双向客流的冲击，施工难度也相对较小，节约成本。铁轨道路则选用无砟轨道，采用刚性较大的黏结硬化材料作为道床板，使得系统的荷载传递、扩散功能显著提高。建筑主体选用玻璃幕墙技术，保证建筑主体安全的同时兼具艺术性与科技性。

According to the relatively flat terrain given and the fact of the road surrounded by buildings, we chose elevated railways for the high-speed railway station to reduce the impact on the original traffic landscape. Besides, the elevated railway also has the advantages of simple engineering and low cost. Commonly used side platforms are adopted to effectively reduce a two-way passenger flow and decrease construction difficulty. The road bed plate of the ballastless track uses a hardened binding material with relatively high rigidness, which significantly improves the load transmission and diffusion functions of the system.

A glass curtain wall is adopted for the main body of the building to ensure safety while embodying artistic values and a sense of technology.

新丝绸之路——郑州航空港站场景设计

New Silk Road—Scene Design of the Zhengzhou Airport Station

参赛学校 University/College
河南城建学院
Henan University of Urban Construction, China

指导教师 Supervisor(s)
王松威 WANG Songwei
王　仪 WANG Yi

参赛学生 Participant(s)
张瑞丹 ZHANG Ruidan
袁彦天 YUAN Yantian
肖林峰 XIAO Linfeng
马凯歌 MA Kaige
何润波 HE Runbo

简介 Description

依据郑州航空港院的实景来设计站房，通过搜索大量的图片和信息确定大致的尺寸和外观，然后利用 Revit 进行翻模，达到大致的外观场景，要求尽量真实和美观，然后将 revit 模型转换为 dae 格式，满足格式要求，接着进行渲染效果图的提取，最终完成作品。

The station was designed according to the actual view of the Zhengzhou Airport Institute. First, we searched for a large number of pictures and information to determine the approximate size and appearance of the object. Then, we used Revit for the rollover to obtain a rough appearance of the scene that is as real and beautiful as possible. Next, Revit was used again to transform the model into dae format required by the contest. The final design was finished after the rendered drawing was extracted.

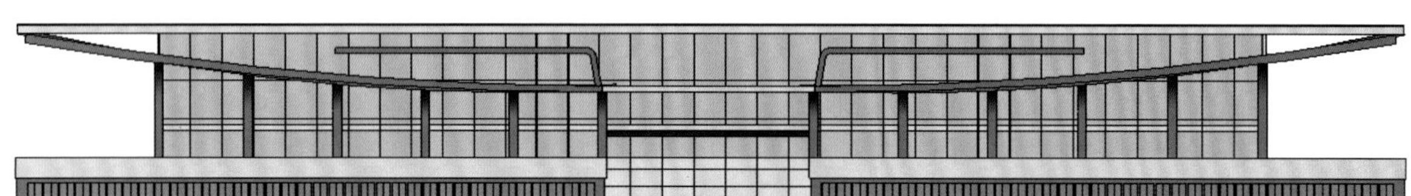

绸缎

Silk

参赛学校 University/College

吉林建筑大学

Jilin Jianzhu University, China

指导教师 Supervisor(s)

宋义坤 SONG Yikun

张广平 ZHANG Guangping

参赛学生 Participant(s)

陈佳宁 CHEN Jianing

陈嘉怡 CHEN Jiayi

李卓十 LI Zhuoshi

汪学林 WANG Xuelin

张显宸 ZHANG Xianchen

简介 Description

铁路是现代交通系统的重要组成部分，高铁时代的到来加速了人流、物流、信息流、资金流的流动，推动了地方产业转移，促进了经贸往来和房地产业、旅游业的长足发展，为社会经济的发展注入了强大动力。本次建模设计选取浙江省嘉兴市的一处新建火车站为对象，旨在探究铁路对于城市发展的意义。

As an important part of the modern transportation system, the arrival of the high-speed railway era has accelerated the flow of people, logistics, information and capital, promoted the transfer of local industries, fueled the rapid development of economic and trade exchanges, the real estate industry and tourism and injected a strong driving force into the development of the social economy. In this modeling design, we selected a new railway station in Jiaxing City, Zhejiang Province in China as the object to explore the significance of railways in urban development.

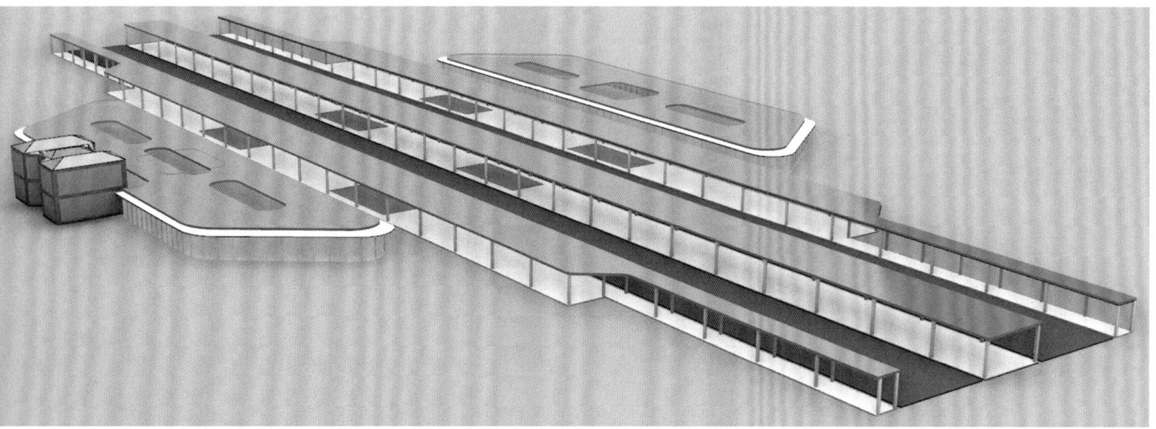

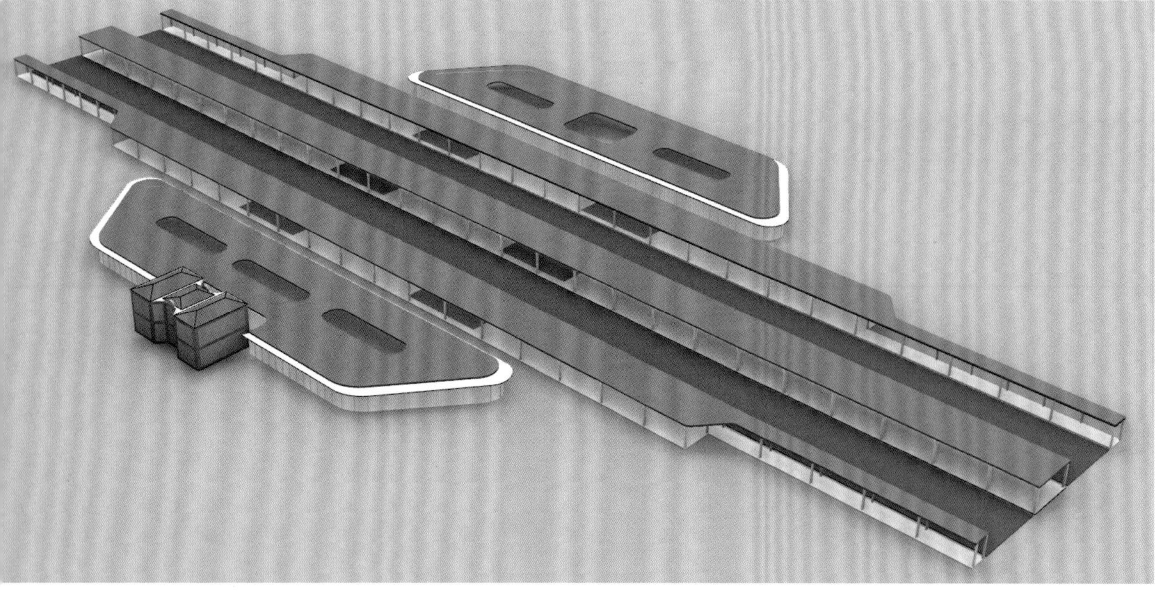

一翩新驿

A New Station

参赛学校 University/College
吉林建筑大学
Jilin Jianzhu University, China

指导教师 Supervisor(s)
宋义坤 SONG Yikun
安　宁 AN Ning

参赛学生 Participant(s)
左一轩 ZUO Yixuan
李　润 LI Run
郭　璐 GUO Lu
范馨穗 FAN Xinsui
张晰然 ZHANG Xiran

简介 Description

铁路建设的不断发展大大缩短了城市之间的距离，铁路在我国的“一带一路”发展中占据了不可或缺的地位。近年来一个个交通项目相继建成和加快推进，新疆对外开放的步伐持续加快。乌鲁木齐正是亚洲的中心，处于“一带一路”规划中丝绸之路的经济带核心区中，是丝绸之路向北向西的窗口。此次建模设计选取乌鲁木齐铁路站台，旨在保留铁路及沿线城市的历史文脉。

The continuous development of railway construction has greatly shortened the distance between cities. Railways have played an indispensable role in the development of China's "Belt and Road" Initiative. In recent years, transportation projects have been completed and accelerated one after another, which helped quicken up the pace of Xinjiang's opening up to the outside world. As the center of Asia, Urumqi, the capital of the Xinjiang Uygur Autonomous Prefecture of China is at the core position of the Economic Belt of the Silk Road under the "Belt and Road" plan and is the window of the Silk Road to the north and west. This modeling design selected the Urumqi railway platform as the object to preserve the history and culture of the railway and the cities along the line.

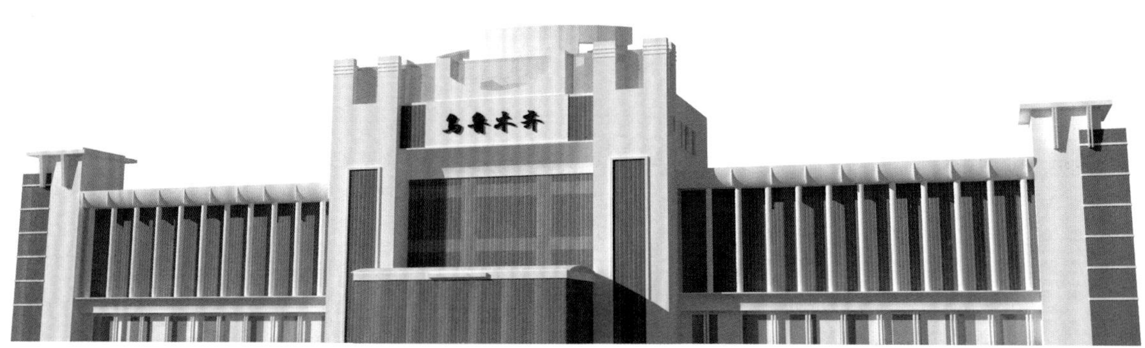
乌鲁木齐

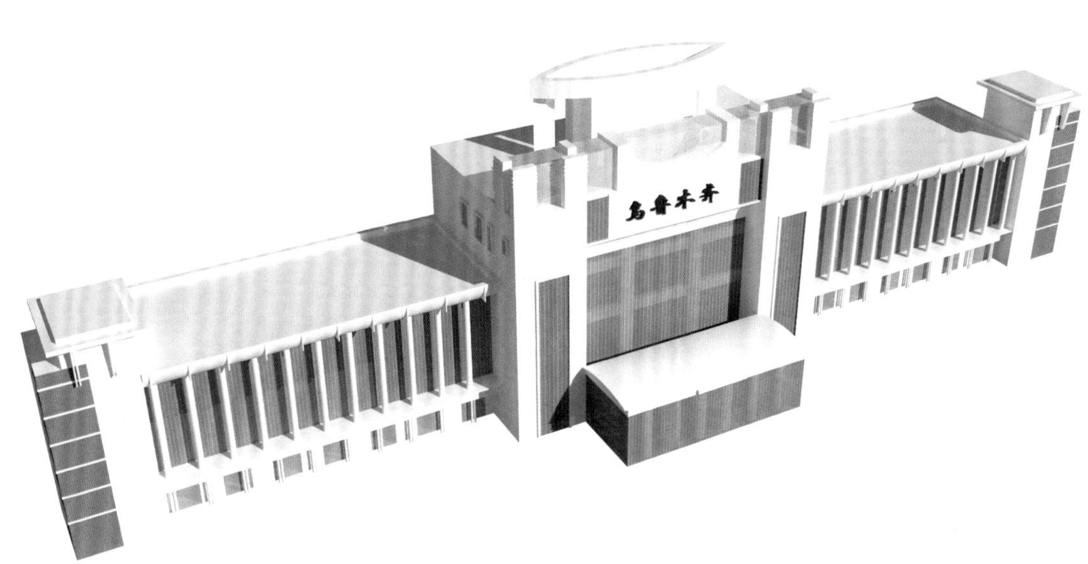
乌鲁木齐

“一带一路”铁路生态园景观设计

The Design of the Belt and Road Railway Ecological Park

参赛学校 University/College

山东建筑大学

Shandong Jianzhu University, China

指导教师 Supervisor(s)

赵学强 ZHAO Xueqiang

张莉莉 ZHANG Lili

参赛学生 Participant(s)

刘玉书 LIU Yushu

潘万里 PAN Wanli

赵一豪 ZHAO Yihao

齐小龙 QI Xiaolong

简介 Description

无论是从历史文化方面，还是现代国际政治经济贸易方面，“一带一路”倡议都有着极其重要的历史意义。随着中国的快速发展,原本交通运输上的主题色“混凝灰”应在新时代赋予新的元素。

建模地点选在“一带一路”中线第二站河南省郑州市某废弃工厂。我们结合“一带一路”倡议对“一带一路”历史文化与现代意义的元素抽象提取进行设计。

在建筑方面，主要功能为城市室内花园和阅读空间的打造。整体采用铁路轨道的元素附着在建筑外表。

在景观上提取“一带”和“一路”进行抽象设计，并且根据历史文化提取丝绸元素用飘带系起每一个国家的心，在经济全球化和人类命运共同体的道路上，手牵手一起走在通往和平、繁荣的未来。

在植物配置上采用各国特色植物贯穿其中，体现国际合作共生的发展理念。

The “Belt and Road” Initiative is of great historical significance in terms of history, culture, modern international politics, economy and trade. With the rapid development of China, the original color of “tuff” that is mainly used in transportation would be endowed new elements to the new era.

The site of the model selected is an abandoned factory in Zhengzhou City in China’s Henan Province, which is also the second station of the central part of the “Belt and Road”. In line with the “Belt and Road” Initiative, our design was based on the abstract elements extracted from its historical culture and modern significance.

The main roles of the architecture itself are urban indoor gardens and reading spaces. Elements of railway tracks are to be attached to the exterior of the building.

The landscape was designed with “one belt” and “one road” in an abstract approach. The ribbon symbolizes silk to connect the heart of every nation. In the road of globalization and the building of a community of shared destiny for mankind, the “Belt and Road” countries join hands to usher in a peaceful and prosperous future.

Representative plants from countries alongside the route were chosen to form the green land of the station, expressing our hope of all countries living together and cooperating with each other.

高铁啤酒博物馆还原设计

Restoration Design for the High-Speed Railway Beer Museum

参赛学校 University/College

山东建筑大学

Shandong Jianzhu University, China

指导教师 Supervisor(s)

张莉莉 ZHANG Lili

参赛学生 Participant(s)

张　秦 ZHANG Qin

宋　莹 SONG Ying

姚子琪 YAO Ziqi

李雅琪 LI Yaqi

苏贝勒 SU Beile

简介 Description

拥有青岛特色品牌的青岛啤酒股份有限公司，正向智能化、科技化、数字化转型。本次高铁啤酒博物馆还原设计以还原青岛高铁原状，可视化呈现啤酒博物馆原貌为目的，融合青岛啤酒元素，展现青岛建筑的建筑形制、景观轴线及节点构成，突出青岛高铁沿线特色建筑。利用犀牛、3Dmax、lumion 等工具，可视化 3D 建模，推进青岛本地企业数字化发展。

Tsingtao Beer Co., Ltd., as the business card of Qingdao, is transforming into an intelligent, scientific and digital enterprise. The restoration design of the high-speed railway beer museum aims to replicate the high-speed railway in Qingdao and visualize the original beer museum. The museum integrates the elements of Qingdao beer and showcases the unique architectural form, landscape axis and node composition in Qingdao architecture with the special buildings along the Qingdao high-speed railway highlighted. Rhinoceros, 3DMAX, lumion and other tools are used to achieve a visualized 3D modeling and promote the digital development of local enterprises in Qingdao.

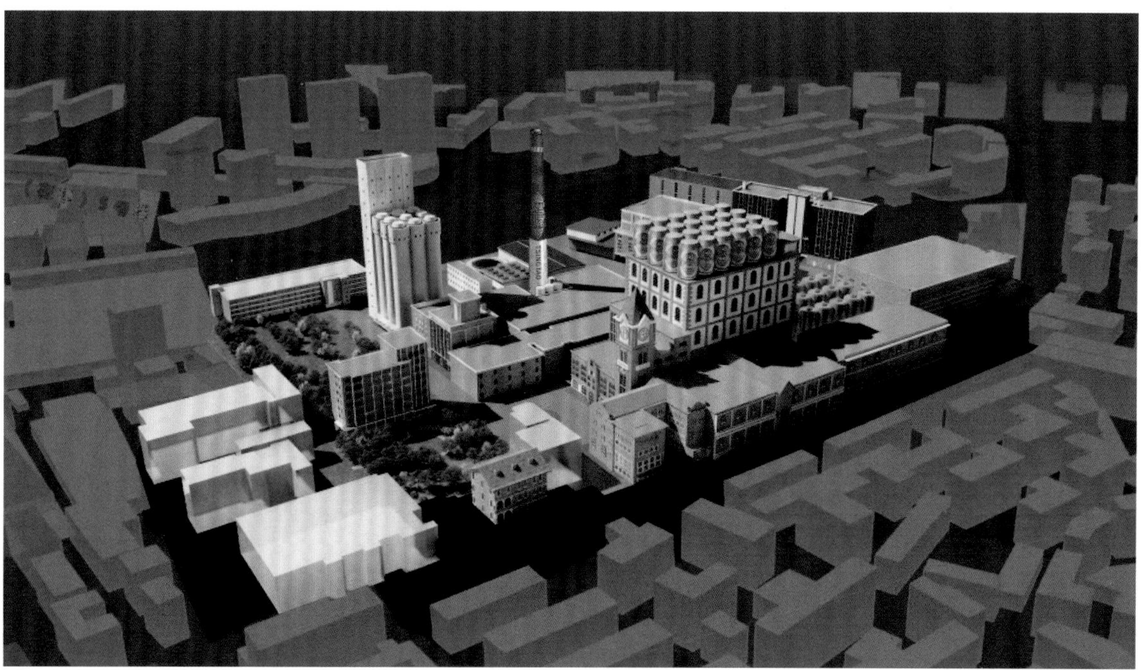

古韵雁塔——基于海量点云数据的大雁塔三维重建

Dayan Pagoda

参赛学校 University/College
西安科技大学
Xi'an University of Science and Technology

指导教师 Supervisor(s)
姚顽强 YAO Wanqiang
蔺小虎 LIN Xiaohu

参赛学生 Participant(s)
黄千翔 HUANG Qianxiang
徐梦雨 XU Mengyu
王欣萌 WANG Xinmeng
张璐爽 ZHANG Lushuang

简介 Description

我们团队考虑到大雁塔本身具有地方特色，且其周围有轻轨路过符合铁路要素的沿线场景这一要求，因此我们选用大雁塔作为本次建模的主体。

团队采用徕卡 ScanStation C10 扫描仪获取大雁塔海量点云数据，联合 Cyclone、AutoCAD 和 3D—Max 平台各自的优点，对大雁塔进行了三维模型重建。本次建模的立意体现在采用三维激光扫描技术，在不接触、无损害的前提下，快速高效地构建了大雁塔真实的横面、剖面、立面模型。

Considering that the modeling scene has requirements for the landscape and can reflect local characteristics, our team chose the Dayan (Wild Goose) Pogoda as the object because it is full of local characteristics and the light rail passing it makes the building meet the requirement that the site of the architecture must have railway-related elements. We combined the massive point cloud data of the Dayan Pagoda obtained by the Leica ScanStation C10 scanner to reconstruct the 3D model of the Dayan Pagoda. In this process, the advantages of the cyclone, AutoCAD and 3D Max platforms were utilized respectively. This modeling intends to use three-dimensional laser scanning technology to quickly and efficiently build the real transverse, section and elevation models of the wild goose pagoda without contacting and damaging the physical building.

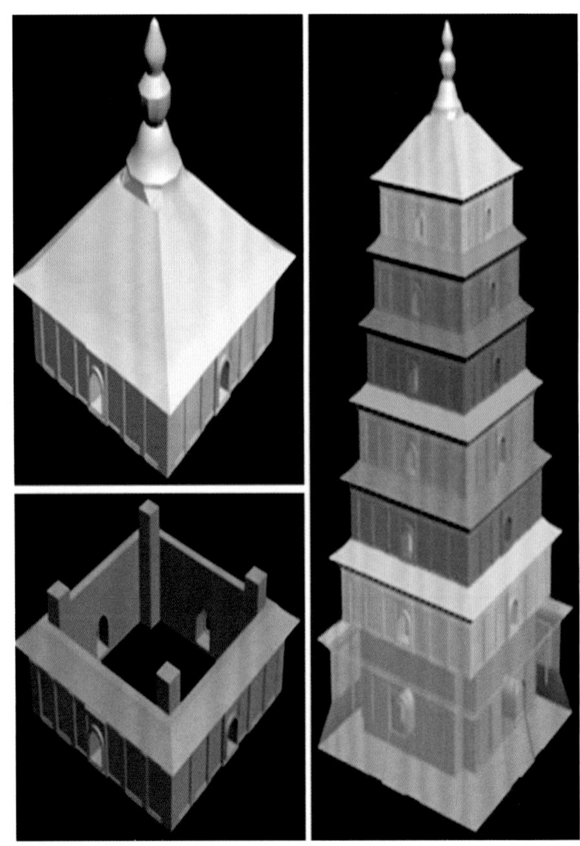

沈丘北站枢纽广场扩建设计

Expansion Design of Shenqiu North Station Hub Square

参赛学校 University/College

中原工学院

Zhongyuan University of Technology, China

指导教师 Supervisor(s)

夏晓敏 XIA Xiaomin

韩文翔 HAN Wenxiang

参赛学生 Participant(s)

陈伟垚 CHEN Weiyao

林勇旗 LIN Yongqi

贾明浩 JIA Minghao

曾治源 ZENG Zhiyuan

简介 Description

该工程位于沈丘北，通过在此处高铁站的升级改造和周围景观设计的深化改造使其重启焕发生机，服务于周边群众，通过加强人流、交通、环境、能源、网络等基础设施的建设和维护，并且使其互联相通，促进经济发展，提高周围人群生活质量，传播当地文化，吸引外来游客，将其打造成一个更深层次的，交流更加广泛的活动平台，打造开放、包容、均衡、普惠的区域经济合作架构，贯彻实现“一带一路”理念。

The project is located in the north of Shenqiu County. Through the upgrading and reconstruction of the high-speed railway station here and the deepening transformation design of the surrounding landscape, the region will be awakened and revitalized to provide services for the people living nearby. By increasing the flow of people and promoting the construction and maintenance of the infrastructure like transportation, environment, energy and internet as well as enhancing their interconnection, the local economic development, people's quality of life, and the local culture will all be promoted with even more tourists attracted here. In that way, the area will become a platform allowing for a wider range of more in-depth activities. Also, we will be able to build a regional economic cooperation framework featuring openness, inclusiveness and balance that is beneficial to all and put into practice the concept of the Belt and Road.

济渎文化展示场景景观场景设计——基于河南省济源站站房服务区及景观设计

Landscape Design of Jidu Culture Exhibition Scene—Based on the Service Area and Landscape Design of Jiyuan Station Building in Henan Province

参赛学校 University/College

中原工学院

Zhongyuan University of Technology, China

指导教师 Supervisor(s)

韩文翔 HAN Wenxiang

夏晓敏 XIA Xiaomin

参赛学生 Participant(s)

谢武城 XIE Wucheng

许允浩 XU Yunhao

王雅慧 WANG Yahui

董恩惠 DONG Enhui

简介 Description

本项目旨在使用符合“一带一路”基本内涵的一种仿古建筑形式（依托于济渎

庙文化）与济源火车站文化产生联系。本项目位置毗邻济源火车站，服务于周边群众，通过加强交通、能源和网络等基础设施的互联互通建设，以促进经济要素有序自由流动、资源高效配置和市场深度融合，开展更大范围、更高水平、更深层次的区域合作，打造开放、包容、均衡、普惠的区域经济合作架构，贯彻实现“一带一路”理念。

This project is committed to using an antique architectural form (relying on the culture of the Jidu Temple) that conforms to the basic connotation of the Belt and Road and has a connection with the culture of the Jiyuan Railway Station. The project site is adjacent to the Jiyuan Railway Station for the convenience of people living nearby. By strengthening the interconnection of infrastructures such as transportation, energy and the internet, economic factors would enjoy greater freedom for flowing, resources will be allocated more efficiently and a deep integration of markets could be achieved. We hope to carry out regional cooperation in a larger scope, at a higher level and in a deeper manner, creating an open, inclusive, balanced and inclusive cooperation framework for the regional economy and put into practice the concept of the Belt and Road.

动脉寻续——集体记忆下的旧铁路驿站沿线景观设计

Arterial Continuity—Landscape Design along the Old Railway Station under Collective Memory

参赛学校 University/College

安徽建筑大学

Anhui Jianzhu University, China

指导教师 Supervisor(s)

孙　升 SUN Sheng

张抗抗 ZHANG Kangkang

参赛学生 Participant(s)

张　墨 ZHANG Mo

戴辰阳 DAI Chenyang

刘正增 LIU Zhengzeng

陈　泽 CHEN Ze

简介 Description

依托淮南丰富的煤和水资源，因铁路运输之便，淮南造纸厂成为中国“一五”计划重点项目，是淮南发展的见证者。随着经济转型和铁路运输萎缩，其早已不复当年荣光。纸厂见证了城市变迁，记录了工业技术轨迹，具有凝结社区记忆、充当城市地标、改善生活品质等重要价值。通过对纸厂进行改造，使其功能上满足品质生活需求，情感上唤醒并塑造老、中、青、幼的集体记忆，最终实现民众、社区和城市在生产、生活、生态等方面的协调发展。

Relying on the Huainan City's abundant coal and water resources as well as the convenient railway transportation, the Anhui Paper Mill of Huainan has become a key project of China's "First Five-Year Plan" and a witness to the development of Huainan. With the economic transformation and the decline of railway transportation, it has lost its glory for long. Paper mills have recorded urban changes and documented the trajectory of industrial technology, which has important values such as refining community memories, existing as city landmarks, and improving the quality of life. Through the transformation of the paper mill, the above functions could be met for quality life, the collective memory and emotional experiences of the old, the middle-aged and the young. The ultimate goal that will be achieved is a coordinated development of people, community and city in production, life, ecological environment and other aspects.

拟改造区域
火车道
拟改造建筑A
拟改造建筑D
拟改造建筑B
拟改造建筑C

无人机鸟瞰——发掘京西煤路新资源，赋予百年老站新生机

Aerial View of Unmanned Aerial Vehicles: Discovering New Resources of Western Beijing Coal Road and Giving New Life to the Century-Old Station

参赛学校 University/College

北京建筑大学

Beijing University of Civil Engineering and Architecture, China

指导教师 Supervisor(s)

黄　鹤 HUANG He

杨军星 YANG Junxing

参赛学生 Participant(s)

朱欣宇 ZHU Xinyu

王丛一 WANG Congyi

赵婧瑶 ZHAO Jingyao

李悦萌 LI Yuemeng

刘羿婷 LIU Yiting

简介 Description

我们选择建模的目标是位于京西的大台站，该车站在2020年被北京市人民政府列为北京市历史建筑，同时与“一带一路”有着密不可分的联系。我们利用所学的专业知识采集数据，利用无人机进行倾斜摄影，然后使用大疆智图建模并导出模型

文件，将导出的模型文件加载到电脑中使用 DpModeler 软件进行后期的精细化处理。同时，我们通过历史资料和当地村民的讲述，挖掘其过去、现在乃至将来的价值，对大台站有了深入的了解。

The object of our modeling is the Datai Station in the west of Beijing in China, which was listed as a historical building in Beijing by the Beijing Municipal Government in 2020 and is inextricable with the "Belt and Road" Initiative. We used our professional knowledge to collect data and utilized drones for oblique photography. Modeling was finished and exported by DJI Maps and the exported file was refined and processed in DpModeler on the computer. We also dived deep into the values of the station in the past, present and future by studying the historical materials and interviewing the elderly about the Datai Station and gained a deeper understanding of it.

大势智慧

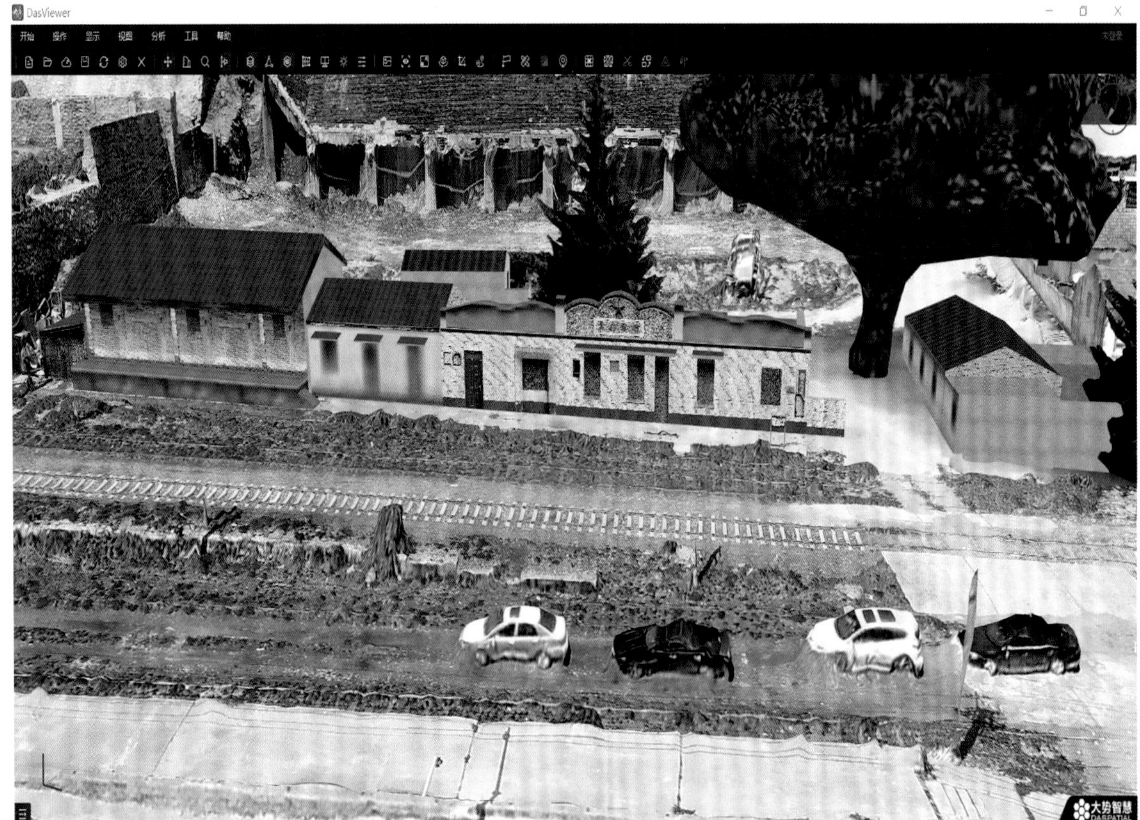
DasViewer
开始
操作
显示
视图
分析
工具
帮助
大势智慧
DASPATIAL

仰视“五岳之首”，俯瞰京沪线上的“泰山”

Taishan Station

参赛学校 University/College
北京建筑大学
Beijing University of Civil Engineering and Architecture, China

指导教师 Supervisor(s)
杨军星 YANG Junxing
刘 星 LIU Xing

参赛学生 Participant(s)
赵嘉仁 ZHAO Jiaren
顾泊桐 GU Botong
李馨悦 LI Xinyue
李邦赫 LI Banghe
何奎志 HE Kuizhi

简介 Description

用无人机进行倾斜摄影后通过 contextcapture 建模并导出模型文件，将导出的模型文件加载到电脑中使用 DP modeler 与 DP manager 进行白模的修建与纹理的映射，再使用 3D MAX 与 PS 进行后期的精细化处理。通过 3D MAX 生成三维数据找出并修复三维瓦片，将选中的瓦片通过 3D MAX 的线面编辑工具与裁剪，对模型的缺失部分进行修整。然后将 DP modeler 与 PS 连接，通过 PS 对道路车辆进行删改呈现整洁道路，最后对破损缺口补洞，完成模型的建立工作。

First, we used drones to conduct tilt photography. Then, the model was built in Contextcapture and exported. Next, DP Modeler and DP Manager were used to construct the white model and map the texture, followed by a later refinement process with 3D MAX and PS. 3D data was generated by 3D MAX to find 3D tiles and repair them. The selected tiles were trimmed by a line-and-surface editing tool in 3D MAX to complete the missing parts. After that, the DP Modeler was connected with PS to delete and modify the vehicles on the road. The final step after the road was tidy was to patch holes for broken gaps and. complete the modeling work.

候车室
进站口
严禁携带
易燃易爆

山止川行——基于参数化的成渝地区高铁站抽象概念转译建模

Wandering in Mountains and Rivers: Modeling Based on Parameterization

参赛学校 University/College
重庆交通大学
Chongqing Jiaotong University, China

指导教师 Supervisor(s)
雷　怡　LEI Yi
刘　锐　LIU Rui

参赛学生 Participant(s)
蒋升望　JIANG Shengwang
陈　漪　CHEN Yi
杜敏翼　DU Minyi
于新月　YU Xinyue

简介 Description

该场地位于中国重庆市永川区，这是重庆地势最平坦的一个区。整个场地的设计理念以永川的水文化为载体，北广场的设计主要是依靠永川的“川”做的场地景观与场地内部的交通，将道路类比成河流从站房流向城市，代表了整个文化的延伸，仿佛一个新鲜的活力注入了城市中。主体建筑的设计理念则是以永川老城经济中心的“三河汇碧”为载体。同时结合现代化的设计，为城市注入新的活力。

The site is located in the Yongchuan District, one of the flattest ones in Chongqing, China. The design concept of the whole site is based on the carrier of the water culture of Yongchuan. The design of the North Square mainly relies on the site landscape representing the “Chuan (Rivers)” in “Yongchuan” and the traffic system inside the site. Compared to a river flowing from the station house to the city, the road embodies the extension of the wider culture, serving as the fresh vitality injected into the city. The design concept of the main building is based on the view named “Three Emerald Rivers Convergence” in the economic center of Yongchuan’s old city, combined with modern design to create new vitality in the city.